Atlas of Structural Geological and Geomorphological Interpretation of Remote Sensing Images

Atlas of Structural Geological and Geomorphological Interpretation of Remote Sensing Images

Edited by

Achyuta Ayan Misra
Reliance Industries Ltd.
Mumbai, Maharashtra, India

Soumyajit Mukherjee
Department of Earth Sciences
Indian Institute of Technology Bombay
Mumbai, Maharashtra, India

WILEY

This edition first published 2023

Registered Offices
John Wiley & Sons, Inc., 111 River Street, Hoboken, NJ 07030, USA
John Wiley & Sons Ltd, The Atrium, Southern Gate, Chichester, West Sussex, PO19 8SQ, UK

Editorial Office
9600 Garsington Road, Oxford, OX4 2DQ, UK

For details of our global editorial offices, customer services, and more information about Wiley products visit us at www.wiley.com.

Wiley also publishes its books in a variety of electronic formats and by print-on-demand. Some content that appears in standard print versions of this book may not be available in other formats.

Library of Congress Cataloging-in-Publication Data applied for
Hardback: 9781119813354

Cover Design: Wiley
Cover Images: © Elena11/Shutterstock.com

Set in 10/12pt Warnock by Straive, Pondicherry, India

We dedicate this book to ***Prof. Ravi P Gupta*** *and* ***Late Prof. Barham Parkash****. They taught us remote sensing and geomorphology, respectively, in our Masters' coursework of Applied Geology in IIT Roorkee (SM: 1999–2002; AAM: 2003–2006). AAM did his Masters' thesis under Prof. Gupta.*

Contents

List of Contributors

DK Aaisyah
Department Geosciences
Universiti Brunei Darussalam
Bandar Seri Begawan, Brunei

Mark Abolins
Department of Geosciences
Middle Tennessee State University
Murfreesboro, TN, USA

Gokul Anand
North Eastern Space Applications Centre
Shillong, Meghalaya, India

Muhammad Awais
Department of Geology
University of Swabi
Swabi, Khyber Pakhtunkhwa, Pakistan
and
Department of Earth, Environmental & Resources Sciences
University of Naples Federico II
Naples, Italy

Ishmohan Bahuguna
Space Application Centre
Ahmedabad, Gujarat, India

Jeff A. Benowitz
Fairbanks, AK, USA

Ankita Biswas
Geological Survey of India
Hyderabad, Telangana, India

Mery Biswas
Department of Geography
Presidency University
Kolkata, West Bengal, India

Jonathan Saul Caine
U.S. Geological Survey
Denver, CO, USA

Laxman S. Chamyal
Department of Geology
The M.S. University of Baroda
Vadodara, Gujarat, India

Jie Chen
State Key Laboratory of Earthquake Dynamics
Institute of Geology, China Earthquake Administration
Beijing, China

Rajarshi Dasgupta
Department of Geography
East Calcutta Girls' College
Kolkata, West Bengal, India

Tanwita Deb
Indian Institute of Science Education and Research
Bhopal, Bhauri, India (former)

Bhushan S. Deota
Department of Geology
The M.S. University of Baroda
Vadodara, Gujarat, India

Chinmay U. Dongare
Department of Geology
The M.S. University of Baroda
Vadodara, Gujarat, India

Arijit Ghosh
Reliance Industries Ltd.
Reliance Corporate Park
Mumbai, Maharashtra, India

Swakangkha Ghosh
Cactus Communications Pvt. Ltd.
Mumbai, Maharashtra, India

Peng Guo
Key Laboratory of Seismic and Volcanic Hazards
China Earthquake Administration
Beijing, China

Suraj Gupta
Department of Geography
Presidency University
Kolkata, West Bengal, India

Ezra Jayasuriya
Department of Geology
Curtin University
Miri, Sarawak, Malaysia

Benjamin R. Jordan
Faculty of Sciences
Brigham Young University – Hawaii
Laie, HI, USA

Aditya U. Joshi
Department of Civil Engineering
Manipal Institute of Technology
Manipal Academy of Higher Education
Manipal, Karnataka, India

Maciej Kania
Faculty of Geography and Geology
Insitute of Geological Sciences
Jagiellonian University
Kraków, Poland

Chuanyou Li
State Key Laboratory of Earthquake Dynamics
Institute of Geology
China Earthquake Administration
Beijing, China

Manoj A. Limaye
Department of Geology
The M.S. University of Baroda
Vadodara, Gujarat, India

Jinrui Liu
Key Laboratory of Seismic and Volcanic Hazards
China Earthquake Administration
Beijing, China

Venkata R. Mandla
CGARD
National Institute of Rural Development and Panchayati Raj
Ministry of Rural Development
Government of India
Hyderabad, Telangana, India

Mudit D. Mankad
Department of Geography
The M.S. University of Baroda
Vadodara, Gujarat, India

Deepak M. Maurya
Department of Geology
The M.S. University of Baroda
Vadodara, Gujarat, India

Achyuta Ayan Misra
Reliance Industries Ltd.
Reliance Corporate Park
Mumbai, Maharashtra, India

Sayoni Mondal
Department of Geography
Presidency University
Kolkata, West Bengal, India

Soumyajit Mukherjee
Department of Earth Sciences
Indian Institute of Technology Bombay
Mumbai, Maharashtra, India

Priyank Pravin Patel
Department of Geography
Presidency University
Kolkata, West Bengal, India

Atul K. Patidar
Department of Petroleum Engineering and Earth Sciences
University of Petroleum and Energy Studies
Dehradun, Uttarakhand, India

Prassana
Department of Geology
Curtin University
Miri, Sarawak, Malaysia

Hongliu Ran
Key Laboratory of Seismic and Volcanic Hazards
China Earthquake Administration
Beijing, China

Nisha Rani
Geological Survey of India
Hyderabad, Telangana, India

Zhikun Ren
Key Laboratory of Seismic and Volcanic Hazards
China Earthquake Administration
Beijing, China

Priyom Roy
National Remote Sensing Centre, ISRO
Hyderabad, Telangana, India

Syaakiirroh Sahari
Department Geosciences
Universiti Brunei Darussalam
Bandar Seri Begawan, Brunei

Ain Said
Department of Geology
Curtin University
Miri, Sarawak, Malaysia

Afroz A. Shah
Department Geosciences
Universiti Brunei Darussalam
Bandar Seri Begawan, Brunei

Amirul Shahbuddin
Department of Petroleum Engineering
Politeknik Brunei
Bandar Seri Begawan, Brunei

Mohamedharoon Shaikh
Department of Geology
The M.S. University of Baroda
Vadodara, Gujarat, India

Tejpal Singh
CSIR – Central Scientific Instruments Organisation
Chandigarh, India

Thota Sivasankar
NIIT University
Neemrana, Rajasthan, India

Gang Su
China Earthquake Disaster Prevention Center
China Earthquake Administration
Beijing, China

Mateusz Szczęch
Faculty of Geography and Geology
Institute of Geological Sciences
Jagiellonian University
Kraków, Poland

Prabhuti Tiwari
Department of Geology
The M.S. University of Baroda
Vadodara, Gujarat, India

Yogi N. Trivedi
Amnex Infotechnologies
Ahmedabad, Gujarat, India

Swarali Vasaikar
Department of Geology
The M.S. University of Baroda
Vadodara, Gujarat, India

Dengyun Wu
Key Laboratory of Seismic and Volcanic Hazards
China Earthquake Administration
Beijing, China

Gongming Yin
State Key Laboratory of Earthquake Dynamics
Institute of Geology
China Earthquake Administration
Beijing, China

Preface

Analyses of geological structure and geomorphology from remote sensing images have been important disciplines in (applied) Earth Sciences. Students have excellent (recent) textbooks on remote sensing, structural geology and geomorphology. However, they lack numerous quality examples to learn how to interpret the images in real-world examples. This atlas fills that gap and benefits both instructors and students. Such a book would be most useful during the tight semester schedule. The book comes out as part of our ongoing efforts to produce atlas and other teaching/lab contents in structural geology and related disciplines (e.g. Mukherjee 2013, 2014, 2015, 2020, 2021; Bose and Mukherjee 2017; Misra and Mukherjee 2018; Mukherjee et al. 2020).

After qualifying their degrees, students will mostly work in industry as geologists and not necessarily as "structural geologists." They would require diverse structural skills, such as image interpretation, which are presently not taught in structural geology courses in most universities. Some amount of remote sensing image interpretation has now become an absolute need in regional structural geological articles that may focus on other issues (e.g. Vanik et al. 2018; Dasgupta and Mukherjee 2017, 2019; Dasgupta et al. 2022; Biswas et al., in press). Instead of using topo-sheets, geologists are now plotting lithocontacts on Google Earth images. Along with the plotted structures obtained from the terrain, interpretation of images for tectonic geomorphology has become essential to strengthen the field findings. International journals are interested in handling articles that approach problems from a multidisciplinary perspective. Having some knowledge in image interpretation therefore has become a "necessary skill."

This edited book provides examples of the process of interpreting remotely sensed images in terms of structural geology and geomorphology. It covers (i) interpretations on remotely sensed images from varied resolutions; (ii) different bands/spectra and their combinations into False Color Composites (FCCs); (iii) diverse criteria used to identify and interpret structural geologic and geomorphologic features; and (iv) providing one uninterpreted and another interpreted image for all examples. The book provides a common platform to look for global examples for anyone interested in remotely sensed images.

In a few years from now and even during the present COVID pandemic, most geologists will be/have been developing virtual field trips. We hope that this book will be useful in that context. We look forward to receiving comments from the readers!

Refer to this book as:

Misra, A.A. and Mukherjee. S. 2022. *Atlas of Structural Geological and Geomorphological Interpretation of Remote Sensing Images*. Wiley. ISBN: 9781119813354.

Refer to individual chapters as:

Misra, A.A. and Mukherjee, S. (2022). Introduction to Atlas of Structural Geological and Geomorphological Interpretation of Remote Sensing Images.
In: *Atlas of Structural Geological and Geomorphological Interpretation of Remote Sensing Images*
A.A. Misra and S. Mukherjee (Eds.) Wiley.
ISBN: 9781119813354.

Achyuta Ayan Misra
Soumyajit Mukherjee

References

Biswas, M., Puniya, M.K., Gogoi, M.P. et al. (2022). Morphotectonic analysis of petroliferous Barmer rift basin (Rajasthan, India). *Journal of Earth System Science* 131, 140.

Biswas, M., Gogoi, M.P., Mondal, B. et al. (in press, 2022) Geomorphic assessment of active tectonics in Jaisalmer basin (western Rajasthan, India). Geocarto International. DOI: 10.1080/10106049.2022.2066726.

Bose, N. and Mukherjee, S. (2017). Map interpretation for structural geologists. In: *Developments in Structural Geology and Tectonics*. Series Editor: Mukherjee, S. Amsterdam: Elsevier. ISBN: 978-0-12-809681-9 ISSN: 2542-9000.

Dasgupta, S. and Mukherjee, S. (2017). Brittle shear tectonics in a narrow continental rift: asymmetric non-volcanic Barmer basin (Rajasthan, India). *The Journal of Geology* 125: 561–591.

Dasgupta, S. and Mukherjee, S. (2019). Remote sensing in lineament identification: examples from western India. In: *Problems and Solutions in Structural Geology and Tectonics*, Developments in Structural Geology and Tectonics Book Series, vol. 5. Series Editor: Mukherjee, S (ed. A. Billi and A. Fagereng), 205–221. Elsevier. ISSN: 2542-9000. ISBN: 9780128140482.

Dasgupta, S., Biswas, M., Mukherjee, S., and Chatterjee, R. (2022). Depositional system, morphological signatures, tectonics and sedimentation pattern along the transform margin- Palar-Pennar basin, Indian east coast. *Journal of Petroleum Science & Engineering* 211: 110155.

Misra, A.A. and Mukherjee, S. (2018). *Atlas of Structural Geological Interpretation from Seismic Images*. Wiley Blackwell. ISBN: 978-1-119-15832-5.

Mukherjee, S. (2013). *Deformation Microstructures in Rocks*, 1–111. Berlin: Springer Geochemistry/ Mineralogy. ISBN: 978-3-642-25608-0.

Mukherjee, S. (2014). *Atlas of Shear Zone Structures in Meso-Scale*, 1–124. Cham: Springer Geology. ISBN: 978-3-319-0088-6.

Mukherjee, S. (2015). *Atlas of Structural Geology*. Amsterdam: Elsevier. ISBN: 978-0-12-420152-1.

Mukherjee, S. (2020). *Teaching Methodologies in Structural Geology and Tectonics*, 1–251. Springer. ISBN: 978-981-13-2781-0.

Mukherjee, S. (2021). *Atlas of Structural Geology*, 2e, 1–260. Amsterdam: Elsevier. ISBN: 978012816802.

Mukherjee, S., Bose, N., Ghosh, R. et al. (2020). *Structural Geological Atlas*. Springer. ISBN: 978-981-13-9825-4.

Vanik, N., Shaikh, H., Mukherjee, S. et al. (2018). Post-Deccan trap stress reorientation under transpression: evidence from fault slip analyses from SW Saurashtra, western India. *Journal of Geodynamics* 121: 9–19.

Acknowledgements

This book comes under the commissioning editorial of **Frank Weinreich** and **Andrew Harrison** (Wiley Blackwell) and under the handling editorial of **Stacey Woods**. We thank **Wiley** for undertaking intense proofreading. **CPDA grant** (IIT Bombay) supported Soumyajit Mukherjee. **Mohit Kumar Puniya** (Survey of India, Dehradun) and Teaching Assistant **Bikramaditya Mondal** (IIT Bombay) assisted us. We thank the **contributing authors** and **reviewers** for participation.

We acknowledge the **anonymous external reviewers** who provided several comments on this book proposal. Soumyajit is thankful to his colleague **Prof. Prabhakar Naraga** who shared the teaching load for the spring semester in 2022. Soumyajit thanks his wife **Payel Mukherjee** for picking up all the household work during the Covid pandemic (2020–ongoing), and allowing uninterrupted free time to finish this book.

Achyuta Ayan Misra and Soumyajit Mukherjee

About the Companion Website

This book is accompanied by a companion website.

www.wiley.com/go/misra/RemoteSensingImages

This website includes:

- Raw (uninterpreted) remote sensing images and instructions for accessing 3D models.

Section A

Background

Introduction to "Atlas of Structural Geological and Geomorphological Interpretation of Remote Sensing Images"

Achyuta Ayan Misra[*,1] *and Soumyajit Mukherjee*[2]

[1] *Reliance Industries Ltd., Reliance Corporate Park, Mumbai, Maharashtra, India*
[2] *Department of Earth Sciences, Indian Institute of Technology Bombay, Mumbai, Maharashtra, India*

Remote sensing has been immensely useful in tectonics, structural geology, and geomorphic studies (e.g. Misra et al. 2014; Dasgupta and Mukherjee 2017, 2019; Shaikh et al. 2020; Dasgupta et al. 2022). This edited book consists of **24 chapters** authored and co-authored by **53 persons** from **21 academic organizations and industries** from **7 countries.**

Misra (2022a; Chapter 1) presents the definition, fundamental processes, and scopes of remote sensing in structural geologic and geomorphologic studies. Misra (2022b; Chapter 2) pens the next introductory chapter, which classifies remote sensing based on data type, source, platform, and imaging media. Gupta and Biswas (2022, Chapter 3) present a morpho-tectonic analysis of a mid-channel bar using remote sensing images from the Jaldhaka river (India). Deota et al. (2022, Chapter 4) present geomorphic indicators of glacier retreat from Jorya-Garang glacier, Baspa Valley (India). They identify different stages of moraines. Jordan (2022, Chapter 5) presents several interesting aerial views of the 2018 Kilauea eruption (U.S.A). Geomorphic features such as lava fissures are documented from images. Awais (2022, Chapter 6) documents different kinds of depositional systems observable in images in Google Earth, viz. different rivers, deltas, lakes, alluvial fans, dunes, and estuaries. Patel et al. (2022, Chapter 7) discuss in detail badland geomorphology from images and several field photos. Their study area was Garbeta (India). Dongare et al. (2022, Chapter 8) discuss the geomorphology of the Indian west coast. The area is important from the perspective of hydrocarbon exploration (Mukherjee et al. 2020) and few recent field data have been available (e.g. Misra and Mukherjee 2017). The present authors focus on the erosional and depositional landforms of Goa beach. Misra (2022c, Chapter 9) works on satellite images and field snaps of the Spiti valley and discusses different fluvial geomorphic features. Sahari et al. (2022, Chapter 10) reports on deformation bands from drone images mainly located in the sandstone outcrops. Field geologists from any part of the globe will find these images interesting and applicable to their terrains. In another contribution, Aaisyah et al. (2022, Chapter 11) present deformation bands, predominantly through meso-scale photographs, from the Lion King Fault Zone (Brunei) that can have far-reaching implications for the petroleum geoscience of the terrain. Patidar et al. (2022, Chapter 12) use DEM and satellite data and document deformation and tectonic geomorphology from a portion of the Kutch basin, particularly the reorganization of the drainage network. Caine and Benowitz (2022, Chapter 13) work on a part of the Alaska-Canada Cordillera for geomorphologic aspects. They also present a historical development of the subject for the terrain, so that the present contribution is well understood as to where it stands. Ghosh et al. (2022, Chapter 14) discuss how image analyses can reveal lineaments from the Indo-Burma Range, which is a humid tropical region. The subject of lineament delineation in mega-scale is of common interest to tectonicians worldwide. Kania and Szczęch (2022, Chapter 15) provide a tectonic interpretation for a part of the Polish Outer Carpathians using airborne-based LiDAR DEM. They identify and interpret faults, joints, overthrusts, bedding planes, and folds. Vasaikar et al. (2022, Chapter 16) work on the Narmada-Tapi interfluve in Gujarat (India) and demonstrate how tectonics, folding, and faulting can alter drainage patterns. Abolins (2022, Chapter 17) utilizes Airborne Visible/Infrared Imaging Spectrometer (AVIRIS) images to study faults in the

*Corresponding Author: achyutaayan@gmail.com

Atlas of Structural Geological and Geomorphological Interpretation of Remote Sensing Images, First Edition.
Edited by Achyuta Ayan Misra and Soumyajit Mukherjee.

Companion website: www.wiley.com/go/misra/RemoteSensingImages

Spring Mountains (U.S.A.). Liu et al. (2022, Chapter 18) study images of coseismic ruptures from Sulawesi Island (Indonesia) after a 2018 earthquake. Several field snaps present the severity of the seismic shock. Rani et al. (2022, Chapter 19) elaborate how ASTER imagery and DEM can be useful in structural and alteration mapping from the Gadak Schist Belt (India). They also present drainage patterns and lineaments from image analyses. Misra et al. (2022d, Chapter 20) present images from the Kutch basin (India), otherwise a favorite spot for paleontologists and sedimentologists, and manifestation of regional faults, scarps, and domes from remote sensing images. Misra (2022d, Chapter 21) shows images from Son River, Central India, and interprets lineaments on various scales. The lineaments are correlated in the field and classified based on the confidence of them being deformation structures. Biswas (2022, Chapter 22) presents regional faulting from the Eastern Dharwar craton using the Sentinel-2A satellite imagery. Biswas and Roy (2022, Chapter 23) study mega-scale fold with faulted limbs from Rajasthan (India). Deb (2022, Chapter 24) study drainage morphology and genesis of bars from Darjeeling Sikkim region (India) using Google Earth images. She also refers to sinuosity magnitudes of several rivers.

Acknowledgements

Wiley Blackwell Acquisition Editors (present: Frank Weinreich, past: Andrew Harrison), Stacey Woods (Handling Editor), and the proofreading team. CPDA grant (IIT Bombay) supported SM.

References

Aaisyah, D., Sahari, S., Shah, A.A. et al. (2022). Disaggregation deformation bands dominate the trapping and sealing process at the Lion King Fault Zone, Brunei, SE Asia. In: *Atlas of Structural Geological and Geomorphological Interpretation of Remote Sensing Images* (ed. A.A. Misra and S. Mukherjee). Wiley Blackwell.

Abolins, M. (2022). Archival Airborne Visible/Infrared Imaging Spectrometer (AVIRIS) image of faults in a mixed carbonate-clastic succession, northwestern Spring Mountains, Nevada, USA. In: *Atlas of Structural Geological and Geomorphological Interpretation of Remote Sensing Images* (ed. A.A. Misra and S. Mukherjee). Wiley Blackwell.

Awais, M. (2022). Depositional systems – an overview via Google Earth. In: *Atlas of Structural Geological and Geomorphological Interpretation of Remote Sensing Images* (ed. A.A. Misra and S. Mukherjee). Wiley Blackwell.

Biswas, A. (2022). Meso scale sinistral shear, eastern Dharwar craton, Telangana, India. In: *Atlas of Structural Geological and Geomorphological Interpretation of Remote Sensing Images* (ed. A.A. Misra and S. Mukherjee). Wiley Blackwell.

Biswas, A. and Roy, P. (2022). Regional polyclinal fold with faulted limbs, Rajasthan, India. In: *Atlas of Structural Geological and Geomorphological Interpretation of Remote Sensing Images* (ed. A.A. Misra and S. Mukherjee). Wiley Blackwell.

Caine, J.S. and Benowitz, J.A. (2022). Tectonics, fault zones, and topography in the Alaska-Canada cordillera with a focus on the Alaska range and Denali fault zone. In: *Atlas of Structural Geological and Geomorphological Interpretation of Remote Sensing Images* (ed. A.A. Misra and S. Mukherjee). Wiley Blackwell.

Dasgupta, S. and Mukherjee, S. (2017). Brittle shear tectonics in a narrow continental rift: asymmetric non-volcanic Barmer basin (Rajasthan, India). *The Journal of Geology* 125: 561–591.

Dasgupta, S. and Mukherjee, S. (2019). Remote sensing in lineament identification: examples from western India. In: *Problems and Solutions in Structural Geology and Tectonics*, Developments in Structural Geology and Tectonics Book Series, vol. 5. Series Editor: Mukherjee, S. (ed. A. Billi and A. Fagereng), 205–221. Elsevier. ISSN: 2542-9000. ISBN: 9780128140482.

Dasgupta, S., Biswas, M., Mukherjee, S., and Chatterjee, R. (2022). Depositional system, morphological signatures, tectonics and sedimentation pattern along the transform margin – Palar-Pennar basin, Indian east coast. *Journal of Petroleum Science & Engineering* 211: 110155.

Deb, T. (2022). Drainage architecture and bar formation of the Rangit tributaries, Darjeeling-Sikkim Himalaya, India. In: *Atlas of Structural Geological and Geomorphological Interpretation of Remote Sensing Images* (ed. A.A. Misra and S. Mukherjee). Wiley Blackwell.

Deota, B.S., Trivedi, Y.N., Bahuguna, I.M. et al. (2022). Geomorphic indicators of glacier retreat from Jorya-Garang glacier of Baspa Valley, Himachal Pradesh, India. In: *Atlas of Structural Geological and Geomorphological Interpretation of Remote Sensing Images* (ed. A.A. Misra and S. Mukherjee). Wiley Blackwell.

Dongare, C.U., Deota, B.S., Joshi, A.U., and Limaye, M.A. (2022). Geomorphology along the west coast of India, through remote sensing. In: *Atlas of Structural Geological and Geomorphological Interpretation of Remote Sensing Images* (ed. A.A. Misra and S. Mukherjee). Wiley Blackwell.

Ghosh, S., Sivasankar, T., and Anand, G. (2022). Use of remote sensing in lineament analysis: exploring its potentials in a humid subtropical and semi-arid environment. In: *Atlas of Structural Geological and Geomorphological Interpretation of Remote Sensing Images* (ed. A.A. Misra and S. Mukherjee). Wiley Blackwell.

Gupta, S. and Biswas, M. (2022). Geodynamic quantification of mid-channel bar morphology-a spatio-temporal study. In: *Atlas of Structural Geological and Geomorphological Interpretation of Remote Sensing Images* (ed. A.A. Misra and S. Mukherjee). Wiley Blackwell.

Jordan, B.R. (2022). Aerial views of the 2018 Kilauea eruption, Hawaii, U.S.A. In: *Atlas of Structural Geological and Geomorphological Interpretation of Remote Sensing Images* (ed. A.A. Misra and S. Mukherjee). Wiley Blackwell.

Kania, M. and Szczęch, M. (2022). Tectonic structures interpretation using airborne-based LiDAR DEM on the examples from the polish outer Carpathians. In: *Atlas of Structural Geological and Geomorphological Interpretation of Remote Sensing Images* (ed. A.A. Misra and S. Mukherjee). Wiley Blackwell.

Liu, J., Wu, D., Chen, J. et al. (2022). Coseismic surface rupture and related disaster during the 2018 Mw 7.5 Palu earthquake, Sulawesi Island, Indonesia. In: *Atlas of Structural Geological and Geomorphological Interpretation of Remote Sensing Images* (ed. A.A. Misra and S. Mukherjee). Wiley Blackwell.

Misra, A.A. (2022a). Introduction to remote sensing. In: *Atlas of Structural Geological and Geomorphological Interpretation of Remote Sensing Images* (ed. A.A. Misra and S. Mukherjee). Wiley Blackwell.

Misra, A.A. (2022b). Classification of remote sensing depending on data type, source, platform and imaging media. In: *Atlas of Structural Geological and Geomorphological Interpretation of Remote Sensing Images* (ed. A.A. Misra and S. Mukherjee). Wiley Blackwell.

Misra, A.A. (2022c). Fluvial geomorphology in a part of the Spiti river basin, Himachal Pradesh, India. In: *Atlas of Structural Geological and Geomorphological Interpretation of Remote Sensing Images* (ed. A.A. Misra and S. Mukherjee). Wiley Blackwell.

Misra, A.A. (2022d). Lineament analysis in a part of the Son River valley, Madhya Pradesh, India. In: *Atlas of Structural Geological and Geomorphological Interpretation of Remote Sensing Images* (ed. A.A. Misra and S. Mukherjee). Wiley Blackwell.

Misra, A.A. and Mukherjee, S. (2017). Dyke-brittle shear relationships in the Western Deccan Strike Slip Zone around Mumbai (Maharashtra, India). In: *Tectonics of the Deccan Large Igneous Province* (ed. S. Mukherjee, A.A. Misra, G. Calvès and M. Nemčok), 269–295. London, Special Publications 445: Geological Society.

Misra, A.A., Bhattacharya, G., Mukherjee, S., and Bose, N. (2014). Near N-S paleo-extension in the western Deccan region in India: does it link strike-slip tectonics with India-Seychelles rifting? *International Journal of Earth Sciences* 103: 1645–1680.

Misra, A.A., Ghosh, A., and Patidar, A. (2022). Identifying subtle deformation structures from satellite images in parts of the Mesozoic Kachchh (Kutch) Basin, Kachchh district, Gujarat, India. In: *Atlas of Structural Geological and Geomorphological Interpretation of Remote Sensing Images* (ed. A.A. Misra and S. Mukherjee). Wiley Blackwell.

Mukherjee, S., Dole, G., Yatheesh, V., and Kale, V. (2020). Tectonics of the Deccan trap: focus on Indian Geoscientists' contribution in the last four years. *Proceedings of the Indian National Science Academy* 86: 237–244.

Patel, P.P., Mondal, S., and Dasgupta, R. (2022). *The Lateritic Badlands of Garbeta*. India: West Bengal.

Patidar, A., Shaikh, M., Tiwari, P. et al. (2022). Surface deformation along Katrol Hill Fault, Kachchh evidenced by satellite and DEM data. In: *Atlas of Structural Geological and Geomorphological Interpretation of Remote Sensing Images* (ed. A.A. Misra and S. Mukherjee). Wiley Blackwell.

Rani, N., Mandla, V.R., and Singh, T. (2022). Structural and alteration mapping using ASTER imagery and DEM for gold mineralization in the Gadag Schist Belt of Karnataka, India. In: *Atlas of Structural Geological and Geomorphological Interpretation of Remote Sensing Images* (ed. A.A. Misra and S. Mukherjee). Wiley Blackwell.

Sahari, S., Aaisyah, D., Shahbuddin, A., and Shah, A.A. (2022). Deformation bands mapped in the Miocene sandstone dominated outcrops, Sengkurong, Brunei Darussalam, SE Asia. In: *Atlas of Structural Geological and Geomorphological Interpretation of Remote Sensing Images* (ed. A.A. Misra and S. Mukherjee). Wiley Blackwell.

Shaikh, M., Maurya, D.M., Mukherjee, S. et al. (2020). Tectonic evolution of the intra-uplift Vigodi-Gugriana-Khirasra-Netra Fault System in the seismically active Kachchh Rift Basin, India: Implications for the western continental margin of the Indian plate. Journal of Structural Geology 140: 104124.

Vasaikar, S., Maurya, D.M., Tiwari, P., and Chamyal, L.S. (2022). Spatial variability of tectonic influences on drainage networks: examples from the Narmada-Tapi interfluve in Gujarat state, western India. In: *Atlas of Structural Geological and Geomorphological Interpretation of Remote Sensing Images* (ed. A.A. Misra and S. Mukherjee). Wiley Blackwell.

1

Remote Sensing Fundamentals

*Achyuta Ayan Misra**

Reliance Industries Ltd., Reliance Corporate Park, Mumbai, Maharashtra, India

1.1 What Is Remote Sensing?

Remote sensing can be formally defined in several ways. The American Society for Photogrammetry and Remote Sensing (ASPRS) defines remote sensing as "the measurement or acquisition of information of some property of an object or phenomenon, by a recording device that is not in physical or intimate contact with the object or phenomenon under study" (Colwell 1983). Fussell et al. (1986) comment that there is no ubiquitous definition of remote sensing and that different definitions exist for diverse uses, and all definitions of the technique are heuristic. They also highlight that the term "remote" here means *indirect* rather than *distant*. Gupta (2017) defines remote sensing as "obtaining information about an object without touching the object itself." It has two facets: the technology of acquiring data through a device, which is located at a distance from the object; and analysis of the data to interpret the physical attributes of the object. These two aspects are intimately linked with each other. As per Campbell and Wynne (2011), remote sensing is "the practice of deriving information about the Earth's land and water surfaces using images acquired from an overhead perspective, using electromagnetic radiation in one or more regions of the electromagnetic spectrum, reflected or emitted from the Earth's surface." Lillesand et al. (2015) provide a wider definition of remote sensing by combing it with "art." They define remote sensing as "the science and art of obtaining useful information about an object, area or phenomenon through the analysis of data acquired by a device that is not in contact with the object, area, or phenomenon under investigation."

To arrive at a holistic definition of remote sensing, it can be put forward as:

Obtaining indirect, systematic, and interpretable information about an area, object, or phenomenon using mostly electromagnetic radiation, avoiding any physical contact with the object itself.

The information collected comprises reflected and emitted energy and must be processed and interpreted depending on the problem at hand to get usable information. Since it is just the radiation that is received, which must be processed to make it usable, the definition has the "indirect" term. The "systematic" term indicates the methodical acquisition of electromagnetic data according and specific to the intended study. The data thus acquired and processed must be "interpretable" by humans or human-developed algorithms to generate usable data. Note that remote sensing includes data acquired by not only by satellite-borne sensors but also any distant/remote sensor.

The definition of remote sensing does not include observing or taking snapshots of distant outcrops with a camera, or even technologies like reflection seismic (e.g. Misra and Mukherjee 2018) or ground penetrating radar (GPR) surveys (e.g. Daniels 2005; Shaikh et al.,2020), which are also imaging technologies. The term remote sensing is applied specifically to acquiring information about the Earth and other objects in the solar system. Currently sensors placed on satellites, manned aerial vehicles (e.g. airplanes and helicopters), and unmanned aerial vehicles (UAVs) are employed to gather data, which is used to identify and analyze objects on the surface or shallow depths. Remote sensing implies

*Corresponding Author: achyutaayan@gmail.com

Atlas of Structural Geological and Geomorphological Interpretation of Remote Sensing Images, First Edition.
Edited by Achyuta Ayan Misra and Soumyajit Mukherjee.

Companion website: www.wiley.com/go/misra/RemoteSensingImages

data acquisition of reflected or emitted electromagnetic radiation (~0.4 μm – 30 cm) from sensors usually tens of meters (e.g. UAVs) to hundreds of kilometers (e.g. satellites) away. The areas of study include the surface, atmosphere, and water bodies, and can be studied based on wavelength of propagated radiation.

In the earliest times, remote sensing was done from platforms like hot air balloons (e.g. Pommereau et al. 1987), which provided great data but lacked a synoptic view of a large portion of the surface. The problem is overcome by the present use of satellites, which provide synoptic information of the surface and environment of not just the Earth but also extraterrestrial objects, e.g. other planets, their natural satellites, and comets in the solar system. Earth-orbiting satellites have remote sensors that provide information about clouds, vegetation, geomorphology, temperature, and wind along with their short-term and long-term variations. They also provide information on mineralogy, gravity, bathymetry, etc. that show very slow to nearly no variations in a long-term time frame. This global, synoptic coverage leads to the analysis of regional to continental/plate scale features like orogenic belts or plate boundaries.

It is not just the Earth that has been studied by remote sensing. Planetary probes, e.g. orbiters, flybys, surface stations, and rovers, have sensors that provide data about the surface and environment of the solar system object (planet/satellite/comet). At the moment, every planet in the solar system has been visited by at least one mission, e.g. Voyager 2 (https://voyager.jpl.nasa.gov) has made a flyby to outer planets. Also, numerous studies have been conducted on comets, even landing on them, e.g. the Philae lander on the Rosetta mission by the European Space Agency (https://rosetta.esa.int). The planetary studies provide information and insight about the solar system, its formation and temporal evolution.

1.2 Fundamental Processes of Remote Sensing

The technique/technology of remote sensing comprises numerous well-defined fundamental processes. Take two simple examples: (i) eyes and (ii) a digital camera.

The eyes of any animal are examples of remote sensors. Most animals observe objects because the solar radiation interacts with the objects and the radiated/reflected light is captured by the retina, comparable to a sensor, in their eyes. The retina sends the signal, which is the data, to the brain, i.e. the processor, which processes the data and converts it into information.

All digital cameras are equivalent to remote sensing instruments. Some are mounted on drones and unmanned aerial vehicles and provide remote sensing data. There is a sensor, usually a charge-coupled device (CCD) or complementary metal oxide semiconductor (CMOS) in the digital camera, which captures the reflected solar radiation (data) and transfers the data to a processor inside the camera to process the data and prepare it in a format that the human eye can read. In low-light conditions, a flash is added with the camera to provide extra source of light for illumination.

The science of remote sensing comprises three main components: (i) acquisition: This encompasses all technologies employed to acquire data by an instrument located at a distance from the object or phenomenon of interest; (ii) processing: The acquired data requires processing to make it usable for the purpose of study; and (iii) analysis: This is the study and interpretation of the acquired and processed data to understand the attributes of the object or phenomenon of interest. The three components are intimately related to each other and the latter two are sometimes grouped into a single component, because they are commonly iterative processes.

These components can be further elaborated into sequential fundamental processes as (Figure 1.1):

i) **Energy sources:** The most important necessity for any remote sensing study is an energy source which provides electromagnetic radiation (EMR) to the study area, which may be the Earth or other celestial objects. They may either be natural, i.e. solar radiation, or artificial, i.e. microwave (Balz and Rocca 2020; Dong et al. 2020). For most remote sensing studies, solar radiation is used as a source of energy.
ii) **EMR–atmosphere interaction:** When the EMR travels from its source to the surface of the Earth (or other celestial objects), it comes in contact with the atmosphere, where present, and interacts with atmospheric constituents (Figures 1.2 and 1.3). This interaction is very strong for the Earth for imaging the surface objects and processes. The EMR reflected from the Earth's surface is received by sensors in the remote sensing instruments. During this process, the EMR once again interacts with the atmosphere. For planets like Jupiter, the atmosphere is composed of ammonia and water vapor cloud. Imaging Jupiter's surface is nearly impossible by current remote sensing systems (e.g. Janssen et al. 2005).The most common interaction mechanisms are (Figure 1.3):
    a) Absorption, where the radiant energy is converted in other forms of energies at different wavelengths and emitted. The absorbed and emitted energy can be measured by remote sensing and important information can be obtained about the atmosphere.

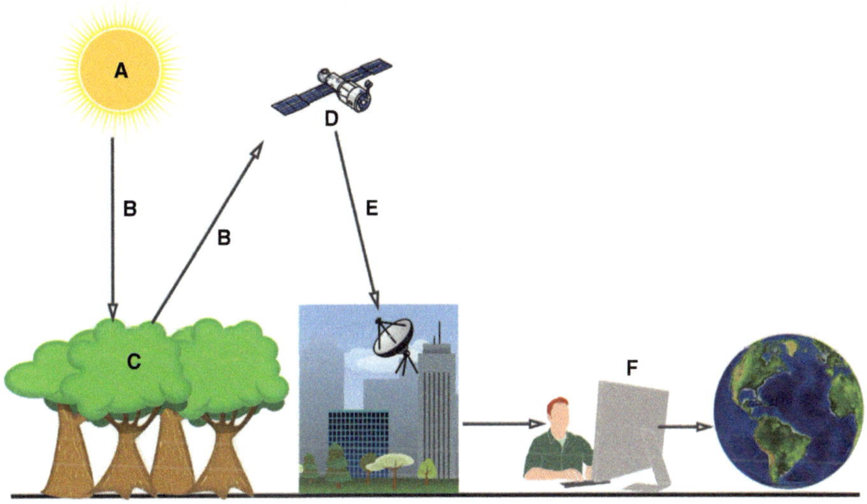

Figure 1.1 Fundamental processes in remote sensing. The different processes are denoted in the figure as A–E. A: Energy source (here, Sun); B: EMR – atmosphere interaction; C: EMR – object interaction; D: Recording and transmission; E: Reception and processing; F: Analysis.

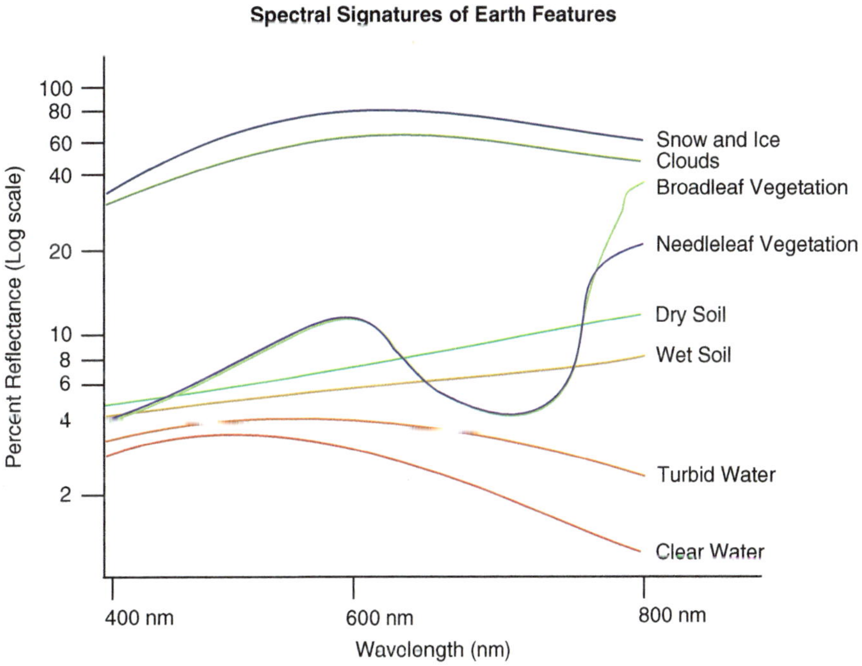

Figure 1.2 Reflectance of various objects and their patterns.
Source: NASA http://missionscience.nasa.gov/ems/09_visiblelight.html
Credited within to: Jeannie Allen.

b) Scattering, where particles, e.g. particulate matter, water vapor, etc., in the atmosphere redirect the EMR from its incident path. Scattering depends on radiation wavelength, size of the interacting particle, and travel distance of the EMR through the atmosphere.

c) Refraction, where the incident EMR passes from one medium into another with a different density, becoming deviated. The entire spectrum of EMR gets affected by refraction. However, the degree of influence depends on the wavelength of the incident radiation along with temperature, pressure, and humidity.

d) Reflection, where the incident radiation is reflected back to the sensor. Reflection occurs very commonly on cloud surfaces.

iii) **EMR–object interaction:** EMR reaching the surface of the Earth or other celestial objects, with a suitable atmosphere transparent to the EMR, through the atmosphere interacts with the surface features (Figure 1.2). This interaction and the result is dependent on characteristic of the EMR and the feature, e.g. water absorbs near infrared (NIR) radiation while healthy vegetation reflects most of the incident NIR radiation. Reflection and absorption are the main interactions of the incident radiation with surface objects.

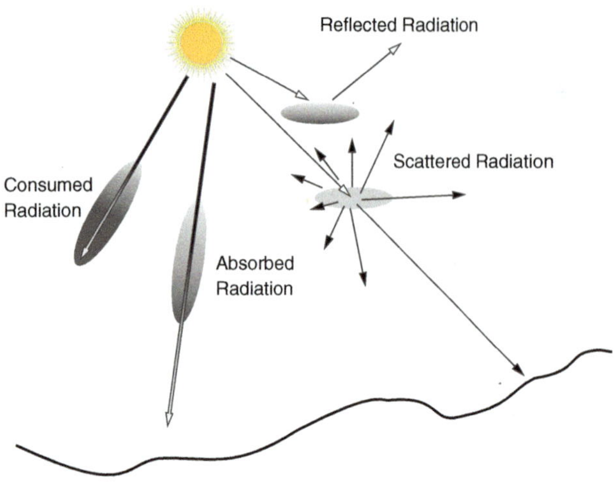

Figure 1.3 Scattering and absorption in the atmosphere. Source: compiled from Lillesand et al. (2015), Gupta (2017).

iv) **Recording and transmission:** After interaction with surface features, the reflected and emitted EMR is transmitted to the sensor, where it is recorded. Sensors are varied and serve different purposes on remote sensing studies (see Chapter 2).The recorded energy from the sensor is transmitted as electronic signals to earth stations. The modes of transmission depend on the type of remote sensing vehicle.

v) **Reception and processing:** The ground-based stations receive the electronic signals transmitted from the remote sensing instrument. These signals are then processed so that they can be read and interpreted by humans or human-developed algorithms. Usually the electronic signals are processed and converted into images.

vi) **Analysis:** The processed image is interpreted and analyzed to extract information about the object of interest.

Surfaces of all terrestrial and extraterrestrial objects reflect, absorb, or transmit variable amount of energy depending on the wavelength. Thus, they have unique spectral signatures or "fingerprints" (Figure 1.2). This information can be used to identify and interpret different objects, features, rocks, and minerals. The number of spectral bands detected by a remote sensing instrument is known as its spectral resolution. This determines how many different types of minerals can be interpreted from the resulting images.

Remote sensing is employed in numerous fields – hydrology, ecology, meteorology, oceanography, glaciology, geology, geography, environment, anthropology, etc. It is also applied in commerce, economic, planning, military, intelligence, etc.

This atlas is restricted to applications of remote sensing in geomorphology and structural geology.

1.3 Advantages of Remote Sensing

There are number of advantages to gathering data through remote sensing compared to ground studies. The most important advantages are:

i) **Time-saving:** Remote sensing analysis is time saving. Nowadays, data is usually present in a repository on the internet. Such datasets can be easily accessed from anywhere in the world. In case an area needs to be analyzed with a specific instrument, data over large parts can be acquired very quickly. Once the data is acquired and processed, it takes a fraction of the time required to undertake a ground-based survey. This also means a lower cost of analysis, involving a smaller workforce for a given study. Remote sensing images can be used to save time in finding sites for drilling and sampling for a variety of studies.

ii) **Cost-effective:** Significant amounts of data and information can be collected at a relatively low cost.

iii) **Beyond visual range data:** Remote sensors provide data in frequencies outside the human visual range, e.g. infrared ranges. This provides information that the human eye cannot detect under normal conditions.

iv) **Unobtrusive:** While collecting data, the EMR, sensor, and (most) instruments do not change the characteristics of the area, object, or phenomenon under study. This is a significant advantage over other means of data collection.

v) **Carbon neutral:** Many current projects must make efforts to reduce their carbon footprint. In near future this will be an important part of every project. Less travel to field areas and/or within the field decreases the carbon footprint. That means the organization undertaking the project will have to offset less CO_2 for the study, minimizing expenditure.

vi) **Better accessibility:** Accessibility to remote locations has improved in the world as new roads have been built. Nevertheless, connectivity is an issue for many field locations that may be dangerous and/or inaccessible, e.g. high mountains or deep inside deserts or forests. Remote sensing images can provide reliable data from those regions and help analyze best routes to possible outcrops, reducing time and effort during field work. Remote sensing analysis can be useful to fill in gaps in a field study. Such gaps may be made due to inaccessibility or may be preconceived to cover larger areas in the ground-based field.

vii) **Comprehensive approach:** Remote sensing helps analyze features/phenomena in conjunction to adjacent features/phenomena. This helps in both understanding regional trends and analyzing relationships of adjacent terrains.

viii) **Multiple usages:** The same dataset can be analyzed for various usages including geology, infrastructure, environment, and land use. This makes remote sensing an economic and easily viable solution to multiple problems.

ix) **Combining products:** Remote sensing products are varied and numerous with their own typical characteristics and advantages. There is the capability in remote sensing analyses to merge different products to form a single product, e.g. Normalized Difference Vegetation Index (NDVI) or Normalized Difference Snow Index (NDSI) to augment interpretations. Other datasets (e.g. topography, bathymetry, geophysical measurements, geochemical surveys) can be compared with remote sensing data to provide new understanding of various properties of natural phenomena.

x) **Variety of scales:** Remote sensing studies can be done at a variety of scales, starting from megaregional (100 km) scales from instruments mounted on satellites to outcrop (cm) scales from UAVs.

xi) **Historical data:** Remote sensing has a very important advantage in the availability of historical data. Such data can be used effectively to detect changes in land use, land cover, infrastructure, etc.

xii) **Sensor/instrument specific advantages:** Most orbital instruments are operational in all weather conditions, have high geometric reliability, are rendered digitally making transmission fast, and are easy to preserve and retrieve over long periods of time. Sensors outside of the visible wavelength range have the benefit of observation on those wavelengths, which helps in interpretation of features invisible/unrecognizable to the human eye.

1.4 Limitations of Remote Sensing

There are a number of limitations to gathering data through remote sensing compared to ground studies. The most important disadvantages are:

i) **Availability of appropriate data:** Appropriate remote sensing data in terms of sensor, resolution, area, cloud cover, etc., may not be immediately available. The tropics have the problem of cloud cover and many sensors cannot transmit data through clouds. Requests for data, acquisition, and processing may take a long time.

ii) **Resolution of sensors:** Resolution of all sensors, especially on orbital platforms, are inadequate for the work in question or the area of interest. However, there are now virtually hundreds of products available, making the resolution issue minimal.

iii) **Requirement of reference data:** Remote sensing studies are never equal to a complete surface-based field study. Thus, there is requirement for reference data, commonly referred to as "ground truth," to support remote sensing analyses.

iv) **Intrusive form:** Some instruments, e.g. drones and other active sensors such as radars and lasers can be intrusive to the object or phenomenon.

v) **Issues on calibration of data:** Remote sensing data depends on calibration of sensors to ground-based calibration points. These calibrations may become un-calibrated with time, resulting in erroneous data and analyses.

1.4.1 Reference Data (Ground Truth)

"Ground truth" is more commonly used for the more appropriate term "reference data" (Lillesand et al. 2015). Another term that is used for reference data is "field data" (Gupta 2017). "Ground truth" is a generalized term for all reference data collected on the ground, in the air or on/in water. The main purposes of obtaining ground truth are (Gupta 2017; Lillesand et al. 2015; Misra et al. 2014):

i) To support, augment, complement, and verify the analyses and interpretations made on remotely sensed data. This is the main purpose for ground truth data collection for the purposes of this book.

ii) To calibrate a sensor.

iii) To verify maps prepared from remote sensing data.

Ground truth data collection must be planned to support remote sensing interpretation either in a deterministic approach, e.g. verifying presence of a fault on the field as interpreted on a satellite image, or a statistical approach, e.g. mean land surface temperature over a period of time at ground stations for a match with remotely sensed data. There are important considerations while collecting ground truth data during a remote sensing project. They are:

i) **Timing:** Timing of ground truth data collection is utmost important. There are two kinds of time-dependent parameters: (i) *time-stable* or *intrinsic* parameters (e.g. those in Boggavarapu and Manoharan 2018 or in Lee et al. 2020), which are parameters that remain constant over long periods of time, e.g. deformation structures, lithology, mineralogy, spectral emissivity, etc.; and (ii) *time-critical* or *time-variant* parameters (e.g. those in Osińska-Skotak et al. 2019 or Wu et al. 2021), are those that change rapidly over time like vegetation, air/water pollution, etc. However, factors such as rainfall, waterlogging, and flooding can hinder collection of ground truth data of time-stable parameters due to accessibility.

ii) **Sampling:** Sampling is usually done depending on the project in hand. Purposive sampling (Gupta, 2017) or deterministic sampling can be done to verify most deformation structures. However, measurements of temperature variations or variation of vegetation due to presence of large fault planes require statistical sampling over time and space and correlate with the remote sensing data. Photographs from the field are also considered a deterministic sampling for verification of deformation and geomorphological features.

iii) **Type of data:** Ground truth data can be collected by field surveys, where sampling, measurements, and observations can be made directly on the ground. Maps of the data collected and field photographs are common products of such surveys. Such photographs have also been provided by modern technologies such as street images acquired and offered by several companies like Street View by Google (Figure 1.4), which is the most widely used service with the widest coverage of street images across 85 countries. There are others like Apple Look Around (Panzarino 2018), Mapillary webpage (https://www.mapillary.com), KartaView webpage (https://kartaview.org), etc. Street images have been used as an effective ground truthing option (e.g. Cao et al. 2018. Kang et al. 2018). Aerial photos from dedicated flights or UAVs are another way to collect ground truth data. UAV images of very high resolution (~1–3 cm) provide ground truth verification nearly equivalent to ground surveys (Figure 1.5). Ground truth data can also be collected from published reports, maps, surveys, records, etc.

Figure 1.4 Google Street View image, looking toward NE, from a road-cut section in the NE–SW trending regional deformation structure named More-Trondelag Fault Complex (see Misra and Mukherjee, 2018 for review). Inset: Location Map. Such images provide great alternative to fieldwork reconnaissance, or may shorten it drastically.

Figure 1.5 Drone imagery from an anonymous location used to map a rocky coast. Note the high-resolution details of the fractures in the image. Such instruments can prepare a "virtual" or "digital" copy of an outcrop in a few hours. Source: Credit :davidfpv02 / 2 images / Pixabay.

Gupta (2017) lists the main types of data that can be collected during ground truth survey. One type is thematic maps: These maps show distribution of features of interest for the study, e.g. faults, fractures, and drainage. They may be older maps prepared in earlier field studies or may be specifically planned to support the remote sensing study. These maps usually contain specific data, e.g. attitudes of planar or linear structures, depths of water bodies, sediment discharge rates, etc. Another type of data collected is spectral data: This data is specifically collected for a particular remote sensing study. Refer to Gupta (2017) and Lillesand et al. (2015) for details on components and instrumentation of spectral data.

iv) **Satellite navigation survey:** Satellite navigation or *SatNav* is a system that uses a constellation of satellites to provide geo-positioning. Small hand-held devices and most smartphones can provide geolocations (latitudes/longitudes) at cm to m accuracy ranges, and elevation (altitude) at few 10s of m accuracy. The United States' Global Positioning System (GPS, since 1979), Russia's Global Navigation Satellite System (GLONASS, since 1983), China's BeiDou Navigation Satellite System (BDS, since 2000), and the European Union's Galileo (since 2016) are fully operational Global Navigation Satellite Systems (GNSSs). Most modern SatNav instruments can read radio signals sent in by visible satellites to provide geo-position and altitude. At least four satellites are required to return a geolocation with high fidelity. Multiple GNSS systems increase the number of visible satellites improving the precise point positioning (PPP) and shorten the average convergence time, thus generating a fast and accurate result. SatNav location data, commonly referred to as GPS data, must be collected with all ground-based surveys and maps prepared with those datasets to have maximum accuracy in the prepared maps.

Acknowledgements

Soumyajit Mukherjee (IIT Bombay) reviewed. Summarized in Misra and Mukherjee (2022). Wiley Blackwell Acquisition Editors (present: Frank Weinreich, past: Andrew Harrison), Stacey Woods (Handling Editor), and proofreading team.

References

Balz, T. and Rocca, F. (2020). Reproducibility and replicability in SAR remote sensing. *IEEE Journal of Selected Topics in Applied Earth Observations and Remote Sensing* 13: 3834–3843.

Boggavarapu, L.N.P. and Manoharan, P. (2018). Classification of hyper spectral remote sensing imagery using intrinsic parameter estimation. In: *International Conference on Intelligent Systems Design and Applications*, 852–862. Cham: Springer.

Campbell, J.B. and Wynne, R.H. (2011). *Introduction to Remote Sensing*. Guilford Press.

Cao, R., Zhu, J., Tu, W. et al. (2018). Integrating aerial and street view images for urban land use classification. *Remote Sensing* 10 (10): 1553.

Colwell, R.N. (1983). *Manual of Remote Sensing*, 2e, 1. Falls Church, VA: American Society of Photogrammetry.

Daniels, D.J. (2005). Ground penetrating radar. In: *Encyclopedia of RF and Microwave Engineering* (ed. K. Chang). New Jersey: Wiley. ISBN: 9780471270539.

Dong, J., Crow, W.T., Tobin, K.J. et al. (2020). Comparison of microwave remote sensing and land surface modeling for surface soil moisture climatology estimation. *Remote Sensing of Environment* 242: 111756.

Fussell, J., Rundquist, D., and Harrington, J.A. (1986). On defining remote sensing. *Photogrammetric Engineering and Remote Sensing* 52 (9): 1507–1511.

Gupta, R.P. (2017). *Remote Sensing Geology*, 3e, 671. New York: Springer. ISBN: 9783642077418.

Janssen, M.A., Hofstadter, M.D., Gulkis, S. et al. (2005). Microwave remote sensing of Jupiter's atmosphere from an orbiting spacecraft. *Icarus* 173 (2): 447–453.

Kang, J., Körner, M., Wang, Y. et al. (2018). Building instance classification using street view images. *ISPRS Journal of Photogrammetry and Remote Sensing* 145: 44–59.

Lee, G., Cheon, J., and Lee, I. (2020). Validation of LIDAR calibration using a LIDAR simulator. *The International Archives of Photogrammetry, Remote Sensing and Spatial Information Sciences* 43: 39–44.

Lillesand, T., Kiefer, R.W., and Chipman, J. (2015). *Remote Sensing and Image Interpretation*, 7e. Wiley. ISBN: 9781118343289.

Misra, A.A. and Mukherjee, S. (2018). *Atlas of Structural Geological Interpretation from Seismic Images*. New Jersey: Wiley. ISBN: 9781119158325.

Misra, A.A. and Mukherjee, S. (2022). Introduction to "atlas of structural geological and geomorphological interpretation of remote sensing images". In: *Atlas of Structural Geological and Geomorphological Interpretation of Remote Sensing Images* (ed. A.A. Misra and S. Mukherjee). Wiley Blackwell. ISBN: 9781119813354.

Misra, A.A., Bhattacharya, G., Mukherjee, S. et al. (2014). Near N-S paleo-extension in the western Deccan region in India: does it link strike-slip tectonics with India-Seychelles rifting? *International Journal of Earth Sciences* 103: 1645–1680.

Osińska-Skotak, K., Radecka, A., Piórkowski, H. et al. (2019). Mapping succession in non-forest habitats by means of remote sensing: is the data acquisition time critical for species discrimination? *Remote Sensing* 11 (22): 2629.

Pommereau, J.P., Fabian, P., Flentje, G. et al. (1987). Intercomparison of stratospheric NO_2 and NO_3 measurements during MAP/GLOBUS 1983. *Planetary and Space Science* 35 (5): 615–629.

Shaikh, M.A., Maurya, D.M., Mukherjee, S. et al. (2020). Tectonic evolution of the intra-uplift Vigodi-Gugriana-Khirasra-Netra fault system in the seismically active Kachchh rift basin, India: implications for the western continental margin of the Indian plate. *Journal of Structural Geology* 140: 104124.

Wu, K., Ryu, D., Nie, L., and Shu, H. (2021). Time-variant error characterization of SMAP and ASCAT soil moisture using triple collocation analysis. *Remote Sensing of Environment* 256: 112324.

Websites

Kartaview webpage: https://kartaview.org/ (accessed 30 October 2021).

Mapillary webpage: https://www.mapillary.com/ (accessed 30 October 2021).

Panzarino, M. (2018). Apple is rebuilding Maps from the ground up. TechCrunchArticle (Date: June 29, 2018). https://techcrunch.com/2018/06/29/apple-is-rebuilding-maps-from-the-ground-up (accessed 01 April 2021).

2

Classification of Remote Sensing Depending on Data Type, Source, Platform, and Imaging Media

*Achyuta Ayan Misra**

Reliance Industries Ltd., Reliance Corporate Park, Mumbai, Maharashtra, India

2.1 Introduction

Currently, remote sensing techniques cover the entire electromagnetic spectrum (EMS) starting from the low-frequency radio waves through the microwave, sub-mm, far infrared, near infrared, visible, ultraviolet, x-ray, and gamma-ray regions of the spectrum (e.g. Elachi and van Zyl 2021; Gupta 2017; Lillesand et al. 2015).The spectrum of the data to be acquired and used for a study depends on the information critical to the analyses.

For capturing spatial information over large surfaces, high-resolution 2D images over limited areas are usually preferred, for example in the analysis of structural geology, geomorphology, and land cover. Two-dimensional images of a wide region or global coverage are used when synoptic data are required, for example in the case of weather studies (e.g. Thies and Bendix 2011; Yang et al. 2013). Specific bands from the entire spectrum may be used for such analysis in passive sensing modes (see Section 2.4). Active sensing (Section 2.4) products, e.g. radar and microwave, are also used for structural geological and geomorphological analyses. Most recently, visual wavelength is used in unmanned aerial vehicle (UAV) (or drone) imagery and street-view photographs (Section 2.5). There are number of examples in this book on the different types of remote sensing data used for structural geological and geomorphological analysis. In the subsequent sections, remote sensing data is classified based on different factors, with emphasis on the interpretation deformation and geomorphological features.

2.2 Data Type

Remote sensing data can be classified based on the most important or the majority of the information needed or used for the particular analysis. Table 2.1 summarizes different types of data acquired in a remote sensing program and the types of sensors that acquire them, along with a few examples of each. There are other types of data, but they are rarely used for structural geological and geomorphological interpretations. They are outside the scope of this book and are avoided here.

2.3 Platforms

There are a variety of platforms that are used in remote sensing applications, depending on the sensors on each platform. The different types of platforms are:

i) **Satellite imagery:** Satellite-based remote sensing started in the early 1960s with military reconnaissance primarily in the US. In 2008, there were more than 150 Earth observation satellites, acquiring and transmitting more than 10 terabytes of data daily (Tatem et al. 2008). Satellite-borne sensors are the most commonly used remote sensing data sources today, mainly because of their numerous advantages over other systems (see Chuvieco 2020; Ferreira et al. 2012; Ide et al. 2011; Liu and Xia 2010), which include: (i) large spatiotemporal coverage; (ii) large spectral coverage possible; (iii) stereo capability using two cameras; and (iv) high-resolution imagery

*Corresponding Author: achyutaayan@gmail.com

Atlas of Structural Geological and Geomorphological Interpretation of Remote Sensing Images, First Edition.
Edited by Achyuta Ayan Misra and Soumyajit Mukherjee.

Companion website: www.wiley.com/go/misra/RemoteSensingImages

Table 2.1 Classification of remote sensing data based on the important type of data required for the intended analysis, with emphasis on identifying and mapping structural geological and geomorphological features.

Important type of information needed	Type of sensor	Examples of sensors
High spatial resolution and wide coverage	Imaging sensors, cameras	Large-format camera (1984); Seasat imaging radar (1978); Magellan radar mapper (1989); Mars Global Surveyor Camera (1996); Mars Rover Camera (2004 and 2014); Cassini Camera (2006)
Limited spectral resolution with high spatial resolution	Multispectral mappers	Landsat multispectral mapper and thematic mapper (1972–1999); SPOT (1986–2002); Galileo NIMS (1989)
High spectral and spatial resolution	Imaging spectrometer	Space-borne imaging spectrometer (1991); ASTER (1999); Hyperion (2000)
High accuracy measurement of location and profile	Altimeters, sounders	Seasat (1978); GEOSAT (1985); TOPEX/Poseidon (1992); Jason (2001) altimeter; Pioneer Venus orbiter radar (1979); Mars orbiter altimeter (1990)
3D topographic mapping	Scanning altimeters and interferometers	Shuttle Radar Topography Mission (2000); Light detection and ranging (Lidar)
Surface displacement mapping	Radar interferometer	Sentinel (2012, 2016); SkyMed (2007); ALOS (2006); TANDEMX (2010); ALOS-2 (2014)
High spatial resolution over limited areas or along track lines	Imaging sensors, cameras (e.g. CMOS)	UAV or drone, street-view imagery

Source: Modified from Elachi and van Zyl (2021).

are also presently available. However, a few disadvantages compared to other systems are: (i) sometimes corrections are difficult; (ii) image processing is time-consuming and involves complex mathematical transformations; and (iii) much of the older data is on film, which needs to be digitized.

ii) **Aerial photography:** Aerial photography was first used in 1858 over Paris, France, by photographer and balloonist Gaspar Felix Tournachon (Professional Aerial Photographers Association web page 2021). Aerial photography (also called airborne remote sensing/imagery) is the technique to acquire data, usually in the visible part of the EM spectrum. Common platforms used for aerial photography include but not limited to aircrafts, helicopters, UAVs (or drones), balloons, kites, parachutes, etc. Vosgerau et al. (2016) provide a great example of helicopter-based photography for delineating reservoir architecture in lava systems. Aerial photography has a number of advantages (see Bhardwaj et al. 2016; Iizuka et al. 2018; Saadatseresht et al. 2015; Yao et al. 2019): (i) very high resolution (~3–20 cm); (ii) surfaces of any slope can be imaged, e.g. cliff faces or quarry walls; and (iii) (re)visit within intended time ranges possible. However, there are a few disadvantages: (i) low coverage area; and (ii) high cost per unit area covered. Aerial photography is used for one-time operations and cover small areas of interest, while satellite imagery is used for large areas with regular, often daily, passes and operations. Along with aerial photography, videography is also becoming a popular choice with bandwidths outside the visible spectrum, e.g. microwave in synthetic aperture radar (SAR).

Aerial photography can be of the following types (Amrullah et al. 2016; Remondino and Gerke 2015; Warrick et al. 2017):

a) Oblique: Aerial images taken at an angle are called oblique photographs, which can be of two types: (i) low-oblique: taken from a low angle relative to the Earth's surface; and (ii) high-oblique: also called steep-oblique, photographs taken from a high angle relative to the Earth's surface.
b) Vertical: Vertical imageries are nadir-looking, i.e. they are taken straight down, and are mainly used in photogrammetry and image interpretation. These images are very important resources in structural geological and geomorphological interpretations. These images are specifically taken with special large-format cameras with calibrated and known geometric properties. Ground-based GPS tagging and georeferencing are essential processes required here.
c) Combined: Aerial imageries are often combined as per purpose. They can be: (i) panoramas, made by stitching several photos taken in a grid pattern achieved either by shooting different angles from one spot, e.g. with a hand-held camera, or from different spots at the same angle, e.g. from a drone; (ii) stereo photography, for the creation of 3D images from two or more

photographs of the same area taken from different angles, e.g. one oblique and another vertical, or two oblique, etc.; or (iii) merged, images from several sensors measuring varied bands of the EM spectrum, and sometimes images taken with different lenses, that are geometrically corrected and combined to produce one image either in the camera or during post-processing.

Detailed analysis on aerial photographs can lead to excellent structural geological (e.g. Lai et al. 2021) and seismic hazard identification (e.g. Jun et al. 2021).

2.4 Energy Source

Remote sensing data sensors are of two types (Figure 2.1):

i) **Passive:** In these systems, reflected sunlight is detected by the sensor. Passive sensors work only when the natural energy (sunlight) illuminates the surface of the area of interest. Since reflected energy from the sun is unavailable at night, most of the passive sensors do not work at that time. Some wavelengths of radiation, e.g. thermal infrared, which is emitted naturally both day and night, can be detected by the passive sensors at night too, provided the amount of energy is sufficient to be sensed. Examples of passive remote sensing systems are Landsat, ASTER, Ikonos, etc.
ii) **Active:** In active systems, the signal is emitted by the instrument on a remote sensing vehicle to the object for illumination and its reflection is detected by the sensor. This resembles making photographs with a camera with an external flash. Such sensors have the natural advantage of acquiring data even during the nighttime. They can also be used to detect wavelengths of radiation mostly absent in solar radiation, e.g. microwave ranges. Examples of active remote sensing systems include LiDAR (e.g. Zhang and Muhammad 2020), microwave (e.g. Dong et al. 2020), SAR (e.g. Balz and Rocca 2020), etc.

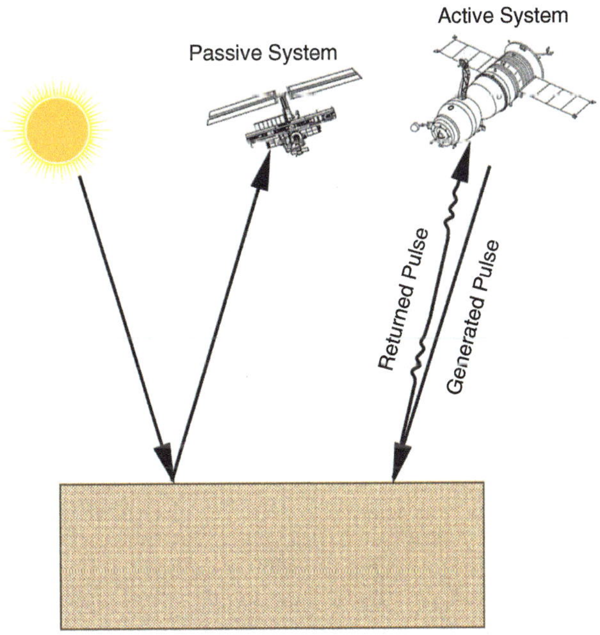

Figure 2.1 Passive and active sensor systems.

2.5 Imaging Media

There are a variety of imaging media used in remote sensing.

i) Cameras: Cameras are the instrument of choice for many remote sensing platforms such as satellite, UAVs, airplanes, helicopters, etc. with overlaps, as required, for stereoscopic imaging. Cameras with photographic sensors are versatile and can acquire images in a large range of bandwidths. These images are associated with vertical exaggeration, spectral and radial distortion, and false topography perception phenomena (Prost 2013 and references therein, Saraf et al. 1996, 2011).
   a) Black and white photography: Black-and-white photography, also known as B/W or panchromatic photography, is preferred in many studies since its cost is much lower than the color images. Structural geological and geomorphological interpretation can be done on B/W images. However, the ability to identify lithologies, soils, and alterations is hampered due to paucity of color. Currently, B/W photos are becoming obsolete owing to the easy availability of low-cost, high-resolution satellite and aerial imaging systems. Notwithstanding, such images in the archives are still used to detect changes in geomorphology, land cover, etc.
   b) Color photography: Digital color photography is the most commonly used satellite data at present, and with the advent of multiple services providing free, color satellite imagery, and in some cases aerial photography too, they have become a product of daily use. Color photos are prepared by combining three bands usually of red, green, and blue. True colors are obtained on an image when blue, green, and red bands of the EM spectrum are combined in blue, green, and red colors. True color images have the obvious advantage of being natural and identifying vegetation, waterbodies, and land cover becomes easier. However, the blue band inherently has a significant haze, which

deteriorates the images. Therefore, False Color Composites (FCC) are used to remove the blue band noise and to accentuate certain features, e.g. vegetation, water bodies, and soil moisture.

c) Thermal scanners: Surface temperatures can be recorded by radiometers, which measure the thermal energy emitted from the surface. There are passive sensors (Section 2.4) that measure thermal radiation from both land and water on the Earth's surface. This method is used very commonly to detect coal fires in subsurface coal mines (Kuenzer and Dech 2013; Prakash and Gupta 1999; Zhang et al. 2004) or to detect oil slicks (e.g. Fingas and Brown 2014 and references therein). Thermal Infrared (TIR) imagery can be used to detect certain thermal anomalies associated with faults (e.g. Wu et al. 2012). TIR imagery is usually acquired at early morning when the thermal emission of all the objects maintains equilibrium.

ii) Radar: Radar (Radio Detection And Ranging) includes all microwave sensors that detect attributes of remote objects and phenomena. Radars are active systems (Section 2.4) that use artificially generated microwaves directed in a particular direction that scatter after colliding with objects. The scattered radiation is received, processed, and analyzed to study the objects. Radar is very sensitive to topography and differences in surface moisture. Advantages (e.g. Lillesand et al. 2015; Casagli et al. 2016; Gupta 2017; Abdelkareem et al. 2020) of radar imagery over other techniques are (i) ability to penetrate clouds, and (ii) ability to acquire images in darkness.

a) Radar altimeters: These are instruments on satellites and aircrafts that radiate radio waves toward the ground and calculate the time taken for the wave to reach the sensor. As the velocity of the radio wave is known, the distance to the surface can be calculated from the two-way travel time. This data can be plotted over an area in 3D to constitute a digital elevation model (DEM) to map the land surface (e.g. Hossain et al. 2021). Significant structural geological and geomorphological interpretations can be done on these images.

b) Radar interferometry: DEMs are sometimes prepared from data collected very close in time and then mapping the difference between the two DEMs. This results in an interferogram that shows minute temporal changes in elevation. This technique has applications in active deformation measurements and also to measure ice elevations with respect to water near the poles.

iii) Multispectral and Hyperspectral Scanners: Multispectral imaging (e.g. Hernandez and Armstrong 2016; Huang et al. 2018) captures images in usually three to fifteen bands (e.g. Landsat, ASTER) in 70–400 nm large bandwidths, whereas hyperspectral images are usually composed of ~100–200 spectral bands in relatively narrow 5–10 nm bandwidth. The advantage of multispectral and hyperspectral imaging are discrimination or identification of certain minerals, soils, vegetation, or other materials like rock and soil alterations, oil seeps, and lithologies. Structural geological and geomorphological interpretation can be done on such images. See Peyghambari and Zhang (2021) for reviews on spectral characteristics of hyperspectral systems.

iv) Lidar: Light detection and ranging (or Lidar) is a high-resolution altimetry technique that uses lasers of ultraviolet (UV), visible, near-infrared (NIR) bandwidths to measure the distance between a surface and the satellite. The Lidar method, much like radar altimetry, can be used to map surface topography. In addition, they can be used to scan outcrops and measure bathymetry, snow cover, etc. Lidar instruments are placed on aircrafts (e.g. aeroplanes, helicopters, and UAVs) and acquired "on demand" in study areas, according to the defined problem. Structural geological and geomorphological features can be well mapped on Lidar data.

v) Potential Fields Systems (gravity and magnetics instruments): Satellite-borne (e.g. ESA's GOCE satellite) or airborne gravity measurements are commonly performed in mineral and hydrocarbon exploration to map anomalies. Satellite-borne potential field data can be of a regional scale (100s–1000s of km), whereas airborne gravity gradiometry data can reach resolutions of a few tens of meters. Satellite gravity data are used to study crustal structure of the Earth and tectonic features, e.g. subduction zones and mobile belts. Airborne high-resolution gravity data can be used to map large faults, volcanic plugs, mineral zones, etc. Magnetometers are placed on spacecraft that measure variations in the strength and direction of the Earth's magnetic field. Such data, usually of regional nature, can be used to map ferrous mineralization zones, volcanic plugs, faults, etc. (detail in Prost 2013).

2.6 Significance in Geomorphology and Structural Geology

Appropriate remote sensing data are very important in augmenting geomorphological and structural geological studies. Such data can identify complicated zones of deformation or fill gaps in inaccessible areas or identify erosion and deposition zones, badlands, and depositional geometries in a synoptic overview at a very low cost of analysis. However, remote sensing can never replace field-based analyses like ground observations, sampling, and measurements. Remote sensing can provide three major advantages: (i) observation of the surface from a different perspective, in a different spectral region or on a different scale; (ii) reduce the area covered in a field work; and (iii) proving information about the region of interest before undertaking the field work. Chapter 1 details the advantages and limitations of remote sensing.

The types of remote sensing data, like those mentioned in the earlier sections in this chapter, are varied and the data to be used in a particular study will depend on the problem definition. The particular geologic analysis to be performed in an area, e.g. identification of features or processes/phenomena, or land use patterns, will guide the type of data that will strengthen the analysis. For example, mapping large-scale crustal deformation potential should use field data. However, to study smaller-scale faulting, images in the EM range are optimal. Resolution is an important factor to be considered. Spatial, spectral, and temporal resolutions are the most important aspects to be covered with respect to the problem in question. Sometimes, vegetation or changes in drainage pattern may be good indicators of structural geological (e.g. Gold 1980) and geomorphological features underneath. In such cases, near infrared or thermal infrared data, or in some cases derived products, e.g. Normalized Difference Vegetation Index (NDVI) images, become ideal for an augmented interpretation. In visible range data, care must be taken to avoid images with remarkable noise due to significant atmospheric and meteorological conditions, e.g. cloud cover and haze.

i) Geomorphology: Geomorphology is the study of landforms (Scheidegger 2012). Landform depends on three major factors: (i) climate and change in climate through time; (ii) structure and lithology of the underlying bedrock; and (iii) the time period involved. Geomorphology is the most widely used application of remote sensing because remote sensing data give a regional view and also directly constrain the surface features and processes. A large number of landforms, e.g. landslides, fluvial systems, glacial systems, volcanic systems, and marine systems are widely studied using remote sensing data.
ii) Structural geology: The deformation nature of an area is the most important factor to influence the surface of an area. Structure controls erosion and denudation of an area's topography, which when observed gives the idea of the deformation process the area has undergone or is presently undergoing. Most commonly, planar discontinuities in the rocks identified by prominent markers observable on remote sensing images are the most evident structures analyzed. However, in some cases deformation is manifested by topography, slope, color, soil, vegetation, etc. In such cases, appropriate remote sensing products can provide insight. Folds and faults can be identified in many cases from the satellite images. Bedding, intrusives, unconformities, etc. can also be inferred from satellite images of appropriate resolution (e.g. Dasgupta et al. 2022; Dasgupta and Mukherjee 2017, 2019; Gogoi et al. 2022; Kaplay et al. 2017; Misra et al. 2014). Morphototectonic analysis is one of the most important analyses for which remote sensing data forms a major part of the dataset required. Usually, topography DEMs and remote sensing imageries in visual to NIR bands of the EM spectrum are used and supported with fieldwork for a morphotectonic study (e.g. Joshi et al. 2013; Radaideh et al. 2016; Pandey and Sharma 2021; Dasgupta et al. submitted).

Acknowledgements

Reviewed by Soumyajit Mukherjee. Summarized in Misra and Mukherjee (2022). Wiley Blackwell Acquisition Editors (present: Frank Weinreich, past: Andrew Harrison), Stacey Woods (Handling Editor) and proofreading team.

References

Abdelkareem, M., Gaber, A., Abdalla, F. et al. (2020). Use of optical and radar remote sensing satellites for identifying and monitoring active/inactive landforms in the driest desert in Saudi Arabia. *Geomorphology* 362: 107197.

Amrullah, C., Suwardhi, D., and Meilano, I. (2016). Product accuracy effect of oblique and vertical non-metric digital camera utilization in UAV-photogrammetry to determine fault plane. *ISPRS Annals of the Photogrammetry, Remote Sensing and Spatial Information Sciences* 3: 41.

Balz, T. and Rocca, F. (2020). Reproducibility and replicability in SAR remote sensing. *IEEE Journal of Selected Topics in Applied Earth Observations and Remote Sensing* 13: 3834–3843.

Bhardwaj, A., Sam, L., Martín-Torres, F.J. et al. (2016). UAVs as remote sensing platform in glaciology: present applications and future prospects. *Remote Sensing of Environment* 175: 196–204.

Casagli, N., Cigna, F., Bianchini, S. et al. (2016). Landslide mapping and monitoring by using radar and optical remote sensing: examples from the EC-FP7 project SAFER. *Remote Sensing Applications: Society and Environment* 4: 92–108.

Chuvieco, E. (2020). *Fundamentals of Satellite Remote Sensing: An Environmental Approach*, 432. Boca Raton: CRC press. ISBN: 9780429506482.

Dasgupta, S. and Mukherjee, S. (2017). Brittle shear tectonics in a narrow continental rift: asymmetric non-volcanic Barmer basin (Rajasthan, India). *The Journal of Geology* 125: 561–591.

Dasgupta, S. and Mukherjee, S. (2019). Remote sensing in lineament identification: examples from western India. In: *Problems and Solutions in Structural Geology and Tectonics.*, Developments in Structural Geology and Tectonics Book Series, vol. 5. (Series Editor: Mukherjee, S.) (ed. A. Billi and A. Fagereng), 205–221. Elsevier. ISSN: 2542–9000. ISBN: 9780128140482.

Dasgupta, S., Biswas, M., Mukherjee, S. et al. (2022). Depositional system, morphological signatures, tectonics and sedimentation pattern along the transform margin – Palar-Pennar basin, Indian east coast. *Journal of Petroleum Science and Engineering* https://doi.org/10.1016/j.petrol.2022.110155.

Dong, J., Crow, W.T., Tobin, K.J. et al. (2020). Comparison of microwave remote sensing and land surface modeling for surface soil moisture climatology estimation. *Remote Sensing of Environment* 242: 111756.

Elachi, C. and Van Zyl, J.J. (2021). *Introduction to the Physics and Techniques of Remote Sensing*. Wiley.

Ferreira, M.A., Andrade, F., Mendes, R.N. et al. (2012). Use of satellite remote sensing for coastal conservation in the eastern African coast: advantages and shortcomings. *European Journal of Remote Sensing* 45 (1): 293–304.

Fingas, M. and Brown, C. (2014). Review of oil spill remote sensing. *Marine Pollution Bulletin* 83 (1): 9–23.

Gogoi, M.P., Gogoi, B., and Mukherjee, S. (2022). Tectonic instability of the petroliferous upper Assam valley (NE India): a geomorphic approach. *Journal of Earth System Science* 131: 18.

Gold, D.P. (1980). Structural geology. In: *Remote Sensing in Geology* (ed. B.S. Siegal and A.R. Gillespie), 419–483. New York: Wiley.

Gupta, R.P. (2017). *Remote Sensing Geology*, 3e, 671. New York: Springer. ISBN: 9783642077418.

Hernandez, W.J. and Armstrong, R.A. (2016). Deriving bathymetry from multispectral remote sensing data. *Journal of Marine Science and Engineering* 4 (1): 8.

Hossain, M.S., Khan Md, S.H., Abdullah, R. et al. (2021). Late Cenozoic transpression at the plate boundary: kinematics of the eastern segment of the Dauki Fault Zone (Bangladesh), and tectonic evolution of the petroliferous NE Bengal basin. *Marine and Petroleum Geology* 131: 106133.

Huang, B., Zhao, B., and Song, Y. (2018). Urban land-use mapping using a deep convolutional neural network with high spatial resolution multispectral remote sensing imagery. *Remote Sensing of Environment* 214: 73–86.

Ide, R., Nakaji, T., Motohka, T. et al. (2011). Advantages of visible-band spectral remote sensing at both satellite and near-surface scales for monitoring the seasonal dynamics of GPP in a Japanese larch forest. *Journal of Agricultural Meteorology* 1104050093–1104050093.

Iizuka, K., Itoh, M., Shiodera, S. et al. (2018). Advantages of unmanned aerial vehicle (UAV) photogrammetry for landscape analysis compared with satellite data: a case study of postmining sites in Indonesia. *Cogent Geoscience* 4 (1): 1498180.

Joshi, P.N., Maurya, D.M., and Chamyal, L.S. (2013). Morphotectonic segmentation and spatial variability of neotectonic activity along the Narmada–son fault, Western India: remote sensing and GIS analysis. *Geomorphology* 180: 292–306.

Jun, C., Jie, C., Jing, L. et al. (2021). Research on oblique aerial photography and its application in detection of hidden geological hazards. *IOP Conference Series: Earth and Environmental Science* 783 (1): 012124. IOP Publishing.

Kaplay, R.D., MD, B., Mukherjee, S. et al. (2017). Morphotectonic expression of geological structures in eastern part of south east Deccan volcanic province (around Nanded, Maharashtra, India). In: *Tectonics of the Deccan Large Igneous Province*, vol. 445 (ed. S. Mukherjee, A.A. Misra, G. Calvès and M. Nemčok), 317–335. Geological Society, London, Special Publications.

Kuenzer, C., Dech, S., and (2013). *Thermal Infrared Remote Sensing: Sensors, Methods, Applications*, vol. 17. Springer Science & Business Media.

Lai, H., Liu, L., Liu, X. et al. (2021). Unmanned aerial vehicle oblique photography-based superposed fold analysis of outcrops in the Xuhuai region, North China. *Geological Journal* 56 (4): 2212–2222.

Lillesand, T., Kiefer, R.W., and Chipman, J. (2015). *Remote Sensing and Image Interpretation*. Wiley.

Liu, D. and Xia, F. (2010). Assessing object-based classification: advantages and limitations. *Remote Sensing Letters* 1 (4): 187–194.

Misra, A.A. and Mukherjee, S. (2022). Introduction to "atlas of structural geological and geomorphological interpretation of remote sensing images". In: *Atlas of Structural Geological and Geomorphological*

Interpretation of Remote Sensing Images (ed. A.A. Misra and S. Mukherjee). Wiley Blackwell. ISBN: 9781119813354.

Misra, A.A., Bhattacharya, G., Mukherjee, S., and Bose, N. (2014). Near N–S paleo-extension in the western Deccan region, India: Does it link strike-slip tectonics with India–Seychelles rifting? *International Journal of Earth Sciences* 103 (6): 1645–1680.

Pandey, P. and Sharma, L.N. (2021). Morphotectonic studies of Ghaggar Basin in the northwestern frontal part of the Himalaya based on remote sensing and GIS techniques. *Journal of the Geological Society of India* 97 (1): 70–78.

Peyghambari, S. and Zhang, Y. (2021). Hyperspectral remote sensing in lithological mapping, mineral exploration, and environmental geology: an updated review. *Journal of Applied Remote Sensing* 15 (3): 031501.

Prakash, A. and Gupta, R.P. (1999). Surface fires in Jharia coalfield, India-their distribution and estimation of area and temperature from TM data. *International Journal of Remote Sensing* 20 (10): 1935–1946.

Prost, G.L. (2013). *Remote Sensing for Geoscientists*, 309–310. New York: CRC Press.

Radaideh, O.M., Grasemann, B., Melichar, R. et al. (2016). Detection and analysis of morphotectonic features utilizing satellite remote sensing and GIS: an example in SW Jordan. *Geomorphology* 275: 58–79.

Remondino, F. and Gerke, M. (2015). Oblique aerial imagery – a review. *Photogrammetric Week* 15 (12): 75–81.

Saadatseresht, M., Hashempour, A.H., and Hasanlou, M. (2015). UAV photogrammetry: a practical solution for challenging mapping projects. *The International Archives of Photogrammetry, Remote Sensing and Spatial Information Sciences* 40 (1): 619.

Saraf, A.K., Das, J.D., Agarwal, B. et al. (1996). False topography perception phenomena and its correction. *International Journal of Remote Sensing* 17 (18): 3725–3733.

Saraf, A.K., Zia, M., Das, J. et al. (2011). False topographic perception phenomena observed with the satellite images of Moon's surface. *International Journal of Remote Sensing* 32 (24): 9869–9877.

Scheidegger, A.E. (2012). *Theoretical Geomorphology*. Springer Science & Business Media.

Tatem, A.J., Goetz, S.J., and Hay, S.I. (2008). Fifty years of earth observation satellites: views from above have lead to countless advances on the ground in both scientific knowledge and daily life. *American Scientist* 96 (5): 390.

Thies, B. and Bendix, J. (2011). Satellite based remote sensing of weather and climate: recent achievements and future perspectives. *Meteorological Applications* 18 (3): 262–295.

Vosgerau, H., Passey, S.R., Svennevig, K. et al. (2016). Reservoir architectures of interlava systems: a 3D photogrammetrical study of Eocene cliff sections, Faroe Islands. *Geological Society, London, Special Publications* 436 (1): 55–73.

Warrick, J.A., Ritchie, A.C., Adelman, G. et al. (2017). New techniques to measure cliff change from historical oblique aerial photographs and structure-from-motion photogrammetry. *Journal of Coastal Research* 33 (1): 39–55.

Wu, W., Zou, L., Shen, X. et al. (2012). Thermal infrared remote-sensing detection of thermal information associated with faults: a case study in Western Sichuan Basin, China. *Journal of Asian Earth Sciences* 43 (1): 110–117.

Yang, J., Gong, P., Fu, R. et al. (2013). The role of satellite remote sensing in climate change studies. *Nature Climate Change* 3 (10): 875–883.

Yao, H., Qin, R., and Chen, X. (2019). Unmanned aerial vehicle for remote sensing applications – a review. *Remote Sensing* 11 (12): 1443.

Zhang, X. and Muhammad, K. (2020). Adaptive enhancement method for multimode remote sensing image based on LiDAR. *Mobile Networks and Applications* 25 (6): 2390–2397.

Zhang, J., Wagner, W., Prakash, A. et al. (2004). Detecting coal fires using remote sensing techniques. *International Journal of Remote Sensing* 25 (16): 3193–3220.

Websites

Professional Aerial Photographers Association. (2021). https://professionalaerialphotographers.com (accessed 10 October 2021).

Section B

Geomorphology

3

Geodynamic Quantification of Mid-Channel Bar Morphology

A Spatio-Temporal Study

*Suraj Gupta and Mery Biswas**

Department of Geography, Presidency University, Kolkata, West Bengal, India

3.1 Introduction

Remote sensing has been a powerful tool to interpret structural geology and geomorphology of terrains (e.g. Misra et al. 2014; Kaplay et al. 2017; Dasgupta and Mukherjee 2017, 2019; Dasgupta et al. 2022; Gogoi et al. 2022). Mid-channel bars are common fluvio-geomorphological features of NE Himalayan alluvial channels. Formation of mid-channel bar is a natural process and their morphodynamics (Sosnowska,= 2020; Tholibon et al. 2016) are interlinked with seasonal flash floods (Paul and Biswas 2019), channel width, channel depth incoming sediment, volume of water discharge, flow velocity, stream power, specific stream power (Adami et al. 2016; Islam and Chandra Das 2015; Lou et al. 2018; Li et al. 2019; Wen et al. 2020), bed roughness, and drop of bed slope with change of slope, etc., in the Jaldhaka River after and before the Nagrakata bridge (26°53′50.49″N/88°53′41.38″E in DMS).

The geomorphological investigations of the active channel shift and change of seasonal active channel are commonly noted near the Nagrakata Bridge of the Jaldhaka River, Jalpaiguri District, West Bengal, India. Therefore, the origin of mid-channel bars and shifting of flow lobes change from season to season. Mean stream power and specific stream power from 2011–2020 vary±11 kW m^{-1} and ± 40 kW m^{-2} (August to January, primary data 2018). Diversion of river in bed channels, channel diversion, and aggradation are associated with water sediment flow of the river Jaldhakaas, which is one of the most dynamic and has the most sediment potential of rivers in North Bengal foothills. Such a geomorphic unit is well quantified and analyzed using satellite data such as Landsat 8 TM data on a spatio-temporal scale. Quantification was performed using the extracted temporal scale data satellite images of three respective years: 2011, 2016, and 2020. Such analysis was verified during field observations (Dec 2018), photo evidence, and primary level survey with the help of Total Station (Figure 3.1d). Sarma and Acharjee (2018) proposed simple indices to infer the braiding characteristics of the river, which are the sum of braid bars and islands of the channel count to the total considered reach of river. There has not been an assessment of the sampling requirements for any of these indices and there has been no systematic study of the equivalence of the indices to each other and their sensitivity to river stage (Khan et al. 2021). Resolution of these issues is essential for progress in studies of braided morphology and dynamics at the scale of the channel network. Duro et al. (2016) and Ghinassi et al. (2021) presented a work that aimed to characterize the channel bar morphology of such braided streams, in order to assess whether the system reaches a steady state under constant flow conditions.

3.1.1 Study Area

For analyzing the braiding pattern, the Jaldhaka River is selected. This river has a total length of 192 km and originates from the Kupup or Bitang Lake in southeastern Sikkim in the eastern Himalayas and flows through Bhutan and the Kalimpong, Jalpaiguri, and Cooch Behar districts of West Bengal, India. However, a small stretch of the Jaldhaka River has been selected for studying the mid-channel bar morphology and braiding pattern with its temporal dynamics. The starting point of the selected stretch has the geographical coordinates (in decimal degrees) of 26.898995°N and 88.894294° E and the stretch ends at 26.175257°N and 88.29485° E (Figure 3.1a).

*Corresponding Author: merybiswas@gmail.com

Atlas of Structural Geological and Geomorphological Interpretation of Remote Sensing Images, First Edition.
Edited by Achyuta Ayan Misra and Soumyajit Mukherjee.

Companion website: www.wiley.com/go/misra/RemoteSensingImages

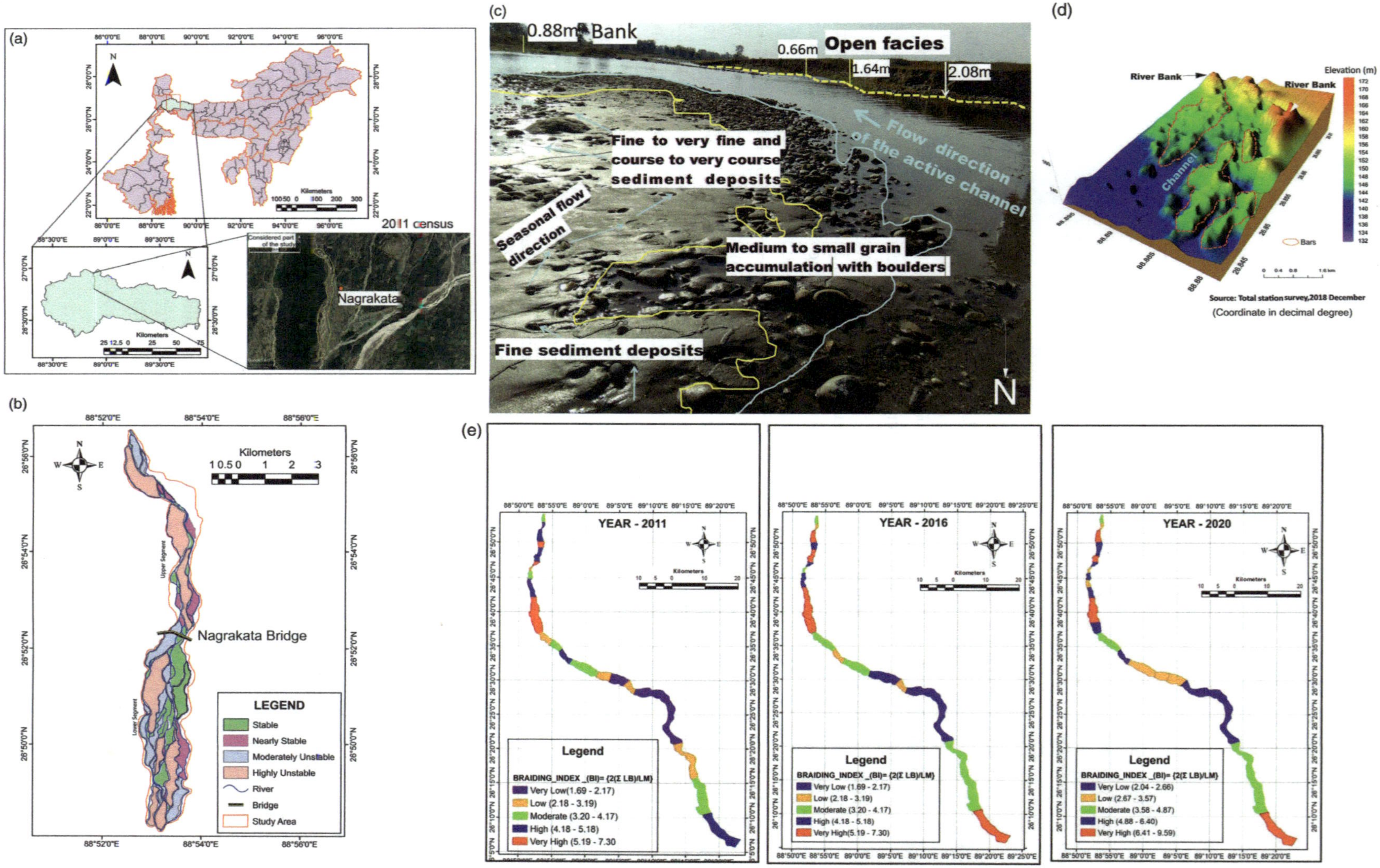

Figure 3.1 (a) Location of the study area with inset of part of Jaldhaka River under consideration (Source: Image courtesy: Google Earth). (b) Classification of stable and unstable bars, based on remote sensing data analysis of 2011, 2016, and 2020. (c) Selective horizontal section of riverbed deposition to observe the sediment aggradation layers in the last two to three years. (d) 3D representation of Total Station data survey of the Jaldhaka riverbed downstream of Nagrakata Bridge, 26.870885°N/88.893267° (Decimal degrees). (e) Comparative representation of braiding index from 2011, 2016, and 2020 with detail classification braiding pattern with magnitude levels.

3.1.2 Methodology

There are two segments of mid-channel bar analysis of Jaldhaka river: one is upstream 18.5 km from the Nagrakata bridge and the other is 17.1 km downstream from the Nagrakata bridge. The quantification process is mainly based on satellite data analysis, which is accompanied by field verification and photographic explanations. The considered study area is divided into two segments: segment 1, upstream from the Nagrakata bridge; and segment 2, downstream from the bridge, respectively. The spatio-temporal analysis is conducted based on a comparative study of 2011, 2016, and 2021. The entire study is coupled up in three phases as satellite image analysis, primary level survey, and photographic analysis. Collectively, the documentation involves application of satellite-based data analysis on a spatio-temporal scale to determine the mid-channel bar morphology. Methodologies are listed in the figure here.

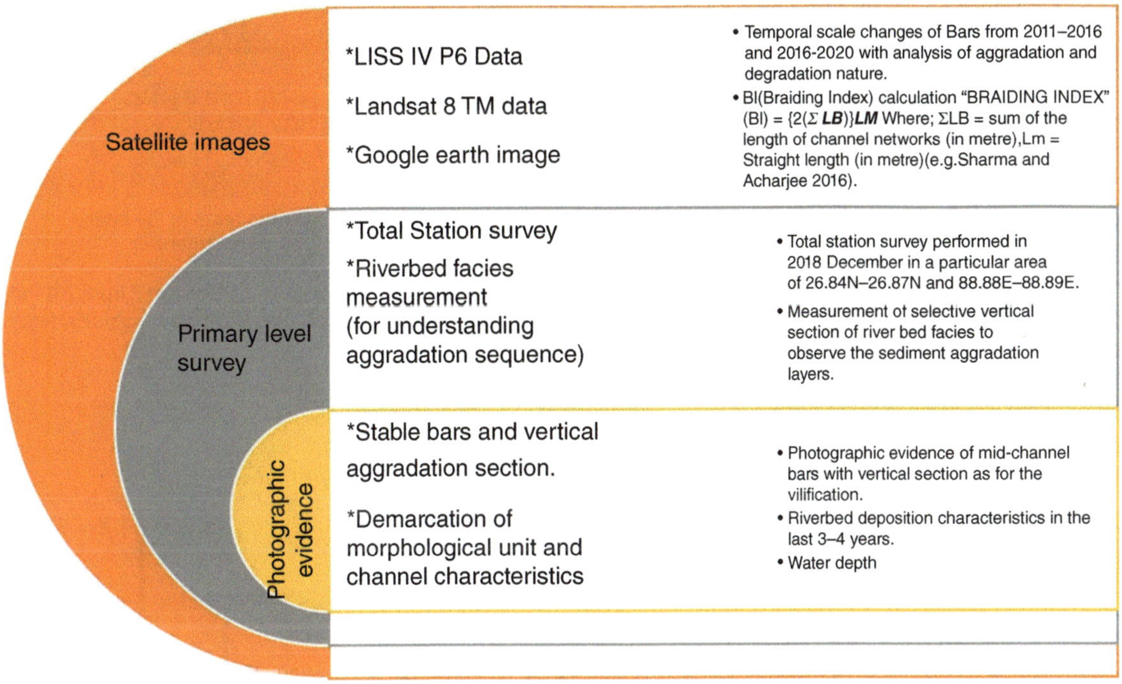

3.2 Discussions

Comparing the upstream and downstream of the bridge, if we analyze the graphical representation, there is a remarkable bifurcation of channels including active channels, seasonally active channels, etc. The 2011, 2016, and 2020 data regarding the channel orientation and intermediate bar dynamics clearly show that the channels are highly bifurcated after the bridge (26°53′30.45"N/88°53′35.53"E) (Coordinate in DMS), reframing the braiding pattern of drainage and stability of bars have increased (Figure 3.1b, d). Firstly, the autogenic causes that signify the slope change in downstream with specific flood tendency. Second, the allogenic causes, e.g. the construction of bridges disturbed the flow and sediment regime of the river, resulting in the channel bifurcation and aggradation process over time (Figure 3.2). From 2011 to 2020, the number of bifurcated channels has increased and seasonal variations every year have induced the sediment aggradation process (Figure 3.3a) to be higher downstream than upstream. To validate the aforementioned illustration of channels orientation, the length to width ratio of up and downstream have been framed, which show that the number of channels is higher downstream than upstream (Figure 3.3b).

The significance of road/rail/bridge construction across the river also plays an important role in mid-channel bar formation. Such construction permanently squeezes up the flow paths and rivers are forced to flow through narrower passages (Biswas and Banerjee 2018). As a result, the river energy increases, along with flow velocity and sediment particle size. This is also accompanied with bridge and river erosion, and increase of river depth just below the flood plain, having a long-term effect on hydrology and sediment volume including the vertical sequence of accumulation with seasonal variation (Biswas and Dhara 2019).

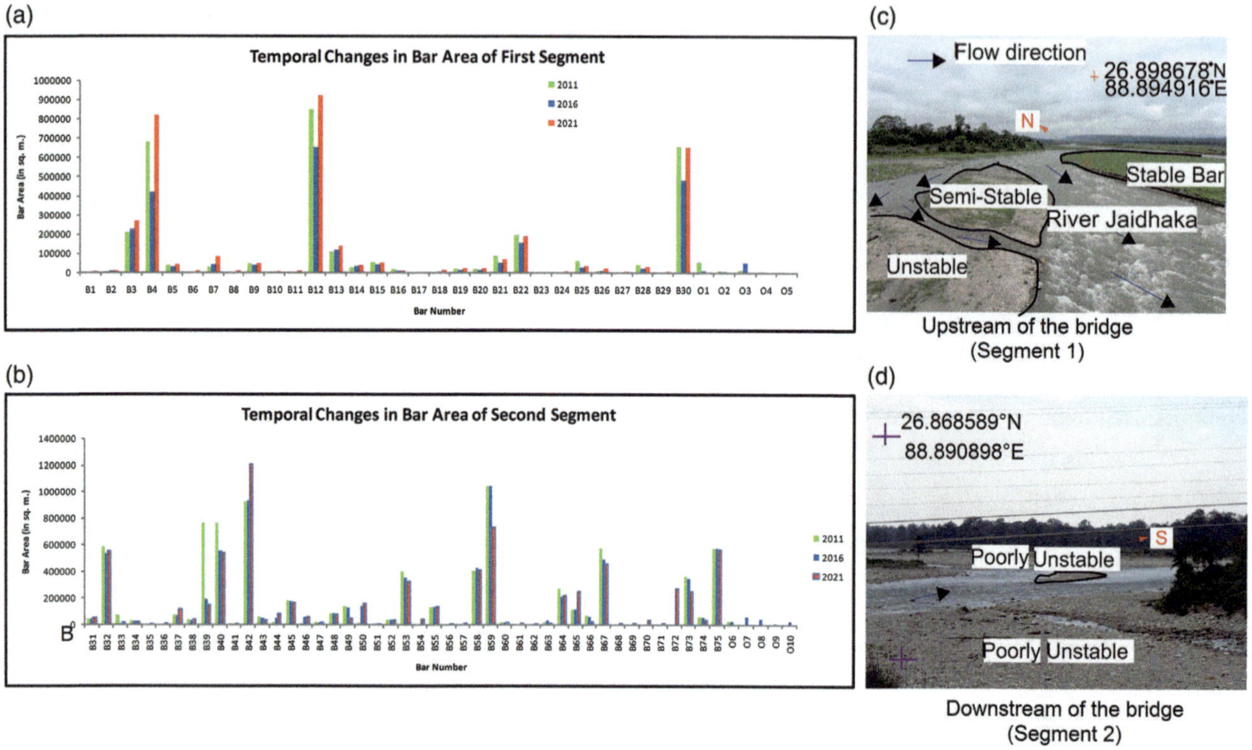

Figure 3.2 (a) Temporal changes of bar area of segment 1 from 2011 to 2020. (b) Temporal changes of bar area of segment 2 from 2011 to 2020. (c) Photographic evidence of stable, semi-stable, and unstable bars upstream from the Nagrakata Bridge. (d) Photographic evidence of stable, poorly unstable bars in the immediate downstream of Nagrakata bridge.

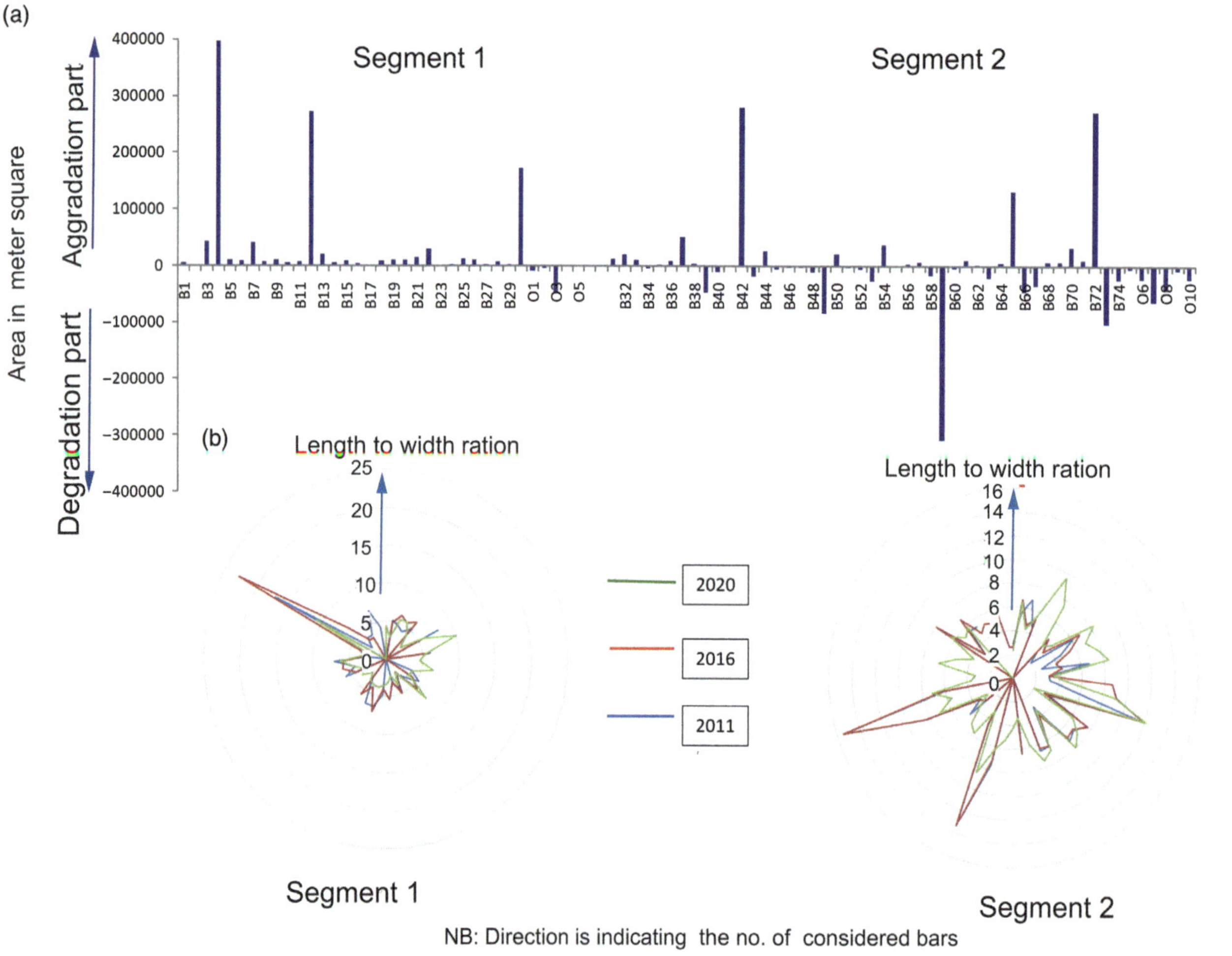

Figure 3.3 (a) Aggradation and degradation sequence of two respective segments in positive and negative manner. (b) Changes in length to width ratio of considered bars in 2011, 2016, and 2020, in two respective segments.

This explanation has been validated using types of trend line analysis: a linear curve fit of radius of curvature of bar lobes in both up and downstream. The decreasing value of R^2 from 2011 to 2020 from 0.038–0.023 to 0.0026 indicate the flattering and widening nature of bar lobes with time that is also accompanied with low gradient and increasing sediment aggradation. The hydro-morphological parameters, both autogenic and allogenic, are responsible for the changing nature of regime, which has become a cause of increased deposition downstream of the bridge, and sudden release of energy has increased the deposition rate with resulting landforms like mid-channel bars (Figure 3.1c, d).

Acknowledgements

Thanks to the Department of Geography, Presidency University (Kolkata) for using the laboratory instruments during the survey. Soumyajit Mukherjee invited to submit an article in the edited book. Achyuta Ayan Misra handled this article. Wiley Blackwell Acquisition Editors (present: Frank Weinreich, past: Andrew Harrison), Stacey Woods (Handling Editor), and proofreading team. Summarized in Misra and Mukherjee (2022).

Appendix

Satellite Data series: Landsat 8 TM data, Spatial resolution: 30 m (visible, NIR, SWIR), Coordinate System: Geographic, WGS84, Sun Angle = 30°, Number of bands: 11.

References

Adami, L., Bertoldi, W., and Zolezzi, G. (2016). Multidecadal dynamics of alternate bars in the alpine Rhine River. *Water Resources Research* 52 (11): 8938–8955.

Biswas, M. and Banerjee, P. (2018). Bridge construction and river channel morphology – a comprehensive study of flow behaviour and sediment size alteration of the river Chel, India. *Arabian Journal of Geosciences* 11: https://doi.org/10.1007/s12517-018-3789-7.

Biswas, M. and Dhara, P. (2019). Correction to: evolutionary characteristics of meander cut-off—a hydro-morphological study of the Jalangi River, West Bengal, India. *Arabian Journal of Geosciences* 12: 739. https://doi.org/10.1007/s12517-019-4971-2.

Dasgupta, S. and Mukherjee, S. (2017). Brittle shear tectonics in a narrow continental rift: asymmetric non-volcanic Barmer basin (Rajasthan, India). *The Journal of Geology* 125: 561–591.

Dasgupta, S. and Mukherjee, S. (2019). Remote sensing in lineament identification: examples from western India. In: *Problems and Solutions in Structural Geology and Tectonics*, Developments in Structural Geology and Tectonics Book Series, vol. 5. Series Editor: Mukherjee, S. (ed. A. Billi and A. Fagereng), 205–221. Elsevier. ISSN: 2542-9000. ISBN: 9780128140482.

Dasgupta, S., Biswas, M., Mukherjee, S., and Chatterjee, R. (2022). Depositional system, morphological signatures, tectonics and sedimentation pattern along the transform margin-Palar-Pennar basin, Indian east coast. *Journal of Petroleum Science & Engineering* https://doi.org/10.1016/j.petrol.2022.110155.

Duro, G., Crosato, A., and Tassi, P. (2016). Numerical study on river bar response to spatial variations of channel width. *Advances in Water Resources* 93: 21–38.

Ghinassi, M., Ielpi, A., Ventra, D. et al. (2021). Distinguishing mid-channel and bank-attached fluvial bars by flow divergence: Implications for the interpretation of stratigraphic records. *Sedimentology* 68 (6): 2783–2797.

Gogoi, M.P., Gogoi, B., and Mukherjee, S. (2022). Tectonic instability of the petroliferous upper Assam valley (NE India): a geomorphic approach. *Journal of Earth System Science* 131: 18.

Islam, A. and Chandra Das, B. (2015). Quantitative indices to measure unit channel bar location: a theoretical and empirical study. *Ethiopian Journal of Environmental Studies and Management* 8 (6): 628–634.

Kaplay RD, Babar Md, Mukherjee S, Kumar, TV. 2017. Morphotectonic expression of geological structures in eastern part of south east Deccan volcanic province (around Nanded, Maharashtra, India) In: Mukherjee S, Misra AA, Calvès G, Nemčok M. (Eds.) *Tectonics of the Deccan Large Igneous Province*. Geological Society, London, Special Publications 445, 317–335.

Khan, M.A., Sharma, N., Pu, J.H. et al. (2021). Experimental observation of turbulent structure at region surrounding the mid-channel braid bar. *Marine Georesources & Geotechnology* 40 (4): 448–461. https://doi.org/10.1080/1064119X.2021.1906366.

Li, D., Lu, X.X., Chen, L., and Wasson, R.J. (2019). Downstream geomorphic impact of the three gorges dam: with special reference to the channel bars in the middle Yangtze River. *Earth Surface Processes and Landforms* 44 (13): 2660–2670.

Lou, Y., Mei, X., Dai, Z. et al. (2018). Evolution of the mid-channel bars in the middle and lower reaches of the Changjiang (Yangtze) river from 1989 to 2014 based on the Landsat satellite images: impact of the three gorges dam. *Environmental Earth Sciences* 77 (10): 1–18.

Misra, A.A. and Mukherjee, S. (2022). Introduction to "atlas of structural geological and geomorphological interpretation of remote sensing images". In: *Atlas of Structural Geological and Geomorphological Interpretation of Remote Sensing Images* (ed. A.A. Misra and S. Mukherjee). Wiley Blackwell. ISBN: 9781119813354.

Misra, A.A., Bhattacharya, G., Mukherjee, S., and Bose, N. (2014). Near N-S paleo-extension in the western Deccan region in India: does it link strike-slip tectonics with India-Seychelles rifting? *International Journal of Earth Sciences* 103: 1645–1680.

Paul, A. and Biswas, M. (2019). Changes in river bed terrain and its impact on flood propagation – acase study of river Jayanti, West Bengal, India. *Geomatics, Natural Hazards and Risk* 10 (1): 1928–1947.

Sarma, J. and Acharjee, S. (2018). A study on variation in channel width and braiding intensity of the Brahmaputra River in Assam, India. *Geosciences (Switzerland)* 8: 1–19. https://doi.org/10.3390/geosciences8090343.

Sosnowska, A. (2020). Dynamics of mid-channel bars in the middle Vistula River in response to ferry crossing abutment construction. *Open Geosciences* 12 (1): 290–298.

Tholibon, D.A., Ariffin, J., Abdullah, J., and Muhamad, N.S. (2016). Centerline Bed Elevation Profile of Sand Bed Channel due to Bar Formation. *IOP Conference Series: Materials Science and Engineering* 136 (1): 012066. IOP Publishing.

Wen, Z., Yang, H., Zhang, C. et al. (2020). Remotely sensed mid-channel bar dynamics in downstream of the three gorges dam, China. *Remote Sensing* 12 (3): 409.

4

Geomorphic Indicators of Glacier Retreat from Jorya Garang Glacier of Baspa Valley, Himachal Pradesh, India

Bhushan S. Deota[1], Yogi N. Trivedi[2], Ishmohan Bahuguna[3], Mudit D. Mankad[4], and Chinmay U. Dongare[*,1]

[1] Department of Geology, The M.S. University of Baroda, Vadodara, Gujarat, India
[2] Amnex Infotechnologies, Ahmedabad, Gujarat, India
[3] Space Application Centre, Ahmedabad, Gujarat, India
[4] Department of Geography, The M.S. University of Baroda, Vadodara, Gujarat, India

4.1 Introduction

Remote sensing has been a powerful tool to interpret structural geology and geomorphology of terrains (Misra et al. 2014; Dasgupta and Mukherjee 2017, 2019; Kaplay et al. 2017; Dasgupta et al. 2022; Gogoi et al. 2022). A glacier is a mass of ice formed by recrystallization of snow, which flowed at some time in the past, under the influence of gravity (Leet and Judson 1969). The Indian Himalayas, popularly known as the third pole, is the world's third largest glacier system after Antarctica and Greenland (Anthwal et al. 2006). However, since 1850 the Himalayan glacier system has been experiencing recession (Mayewski and Jeschke 1979) and the rate of recession has increased in recent times due to climate change (Jangpang and Vohra 1962; Leet and Judson 1969; Kurien and Munshi 1972; Srikantia and Padhi 1972; Vohra 1981; Kulkarni and Alex 2003). Glaciers have a unique way of eroding, transporting, and depositing earth materials, which give rise to characteristic landforms of erosional and depositional nature. These landforms significantly reflect the processes responsible for their carving and also act as indicators of climate change. A notable characteristic of modern glaciers is that they follow same general pattern of growth and wastage around the globe and serve as excellent indicators of paleo-glacial activities (Leet and Judson 1969).

4.2 Geomorphic Characteristics of the Jorya Garang Glacier

The characteristic geomorphic indicators of the Jorya Garang glacier, Baspa valley, Himachal Pradesh, India (Figures 4.1 and 4.2), viz. accumulation zone, ablation zone, snout, deglaciated valley, terminal moraine, lateral moraine, and medial moraine are mapped from IRS-P6, LISS-III satellite data (28 August 2005, resolution 23.5 m) (Figure 4.3) (Trivedi et al. 2007; Deota et al. 2011; Deota et al. 2018). The northeast facing accumulation zone of Jorya Garang glacier is significantly observed with cyan color and fine texture (Figure 4.3). Below the accumulation zone, the east–west trending blue colored and fine textured exposed ablation zone is clearly demarcated, while the moraine-covered ablation zone trending north represents with a pinkish brown color and medium texture (Figure 4.3). The snout of JoryaGarang marks the end of the ablation where the glacial system ends and fluvial system initiates (Figure 4.5b). It usually appears as a black color in images on account of the shadow of the ice-wall (Figures 4.3 and 4.4). Two stages of lateral moraines are inferred for the Jorya Garang glacier (Deota et al. 2006). Stage 1 lateral moraines are marked at the outer edge of the glacier and are longer compared to stage 2 lateral moraines (Figure 4.3). Stage 1 lateral moraines exhibit brown color on satellite images (Figure 4.3), generally

*Corresponding Author: chinmayd.avs@gmail.com

Atlas of Structural Geological and Geomorphological Interpretation of Remote Sensing Images, First Edition.
Edited by Achyuta Ayan Misra and Soumyajit Mukherjee.

Companion website: www.wiley.com/go/misra/RemoteSensingImages

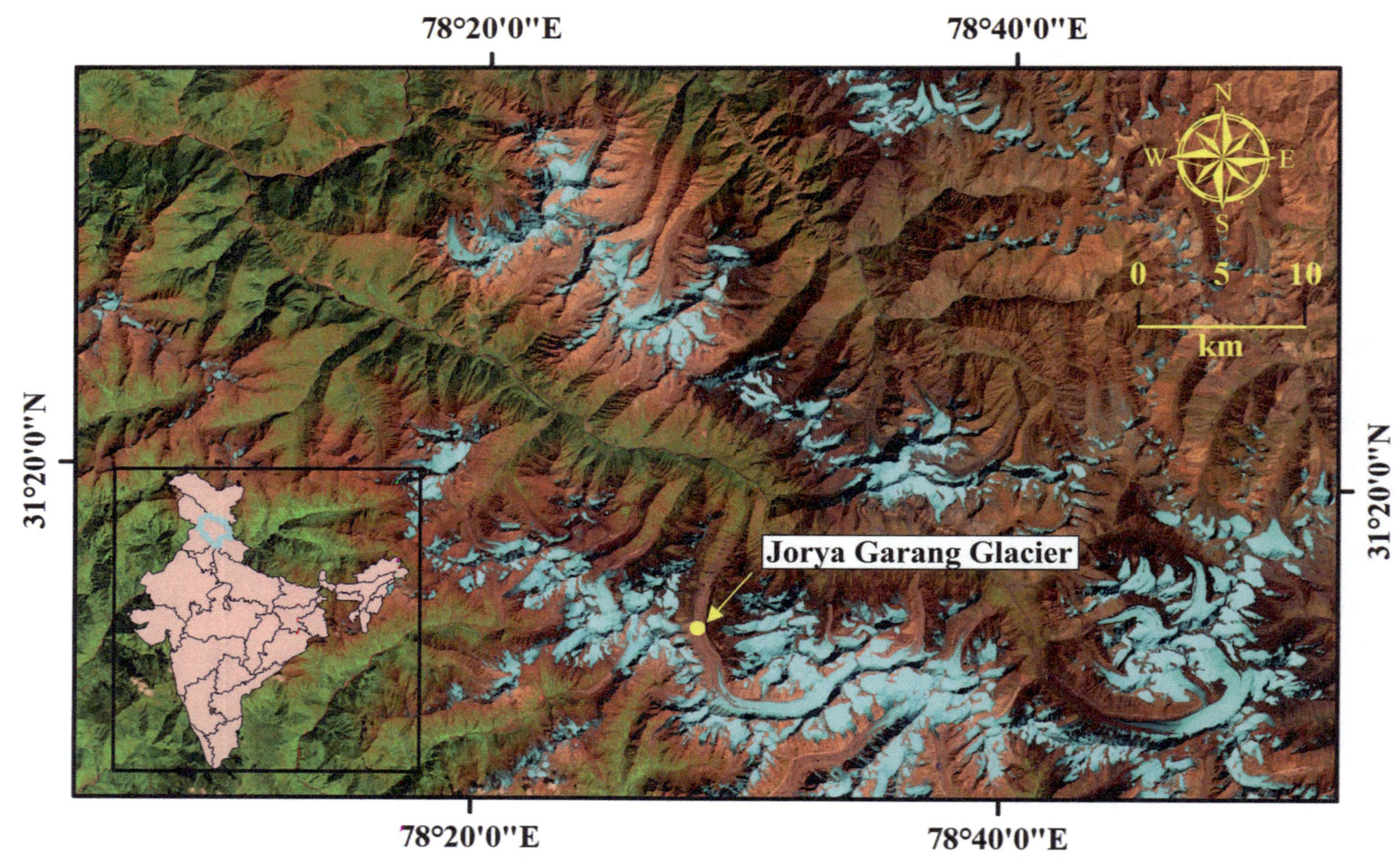

Figure 4.1 Location map of Jorya Garang Glacier. False Color Composite (FCC) (SWIR, NIR, and Green band) of Landsat-8 OLI-TIRS image, September 2017.

Figure 4.2 Uninterpreted False Color Composite (FCC) (SWIR, NIR, and Green band) of IRS-P6, LISS-III Satellite image – August 2005, Jorya Garang glacier. (Deota et al. 2018).

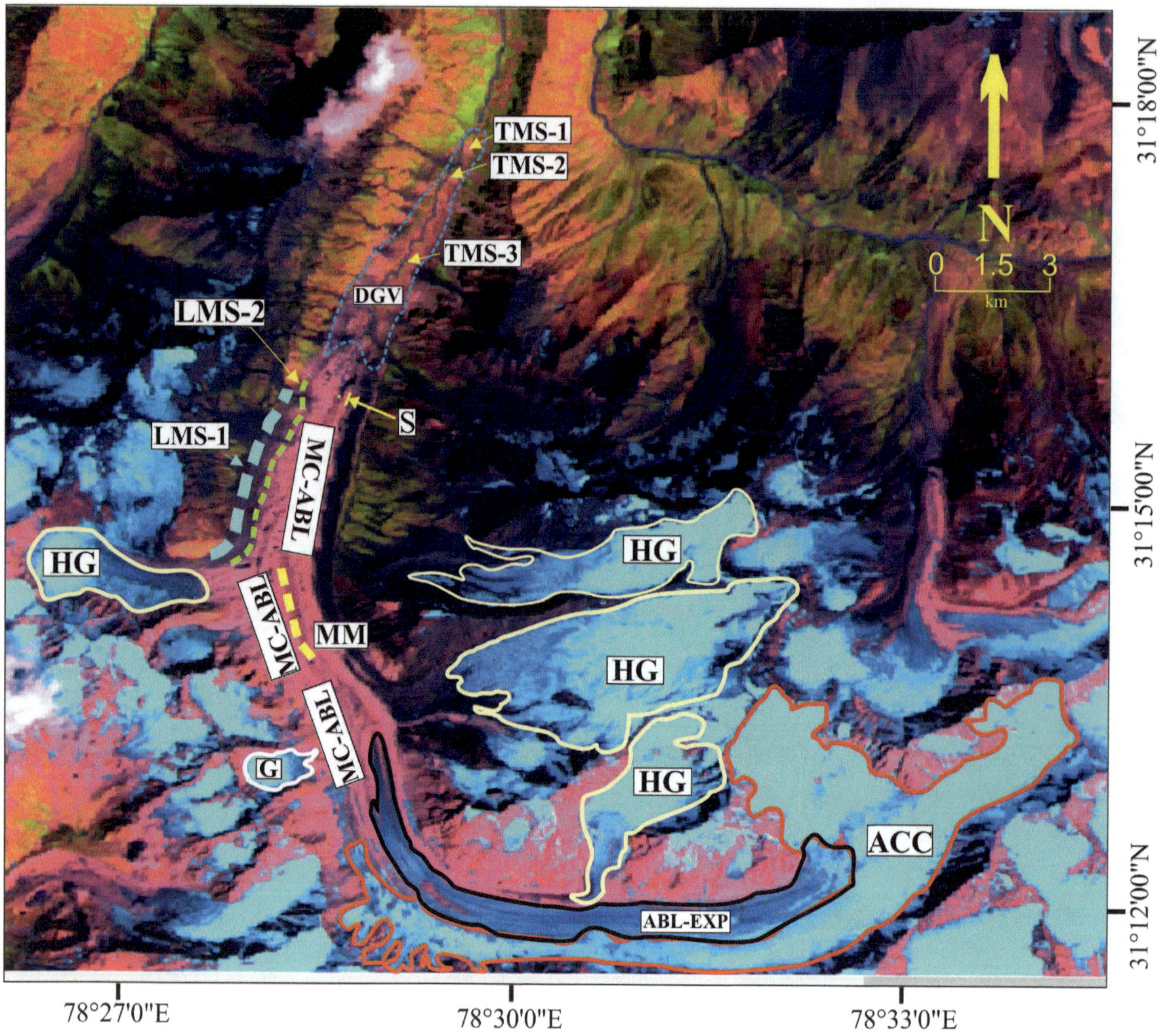

Figure 4.3 Interpreted geomorphic features of Jorya Garang Glacier from FCC (SWIR, NIR, and Green band) of IRS-P6, LISS-III satellite image – August 2005 (Deota et al. 2018). **ACC**: Accumulation Zone; **ABL-EXP**: Exposed Ablation Zone; **MC-ABL**: Moraine Covered Ablation Zone; **HG**: Hanging Glacier; **G**: Glacieret; **MM**: Medial Moraine; **LMS-1**: Lateral Moraine Stage-1; **LMS-2**: Lateral Moraine Stage-2; **TMS-1**: Terminal Moraine Stage-1; **TMS-2**: Terminal Moraine Stage-2; **TMS-3**: Terminal Moraine Stage-3; **S**: Snout; **DGV:** Deglaciated Valley.

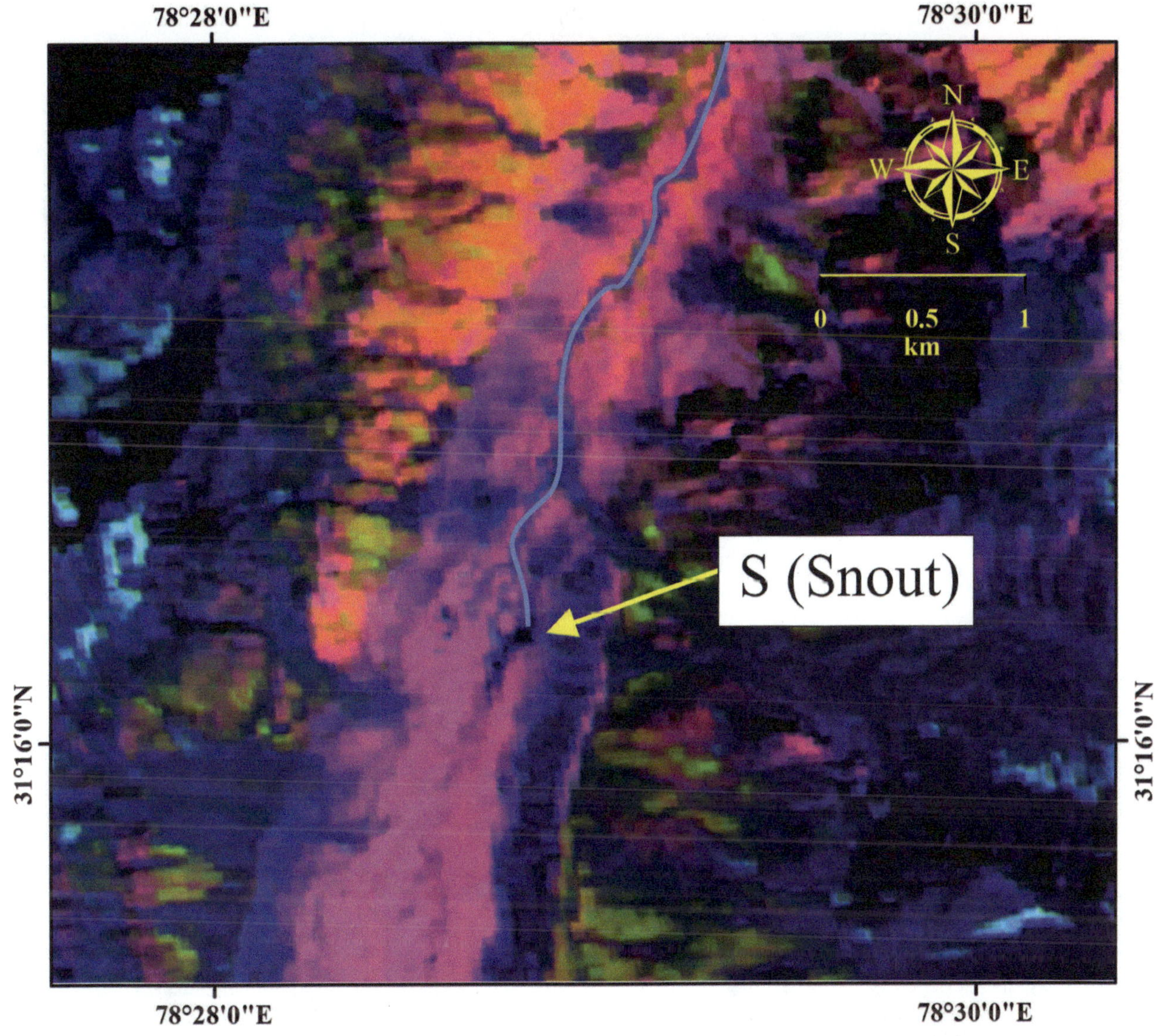

Figure 4.4 A closer look to Snout of Jorya Garang Glacier on FCC (SWIR, NIR, and Green band) of IRS-P6, LISS-III satellite image – August 2005.

Figure 4.5 (a) Photograph displaying Accumulation Zone, Moraine Covered Ablation Zone, and Lateral Moraines of Jorya Garang Glacier. (b) Photograph displaying Snout of Jorya Garang Glacier.

Figure 4.6 (a) Photograph displaying stage 1 Lateral Moraine of Jorya Garang Glacier. (b) Photograph displaying stage 1 Terminal Moraine of Jorya Garang Glacier.

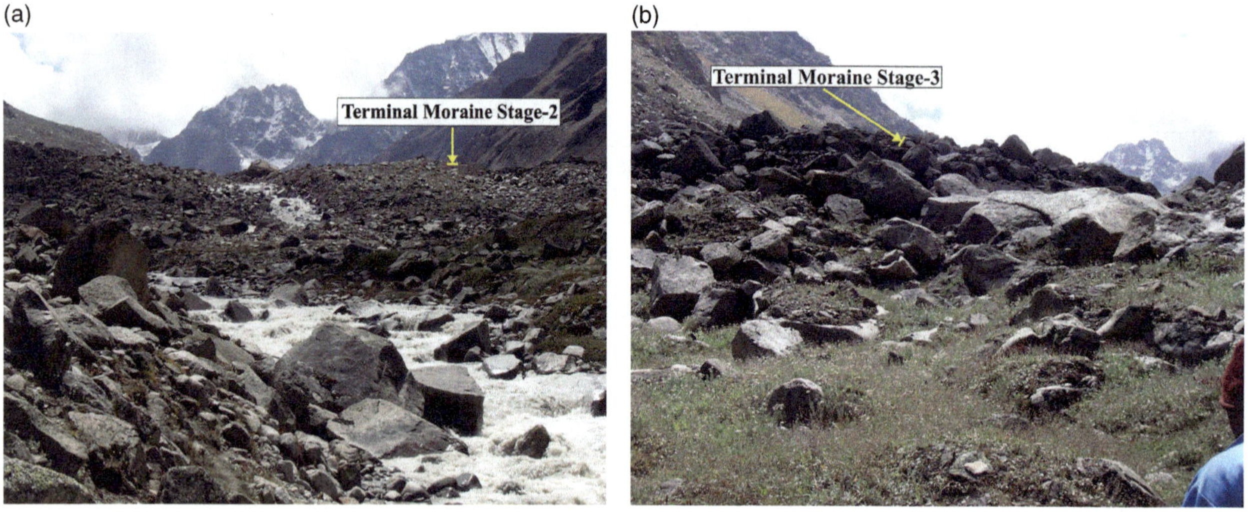

Figure 4.7 (a) Photograph displaying stage 2 Terminal Moraine of Jorya Garang Glacier. (b) Photograph displaying stage 3 Terminal Moraine of Jorya Garang Glacier.

possess vegetation, and are overlapped with glacio-fluvial fans and rockfalls (Figure 4.6a). Stage 2 lateral moraines exhibit pinkish-brown color on satellite images (Figure 4.3) and are composed of fresh angular boulders. A linear ridge-like feature, oriented in the northwest direction displaying pinkish-brown color and medium texture on satellite image, is identified as a medial moraine (Figure 4.3) (Deota et al. 2018), which represents the active glacier. Further, three stages of terminal moraines (Figures 4.6b, 4.7a, 4.7b), exhibiting crescent form, with convex side down the valley are inferred by pinkish red color and medium texture from satellite image (Figure 4.3), (Deota et al. 2018). Below the present-day ablation zone, the U-shaped long deglaciated valley is marked by pinkish orange color with medium texture (Figures 4.3 and 4.8). Four hanging glaciers and one glacieret nourished on the flanks of high mountains, flowing down to join the Jorya Garang glacier, are conspicuously seen on satellite data (Figure 4.3).

Remote sensing data has proved to be economical and reliable in obtaining information about the excessive rugged terrains with inclement weather conditions. IRS-P6, LISS-III data acquired between mid-August to the first week of October is preferred to map glacial geomorphic features, as glaciers are exposed to their maximum during this time period. The presence of two stages of lateral and three stages of terminal moraines suggest glacial retreat along its length and width during de-glaciation (Figure 4.3).

Figure 4.8 Photograph displaying Deglaciated Valley (DGV) of Jorya Garang Glacier.

Acknowledgements

Soumyajit Mukherjee invited submission of article. Handled by Achyuta Ayan Misra. Wiley Blackwell Acquisition Editors (present: Frank Weinreich, past: Andrew Harrison), Stacey Woods (Handling Editor), and proofreading team. Summarized in Misra and Mukherjee (2022).

References

Anthwal, A., Joshi, V., Sharma, A., and Anthwal, S. (2006). Retreat of Himalayan glaciers–indicator of climate change. *Nature and Science* 4 (4): 53–59.

Dasgupta, S. and Mukherjee, S. (2017). Brittle shear tectonics in a narrow continental rift: asymmetric non-volcanic Barmer basin (Rajasthan, India). *The Journal of Geology* 125: 561–591.

Dasgupta, S. and Mukherjee, S. (2019). Remote sensing in lineament identification: examples from western India. In: *Problems and Solutions in Structural Geology and Tectonics*, Developments in Structural Geology and Tectonics Book Series, vol. 5. Series Editor: Mukherjee, S. (ed. A. Billi and A. Fagereng), 205–221. Elsevier. ISSN: 2542-9000. ISBN: 9780128140482.

Dasgupta, S., Biswas, M., Mukherjee, S. et al. (2022). Depositional system, morphological signatures, tectonics and sedimentation pattern along the transform margin Palar-Pennar basin, Indian east coast. *Journal of Petroleum Science and Engineering* 211: 110155.

Deota, B.S., Trivedi, Y.N., Bahuguna, I.M., et al. (2006). Geomorphic records of paleoglacial activity from theupper reaches of Baspavalley, H.P. In: *Proceedings of SPIE 6412*, Disaster Forewarning Diagnostic Methods and Management, 64121I. https://doi.org/10.1117/12.700791.

Deota, B.S., Trivedi, Y.N., Kulkarni, A.V. et al. (2011). RS and GIS in mapping of geomorphic records and understanding the local controls of glacial retreat from the Baspa Valley, Himachal Pradesh, India. *Current Science* 1555–1563.

Deota, B.S., Trivedi, Y.N., Kulkarni, A.V., and Mankad, M.D. (2018). Reconstructing glacial history of JoryaGarang glacier from little ice age to present. *Journal of Geosciences* 3 (2): 163–170.

Gogoi, M.P., Gogoi, B., and Mukherjee, S. (2022). Tectonic instability of the petroliferous upper Assam valley (NE India): a geomorphic approach. *Journal of Earth System Science* 131: 18.

Jangpang, B.S. and Vohra, C.P. (1962). The retreat of the Skunkulpa (Ralam) glacier in the central Himalaya, Pithoragarh District, UP, India. *The International Association of Hydrological Sciences* 58: 234–238.

Kaplay, R.D., Md, B., Mukherjee, S., and Kumar, T.V. (2017). Morphotectonic expression of geological structures in eastern part of south east Deccan volcanic province (around Nanded, Maharashtra, India). In: *Tectonics of the Deccan Large Igneous Province*, vol. 445 (ed. S. Mukherjee, A.A. Misra, G. Calvès and M. Nemčok), 317–335. Geological Society, London, Special Publications.

Kulkarni, A.V. and Alex, S. (2003). Estimation of recent glacial variations in Baspa Basin using remote sensing technique. *Journal of the Indian Society of Remote Sensing* 31 (2): 81.

Kurien, M.N. and Munshi, M.N. (1972). A survey of Sonapani glacier, Lahaul District, Punjab. *Geological Survey of India*, Miscellaneous Publication 15: 83–88.

Leet, L.D. and Judson, S. (1969). *Physical Geology*, 3e, 404. Wiley.

Mayewski, P.A. and Jeschke, P.A. (1979). Himalayan and trans-Himalayan glacier fluctuations since AD 1812. *Arctic and Alpine Research* 11 (3): 267–287.

Misra, A.A. and Mukherjee, S. (2022). Introduction to "atlas of structural geological and geomorphological interpretation of remote sensing images". In: *Atlas of Structural Geological and Geomorphological Interpretation of Remote Sensing Images* (ed. A.A. Misra and S. Mukherjee). Wiley Blackwell.

Misra, A.A., Bhattacharya, G., Mukherjee, S., and Bose, N. (2014). Near N-S paleo-extension in the western Deccan region in India: does it link strike-slip tectonics with India-Seychelles rifting? *International Journal of Earth Sciences* 103: 1645–1680.

Srikantia, S.V. and Padhi, R.N. (1972). Recession of the Barashigri glacier, Lahaul-Spiti District, Punjab, Miscellaneous Publication. *Geological Survey of India* 15: 97–100.

Trivedi, Y.N., Deota, B.S., Rathore, B.P. et al. (2007). IRS images for glacial geomorphological studies of Baspa Valley. *Journal of Indian Geomorphology* 11: 12.

Vohra, C.P. (1981). Note on recent glaciological expeditions in Himachal Pradesh. *Geological Survey of India Special Publication* 6: 26–29.

5

Aerial Views of the 2018 Kilauea Eruption, Hawaii, U.S.A.

*Benjamin R. Jordan**

Faculty of Sciences, Brigham Young University-Hawaii, Laie, HI, USA

5.1 Introduction and Start of Eruption

Remote sensing has been a powerful tool to interpret structural geology and geomorphology of terrains (e.g. Misra et al. 2014; Kaplay et al. 2017; Dasgupta and Mukherjee 2017, 2019; Dasgupta et al. 2022; Gogoi et al. 2022). On 30 April 2018, the ongoing eruption of Kilauea Volcano (Figure 5.1), which had been erupting continuously since 1983 (Babu and Kumar 2019; Williams et al. 2020), underwent a dramatic change in its behavior. At the time, the volcano had two active volcanic centers where lava was present: the Halema'uma'u summit crater and the Pu'u O'o cone on the southeastern flank of the volcano (Liu et al. 2018; Williams et al. 2020). Lava lakes existed in Halema'uma'u and Pu'u O'o, with relatively small volume flows from Pu'u O'o reaching the ocean through a series of lava tubes (Figure 5.2) and forming a series of lava deltas (Figure 5.3). On 30 April and 1 May, several earthquakes occurred, and the lava lake levels within Halema'uma'u and Pu'u O'o dropped dramatically. For two days, earthquake swarms indicated that magma was moving downslope and southeastward beneath the Southeast Rift Zone of the volcano. On 3 May, a large rift eruption, consisting of a long, linear fracture in the Earth's crust that erupts lava in a line, began within the populated neighborhood of Leilani Estates (Williams et al. 2020) (Figure 5.4). Similar eruptions had occurred in this same area in 1955 and 1960 (Gregg et al. 2008; Morales et al. 2010; Liu et al. 2018). However, despite the area being directly over the southeast rift zone of Kilauea Volcano and designated as lava hazard zones 1 and 2 (the highest zone designations) (Wright et al. 1992), approval was given for the development and construction of homes.

5.2 Lava Behavior

As the eruption progressed, a total of 24 lava fissures eventually formed (Figure 5.5). The eruptive activity became concentrated at Fissure 8, which had lava fountains that were estimated to reach heights of 80 m (Williams et al. 2020). The ejected lava flowed downslope to the ocean in multiple coalescing channels (Figure 5.6). The channels developed large levees. These levees consisted of solidified rock that formed natural walls or banks along the lava channel (Figure 5.7). These levees extended the length of the flows to the sea (Figure 5.8). The erupted lava was comparatively viscous basalt lava, solidifying with thick crusts of a'a lava (Figure 5.9), with a smaller amount of pahoehoe (Figure 5.7). Upon entering the ocean, phreatomagmatic steam explosions quenched and shattered the lava and filled the local coastal waters with suspended particles (Figure 5.10). High concentrations of these particulates created a distinctive color change within the sea (Figure 5.11).

5.3 Eruption End

The eruption rate of the lava rapidly fell on 4 August and by 21 August lava was no longer reaching the ocean. Final observed eruptive activity occurred on 4 September. The eruption lasted 107 days. Despite the short period of time

*Corresponding Author: ben.jordan@byuh.edu

Atlas of Structural Geological and Geomorphological Interpretation of Remote Sensing Images, First Edition.
Edited by Achyuta Ayan Misra and Soumyajit Mukherjee.

Companion website: www.wiley.com/go/misra/RemoteSensingImages

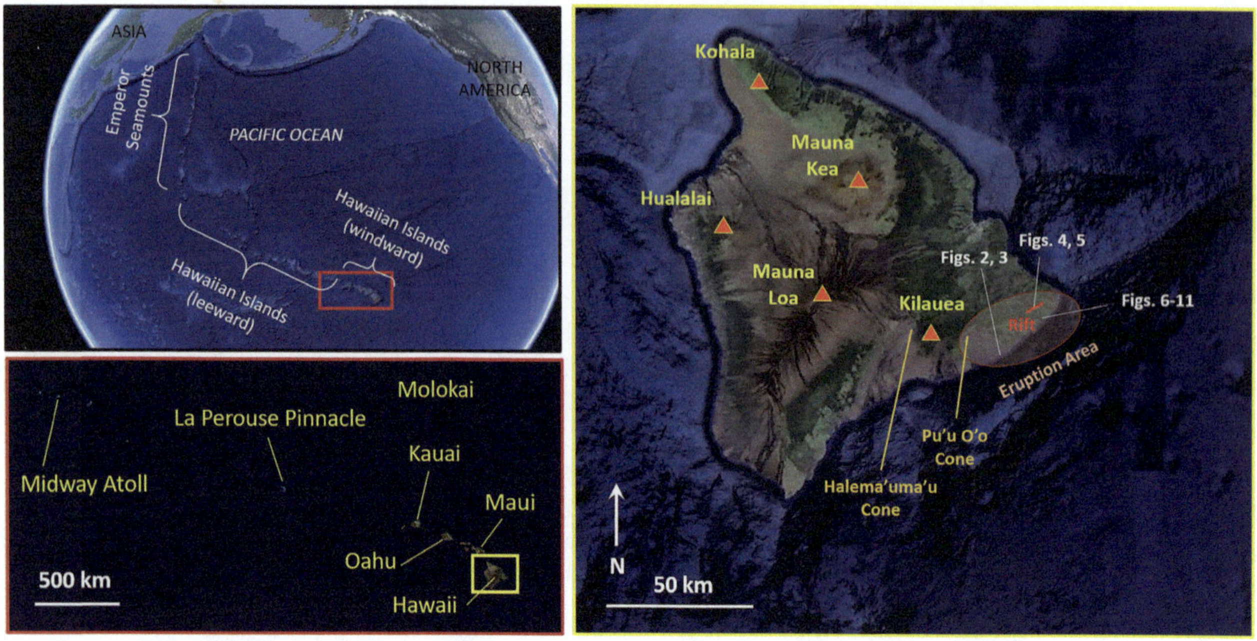

Figure 5.1 Map showing the location of the 2018 eruption of Kilauea Volcano in Hawaii, including the rift where the eruption occurred (red line in right figure) and the locations of the images in this atlas. Base maps for the right and upper left figures from Google Earth. Lower left base map image from NASA satellite imagery.

Figure 5.2 Location of lava entry into the ocean for lava flows erupted from the Pu'u O'o cone of Kilauea Volcano. This image clearly shows a skylight through which the flowing lava is exposed. A skylight is an opening into the lava tube caused by collapse of the overlying, cooled lava (Bunnell, 2008). Just prior to this image being taken there was a collapse of a lava delta, which was similar to that shown in Figure 5.3. A new delta is in the process of forming. This image taken from a helicopter on 31 July 2008 using a handheld Nikon Coolpix P50 (F-stop: f/5.6, exposure time: 1/141 s, focal length: 17 mm, ISO speed: ISO-64, image resolution: 300 PPI, image uncropped).

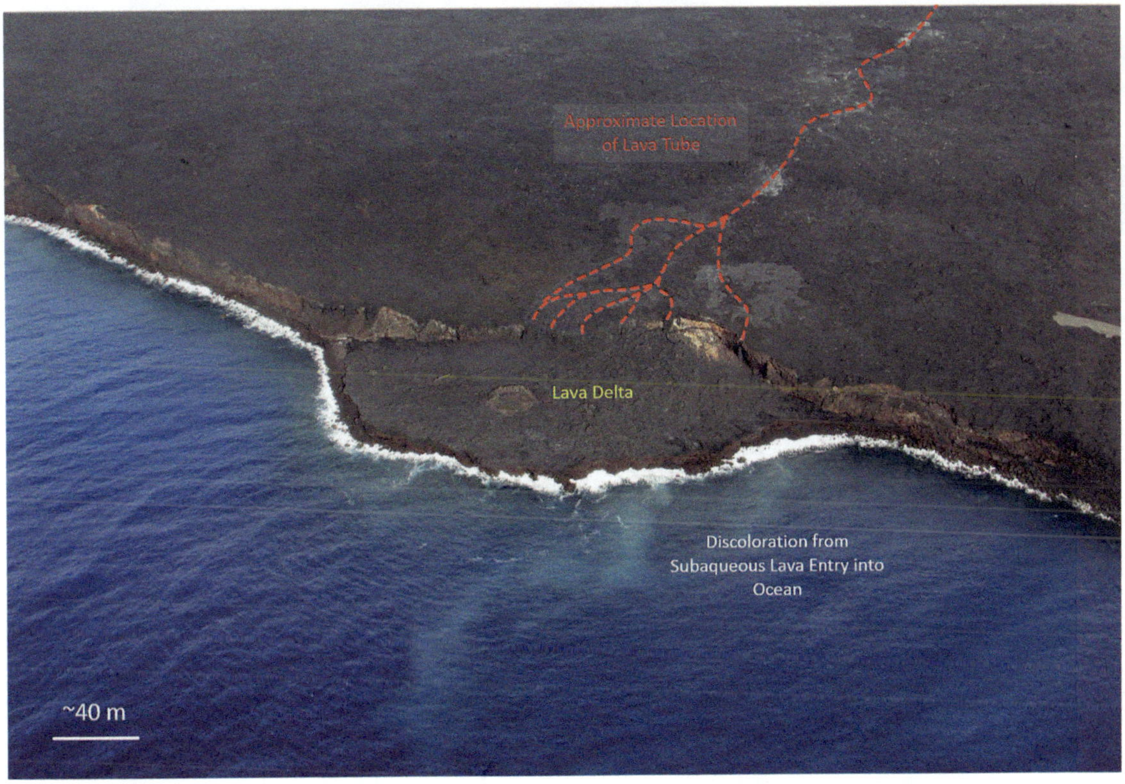

Figure 5.3 A view of one of the lava deltas formed by flows from the Pu'u O'o cone of Kilauea Volcano. Image taken on 10 March 2018, a few weeks before the collapse of the Pu'u O'o lava lake. Image taken from a helicopter using a handheld Canon PowerShot SX500 IS (F-stop: f/5, exposure time: 1/125 s, focal length: 8 mm, ISO speed: ISO-100, image resolution: 180 PPI, image uncropped).

Figure 5.4 Image taken of the 2018 Kilauea East Rift eruption shortly after it began. The East Rift extends through the neighborhood of Leilani Estates. Erupted lavas eventually inundated and destroyed most of the area in the image (red area). This image is digitally stitched together, using *ArcSoft Panorama Maker 6* (2011) from two images taken from a helicopter on 11 May 2018 using a handheld Pentax K-70 (both images used F-stop: f/9, exposure time: 1/160 s, focal length: 55 mm, ISO speed: ISO-400, image resolution: 300 PPI, images cropped after stitching).

Figure 5.5 Image of Fissure 22, one of the more active fissures during the 2018 Kilauea East Rift eruption. This fissure was near the Puna Geothermal Venture power plant (just out of view beyond the bottom of the image), but did not destroy the plant. Image taken from a helicopter on 24 May 2018 using a handheld Pentax K-70 (F-stop: f/11, exposure time: 1/250, focal length: 60 mm, ISO speed: ISO-200, image resolution: 300 PPI, image uncropped).

Figure 5.6 View of two of the major lava channels formed during the 2018 Kilauea East Rift eruption, illustrating the large area destroyed by lava flows. Image taken from a helicopter on 24 May 2018 using a handheld Pentax K-70 (F-stop: f/13, exposure time: 1/125 s, focal length: 18 mm, ISO speed: ISO-200, image resolution: 300 PPI, image uncropped).

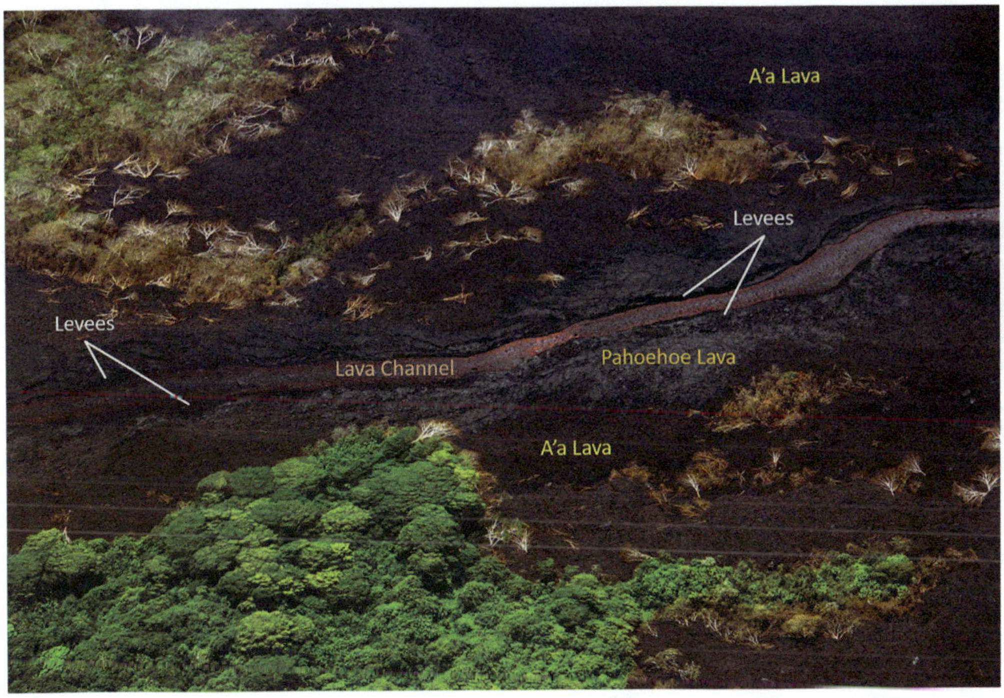

Figure 5.7 Closer view of the eastern lava channel from Figure 5.5. This image shows the channelized flow resulting from the formation of levees on the sides of the flow as the lava cools. Also evident are the contrasting lava types of a'a, with its rough and jagged surface, and pahoehoe, with its much less-fractured surface reflecting more sunlight, giving it a shiny appearance. Image taken from a helicopter on 24 May 2018 using a handheld Pentax K-70 (F-stop: f/10, exposure time: 1/250 s, focal length: 115 mm, ISO speed: ISO-400, image resolution: 300 PPI, image cropped).

Figure 5.8 View of the ocean entry of the lava channel in Figure 5.7. Image taken from a helicopter on 24 May 2018 using a handheld Pentax K-70 (F-stop: f/10, exposure time: 1/200 s, focal length: 88 mm, ISO speed: ISO-200, image uncropped).

Figure 5.9 Ocean-level view of the 2018 Kilauea East Rift eruption lava flows entering the ocean. Lavas consisted of a'a basalt. Images taken from a boat using a handheld Pentax K-70 (F-stop: f/10, exposure time: 1/200 s, focal length: 100 mm, ISO speed: ISO-1600, image cropped).

Figure 5.10 Ocean level view of the ocean entry point for some of the 2018 Kilauea East Rift eruption lava flows. Explosive fragmentation and seawater flashing to steam produces the large clouds and discoloration of the seawater. Images taken from a boat using a handheld Pentax K-70 (F-stop: f/8, exposure time: 1/640 s, focal length: 18 mm, ISO speed: ISO-200, image cropped).

relative to the previous eruptive event at the Pu'u O'o cone, the 2018 eruption is considered the most destructive volcanic event in Hawaii since at least 1790 CE. It is also one of the most expensive volcanic disasters in U.S. history (Williams et al. 2020). In total, it is estimated that the eruption caused approximately $800 million in damages including lost homes/property, population displacement, and local financial impacts (Dayton 2018). This article has been summarized in Misra and Mukherjee (2022).

Figure 5.11 Aerial view of the discoloration of the ocean due to suspended particles produced by fragmentation of lava from steam explosions at the ocean entry of the 2018 Kilauea East Rift eruption lava flows. Image taken from a helicopter on 24 May 2018 using a handheld Pentax K-70 (F-stop: f/13, exposure time: 1/125 s, focal length: 18 mm, ISO speed: ISO-200, image cropped).

Acknowledgements

Wiley Blackwell Acquisition Editors (present: Frank Weinreich, past: Andrew Harrison), Stacey Woods (Handling Editor), and proofreading team. Summarized in Misra and Mukherjee (2022).

References

Babu, A. and Kumar, S. (2019). SBAS interferometric analysis for volcanic eruption of Hawaii island. *Journal of Volcanology and Geophysical Research* 370: 31–50. https://doi.org/10.1016/j.jvolgeores.2019.01.011.

Bunnell, D. (2008). *Caves of Fire: Inside America's Lava Tubes*. Hunstville, Alabama: National Speleological Society 124 pp.

Dasgupta, S. and Mukherjee, S. (2017). Brittle shear tectonics in a narrow continental rift: asymmetric non-volcanic Barmer basin (Rajasthan, India). *The Journal of Geology* 125: 561–591.

Dasgupta, S. and Mukherjee, S. (2019). Remote sensing in lineament identification: examples from western India. In: *Problems and Solutions in Structural Geology and Tectonics*, Developments in Structural Geology and Tectonics Book Series, vol. 5. Series Editor: Mukherjee, S. (ed. A. Billi and A. Fagereng), 205–221. Elsevier. ISSN: 2542-9000. ISBN: 9780128140482.

Dasgupta, S., Biswas, M., Mukherjee, S., and Chatterjee, R. (2022). Depositional system, morphological signatures, tectonics and sedimentation pattern along the transform margin- Palar-Pennar basin, Indian east coast. *Journal of Petroleum Science & Engineering*. https://doi.org/10.1016/j.petrol.2022.110155.

Dayton, K. (2018). Recovery from Kilauea eruption might cost $800M. *Honolulu StarAdvertiser*. https://www.staradvertiser.com/2018/09/01/hawaii-news/recovery-from-kilauea-eruption-might-cost-800m.

Gogoi, M.P., Gogoi, B., and Mukherjee, S. (2022). Tectonic instability of the petroliferous upper Assam valley (NE

India): a geomorphic approach. *Journal of Earth System Science* 131: 18.

Gregg, C.E., Houghton, B.F., Paton, D. et al. (2008). Hawaiian cultural influences on support for lava flow hazard mitigationmeasures during the January 1960 eruption ofKīlauea volcano, Kapoho, Hawai'I. *Journal of Volcanology and Geophysical Research* 172: 300–307. https://doi.org/10.1016/j.jvolgeores.2007.12.025.

Kaplay, R.D., Md, B., Mukherjee, S., and Kumar, T.V. (2017). Morphotectonic expression of geological structures in eastern part of south East Deccan volcanic province (around Nanded, Maharashtra, India). In: *Tectonics of the Deccan Large Igneous Province*, vol. 445 (ed. S. Mukherjee, A.A. Misra, G. Calvès and M. Nemčok), 317–335. Geological Society, London, Special Publications.

Liu, C., Lay, T., and Xiong, X. (2018). Rupture in the 4 may 2018M_W6.9 earthquake seawardof the Kilauea east rift zone fissure eruption in Hawaii. *Geophysical Research Letters* 45: 9508–9515. https://doi.org/10.1029/2018GL079349.

Misra, A.A. and Mukherjee, S. (2022). Introduction to "atlas of structural geological and geomorphological interpretation of remote sensing images". In: *Atlas of Structural Geological and Geomorphological Interpretation of Remote Sensing Images* (ed. A.A. Misra and S. Mukherjee). Wiley Blackwell. ISBN: 9781119813354.

Misra, A.A., Bhattacharya, G., Mukherjee, S., and Bose, N. (2014). Near N-S paleo-extension in the western Deccan region in India: does it link strike-slip tectonics with India-Seychelles rifting? *International Journal of Earth Sciences* 103: 1645–1680.

Morales, J., Zhao, X., and Goguitchaichvili, A. (2010). Geomagnetic field intensity from Kilauea 1955 and 1960 lava flows: towards a better understanding of paleointensity. *Studia Geophysica et Geodaetica* 54: 561–574. https://doi.org/10.1007/s11200-010-0034-6.

Williams, D.M., Avery, V.F., Coombs, M.L., et al. (2020). U.S. geological survey 2018 Kilauea volcano eruption response in Hawai'I – after-action review: U.S. Geological Survey Open-File Report 2020-1041, 56 pp. http://dx.doi.org/10.3133/ofr20201041.

Wright, T.L., Chun, J.Y.F., Esposo, J., et al. (1992). Map showing lava-flow hazard zones, Island of Hawaii, *USGS Miscellaneous Field Studies*, Map MF-2193.

6

Depositional Systems – An Overview Via Google Earth

*Muhammad Awais**

Department of Geology, University of Swabi, Swabi, Khyber Pakhtunkhwa, Pakistan
Department of Earth, Environmental & Resources Sciences, University of Naples Federico II, Naples, Italy

6.1 Introduction

Remote sensing has been a powerful tool to interpret structural geology and geomorphology of terrains (e.g. Misra et al. 2014; Kaplay et al. 2017; Dasgupta and Mukherjee 2017, 2019; Dasgupta et al. 2022; Gogoi et al. 2022). There are different types of depositional environments and settings such as alluvial, deltaic, eolian, fluvial, lacustrine, transitional, and marine, which can be studied using satellite images. Google Earth is quite useful in the teaching and studying of geological and geomorphological features. It allows us to visualize and study satellite images portraying different geologic and geomorphologic features of the planet Earth. Google Earth shows high-resolution and updated true color satellite imagery (the majority of land areas are covered in >15 m per pixel resolution satellite imagery or by small-scale aerial photographs) (Tooth 2015). Any terrain can be analyzed in Google Earth using a variety of scales and perspective, although sometimes image quality is variable within and between regions (Tooth 2013). In Google Earth, some basic quantitative measurements such as distance, length, and slope can be seen using built-in tools, but there are certain limitation(s) in slope measurement (Tooth 2013). It can be considered as a *gateway* to try working with other satellite imageries and specialized softwares like Global Mapper and Geographical Information System (GIS) packages (Tooth 2013). This high-resolution archive is open source, cost-free, instantly accessible, and available for multiple types of scientific studies. On the contrary, other satellite imageries provide images at a wide range of electromagnetic spectrum but at high cost and procurement; processing and interpretation are lengthy processes and need specialized human resource(s).

The present chapter considers the demonstration of different depositional settings from different parts of the world such as Indus River (Pakistan); Satpara Lake and alluvial fans in Skardu (Pakistan); Meandering River System (Canada); Horton River System and Horton Delta (Canada); Nile River and Nile Delta (Egypt); Lake Ayakum (China); alluvial fans in China; Dunes in Rub al-Khali (southern Arabian Peninsula); Star Dunes in Algeria; and Musa Bay (Estuary) in Iran (Figures 6.1–6.12).

6.2 Indus River (Pakistan)

The Indus River is one of the largest river systems in the world originating from Mount Kailas in the Gangdese Range of southern Tibet (Kumar and Srivastava 2018); based on the Indus River catchment area from source to sink, it is placed 12th (Kumar and Srivastava 2018). The Indus River has meandering, braided, and anastomosing patterns in different areas throughout its course (Kumar and Srivastava 2018). Starting from the vicinity of the Tarbela dam to Attock Khurd area, the Indus River has a well-developed braided river system, as shown in Figure 6.1. The overall course of Indus River in the aforementioned area is east–west (Figure 6.1).

6.3 Meandering River System (Alberta, Canada)

In the northwestern part of Alberta (Canada), there is very well-developed meandering river system (Figure 6.2). In the northeast and southwest directions, the meandering river system is normal, however, in the

*Corresponding Author: awais.geo89@gmail.com

Atlas of Structural Geological and Geomorphological Interpretation of Remote Sensing Images, First Edition.
Edited by Achyuta Ayan Misra and Soumyajit Mukherjee.

Companion website: www.wiley.com/go/misra/RemoteSensingImages

Table 6.1 River types in relation to Sinuosity Index (SI) (Charlton, 2008; Rhoads, 2020).

S. No.	Sinuosity Index (SI)	River type
1	~1.0	Straight
2	1.1–1.5	Sinuous
3	>1.5	Meandering
4	>2.5	Highly meandering

middle part of this area the active meandering river system is associated with many oxbow lakes and abandoned channels (Figure 6.2). This meandering river system has high sinuosity (i.e. > 1.5–2.5) (Figure 6.2). According to Charlton (2008), different river patterns/types have different sinuosity index (Table 6.1). The sinuosity index increases as the river departs from the straight course (Table 6.1; Charlton 2008; Rhoads 2020).

6.4 Horton River System and Horton Delta (Canada)

In Canada, the Horton River flows into the Franklin Bay and a delta called Horton Delta is formed where the river discharges water into the marine environment (Mackay and Slaymaker 1989). Earlier, the Horton River discharged into the Harrowby Bay in the northwest direction (Mackay 1958, 1981). In the Horton River, meanders are well developed (Figure 6.3). The Horton River (old) to the northwest of the Horton Delta consists of abandoned meanders (Figure 6.3).

6.5 Nile River and Nile Delta (Egypt)

The Nile River and Nile Delta are one of the most spectacular geological-geomorphological features in the world (Figure 6.4). The Nile Delta was formed during the Upper Miocene to Recent and was formed by the alluvium carried by the earlier seven active branches of the Nile River (Fishar 2018). Now these branches have been silted and hence at present only two branches are active: Rosetta and Damietta (Stanley and Warne 1993; Fishar 2018). The Nile River System has a meandering flow pattern between the western and eastern deserts of Egypt (Figure 6.4). The Nile Delta consists of different lakes from west to east including Manzalah, Burullus, Edku, and Mariut (Fishar 2018).

6.6 Lake Ayakum, Tibet (China)

Lake Ayakum is present in the vicinity of the northern periphery of the Tibetan Plateau, China (Figure 6.5; NASA 2021). Many small-sized glacier and snowmelt-fed streams/rivers discharge and empty into Lake Ayakum (NASA 2021). In the lower left (southwestern) portion of Lake Ayakum, there are couple of river-deltas (Figure 6.5). The deltas have both active deltas and older surfaces with many distributaries (Figure 6.6; NASA 2021). It is interesting to note that the overall shape of the lake looks like that of a giant animal, perhaps a dinosaur (Figure 6.5).

6.7 Satpara Lake and Alluvial Fans in Skardu (Pakistan)

Satpara Lake is high altitude freshwater lake in Skardu (Gilgit-Baltistan), Pakistan. The Satpara Lake is fed by the Satpara Stream originating from Deosai Plains. Satpara Lake's longer dimension is roughly parallel to north–south direction (Figure 6.7). Alluvial fans are also formed in the vicinity of Satpara Lake (Figure 6.7 and 6.8). The alluvial fans are of different sizes, from small to large (Figure 6.7 and 6.8). Likewise, the physical forms/shapes of alluvial fans are also many, varying from perfectly fan shaped to sheet-shaped (Figure 6.8). A dam called the Satpara Dam is located in the north-northeast part of the Satpara Lake (Figure 6.7).

6.8 Alluvial Fans in China

In the vicinity of Taiyang Lake (Central China), well-developed alluvial fans are noticed (Figure 6.9). In this area, there are higher altitude glacier-filled valleys, which are involved in the formation of different alluvial fans at the foot of the mountains (Figure 6.9; Scheffers et al. 2015). The glacier-filled valleys are present in the north and south directions and hence alluvial fans are formed in both directions, such that the opposite alluvial fans are facing each other(s), as shown in Figure 6.9.

6.9 Dunes in Rub al-Khali (Southern Arabian Peninsula)

The Rub al-Khali desert, occupying most of the central and southern Arabian Peninsula, is one of the largest sand deserts in the world (Searle 2019; Figure 6.10). It stretches from northern UAE, i.e. Ras al Khaimah in north–southwards, and crosses central and south Arabia to the mountain ranges of Yemen and Dhofar (Searle 2019). The Rub al-Khali sand dunes have an elevation of 1200 m and individual dunes heights reach up to 250 m. The Rub al-Khali desert consists of sand dunes and scattered patches of sabkha, white gypsum, and salt flats (Searle 2019; Figure 6.10).

6.10 Star Dunes in Algeria

The Ouargla province in Algeria contains a desert (part of the famous Sahara Desert) full of well-developed star dunes (Figure 6.11). The star dunes have a pyramidal shape and large size (height may reach >300 m) with radiating arms (Lancaster 1989a). The star dunes are also called the Stellate dunes (Glennie 1970). The star dunes are one of the largest eolian bedforms in the various sand seas of the world (Lancaster 1989a, 1989b). The radial symmetry and scale of the star dunes reflects that they are formed as a result of complicated interactions between multidirectional wind systems and topography (Lancaster 1989b; Zhang et al. 2000).

6.11 Musa Bay (Estuary) in Iran

The Musa Bay is a shallow estuary and is positioned at the northwestern site of the Persian Gulf (Figure 6.12; ESA 2017). In Musa Bay, the water depth varies from 6 to 40 m in many parts with maximum depth of ~75 m at certain locations (Yeknami et al. 2016). The northernmost part of Musa Bay has a dendritic pattern (Figure 6.12).

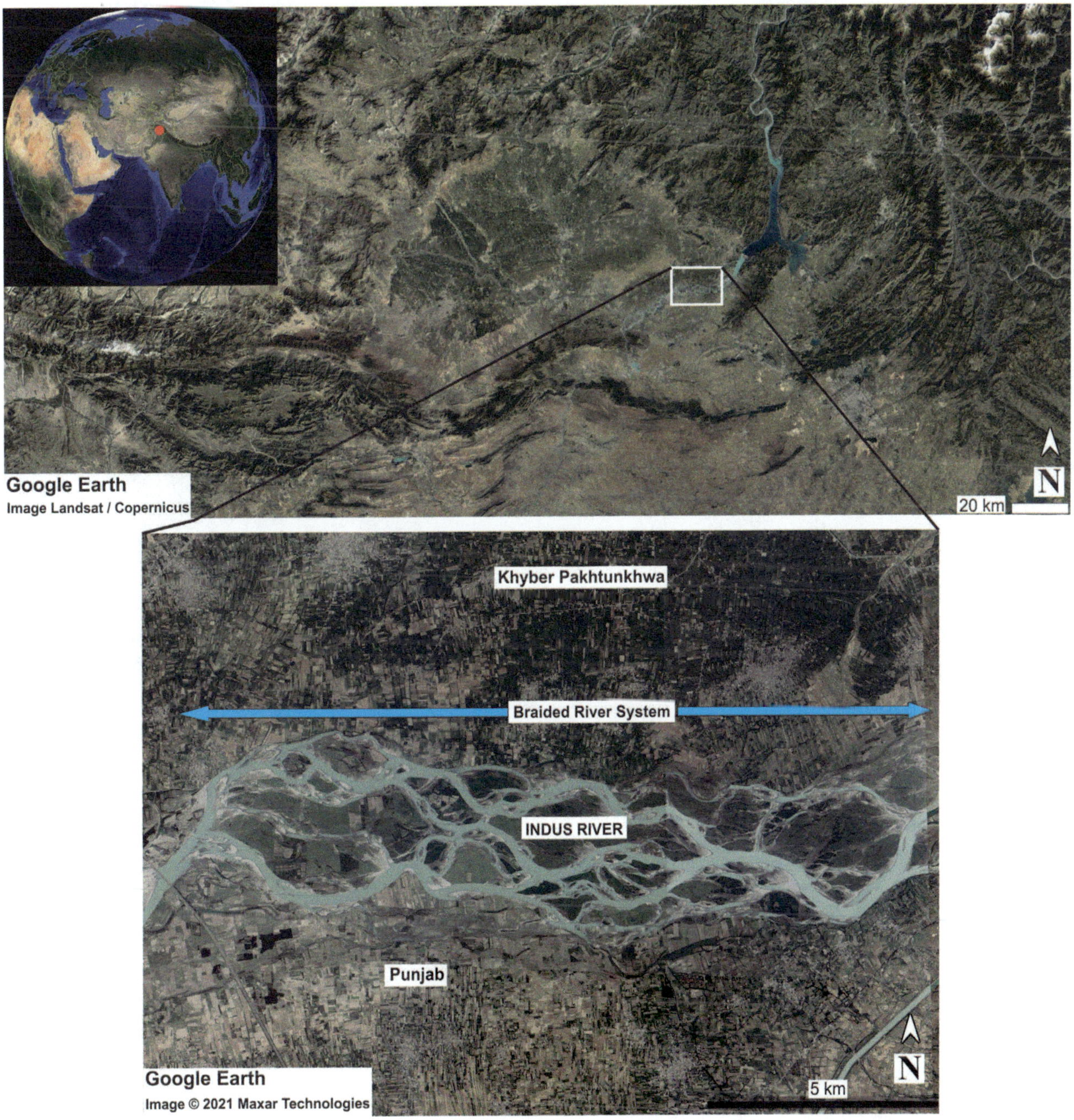

Figure 6.1 Indus River system near Tarbela Dam, Khyber Pakhtunkhwa, Pakistan. The river system is a braided river. Zoomed-in version of the image of the braided river acquired from Google Earth on 20 February 2021, and the regional areas on 12 September 2021. Actual date of acquisition of the image (zoomed-in version) by the satellite system is 14 September 2020. Source: Image © 2021 Maxar Technologies. Regional area (top) image is Landsat/Copernicus. On the world globe, the red dot indicates the area of Figure 6.1.

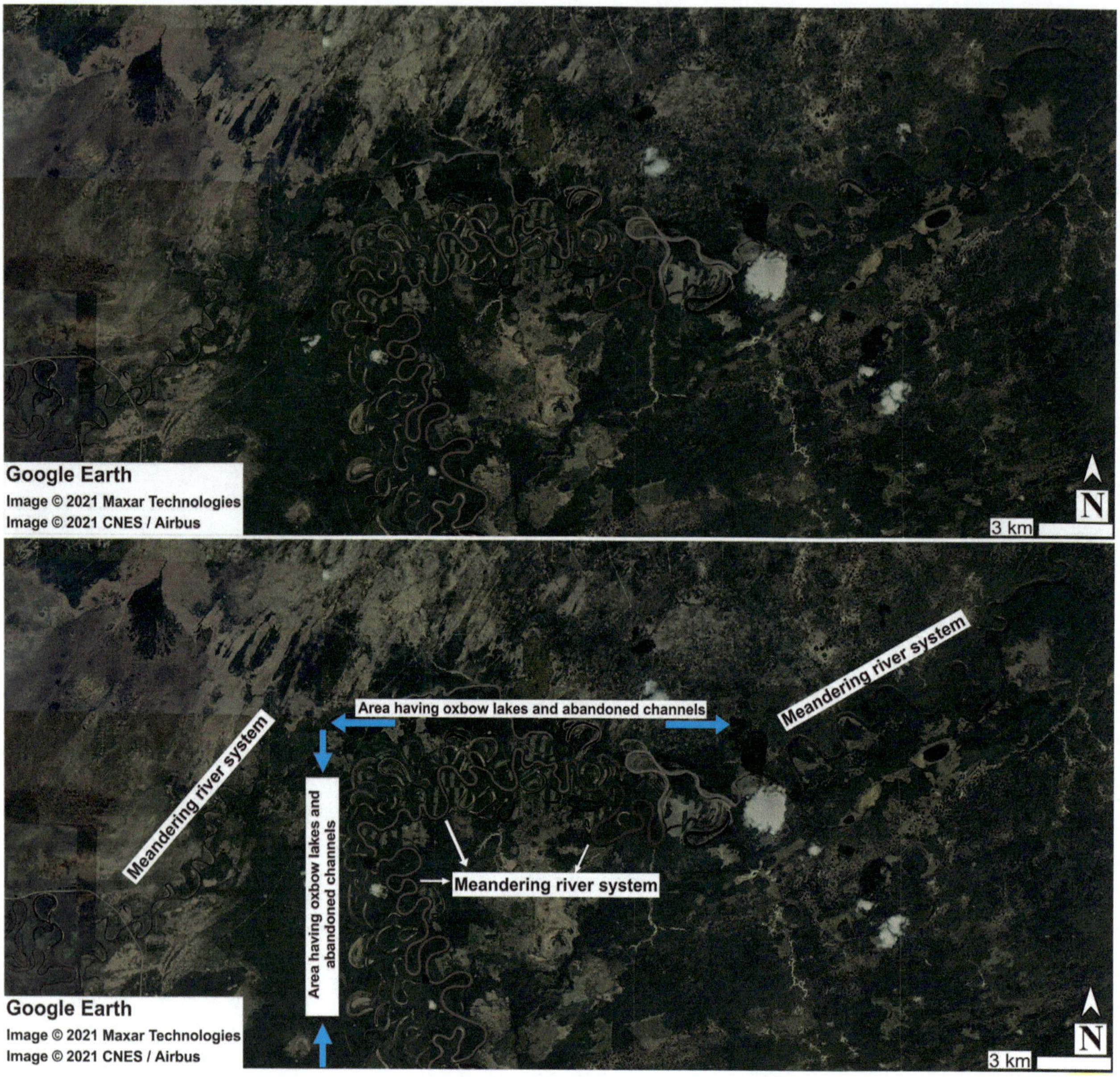

Figure 6.2 Meander River system having well-developed meanders, oxbow lakes/abandoned channels in Alberta, Canada. Source: Image acquired from Google Earth on 13 September 2021. Actual date of acquisition of the image by the satellite system is 27 June 2021. Image © 2021 CNES/Airbus and Maxar Technologies.

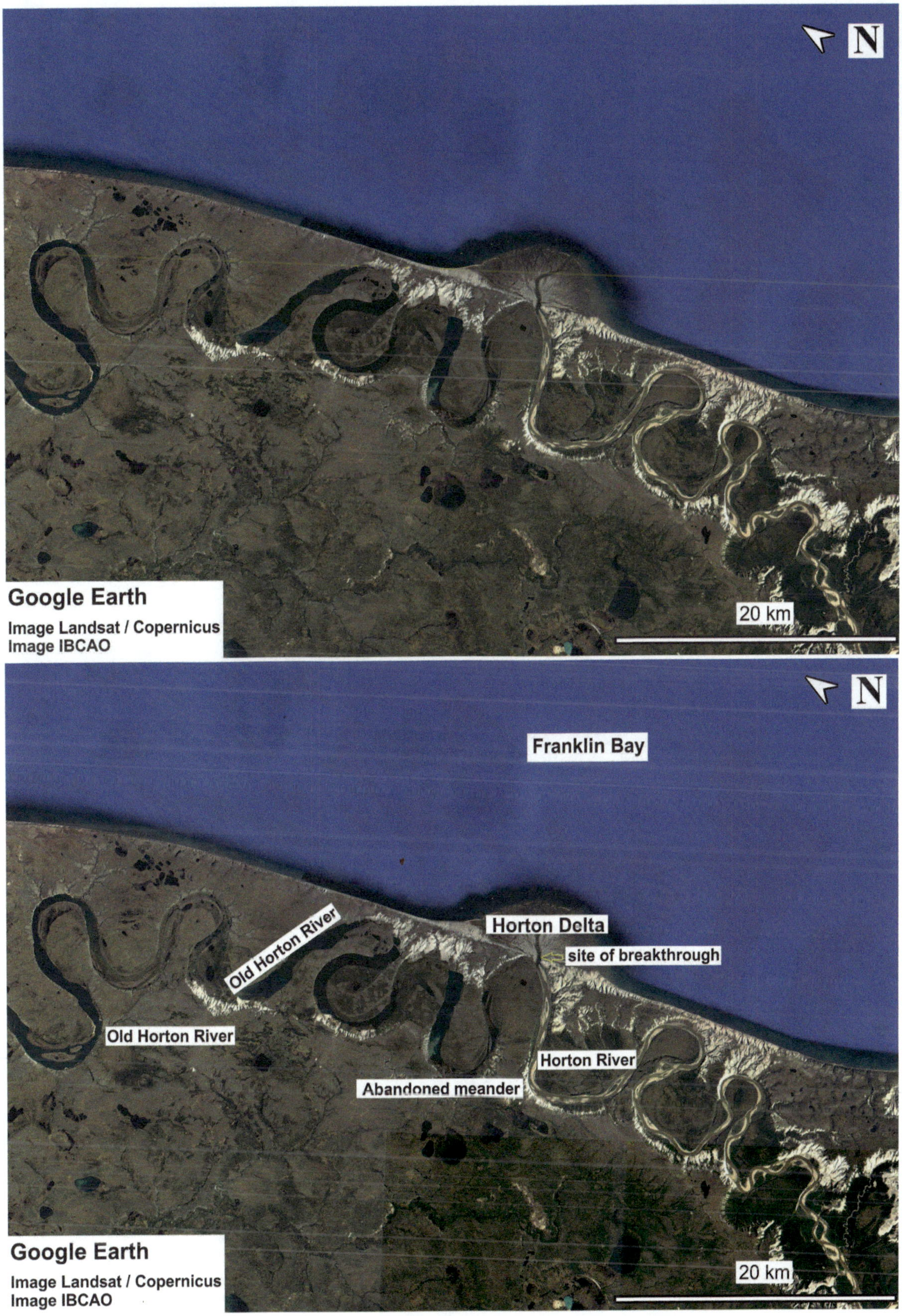

Figure 6.3 Horton River system and Horton Delta. Source: Image acquired from Google Earth on 19 February 2021. Actual date of acquisition of the image by the satellite system is 26 July 2006. Image Landsat/Copernicus and IBCAO.

Figure 6.4 Nile River (meandering river) and Nile Delta, Egypt. The branches and associated lakes of the Nile River are also shown. Identification of different features labeled in Figure 6.4 are based on Fishar (2018) and Stanley (2019). Source: Image acquired from Google Earth on 21 February 2021. Actual date of acquisition of the image by the satellite system is 14 December 2015. Image Landsat/Copernicus; Data SIO, NOAA, U.S. Navy, NGA, GEBCO.

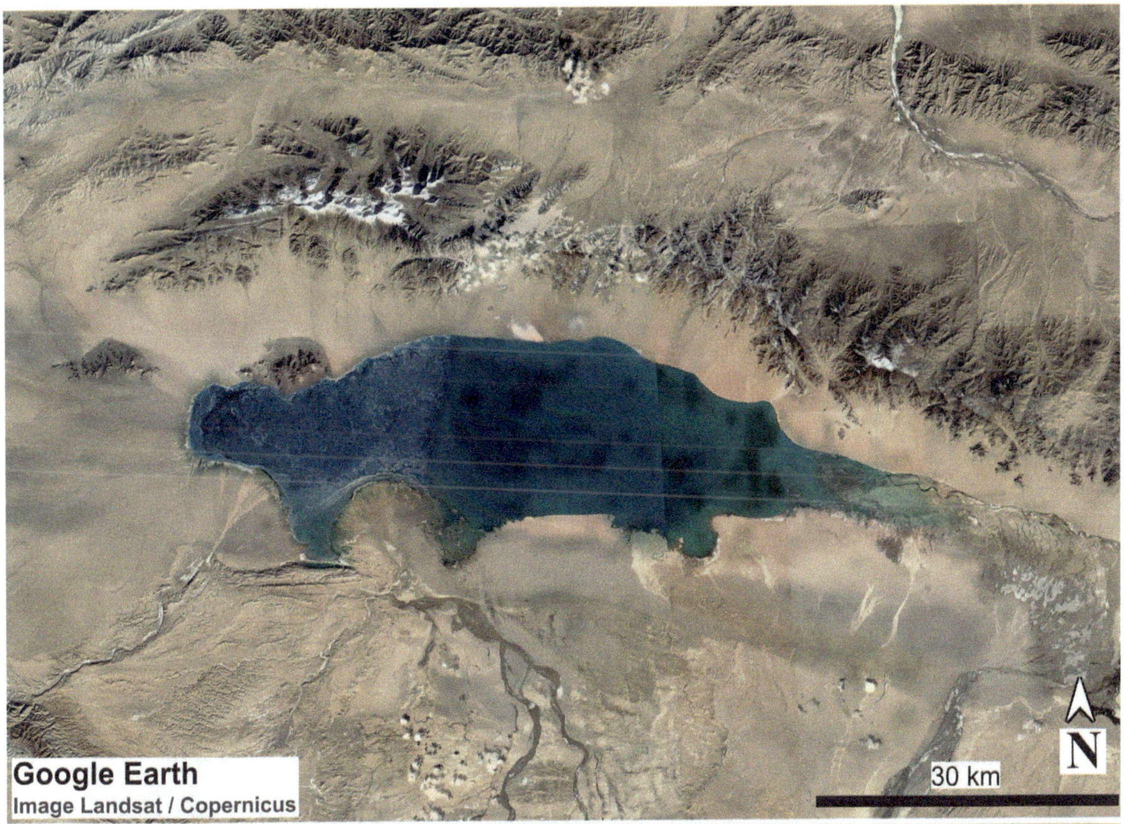

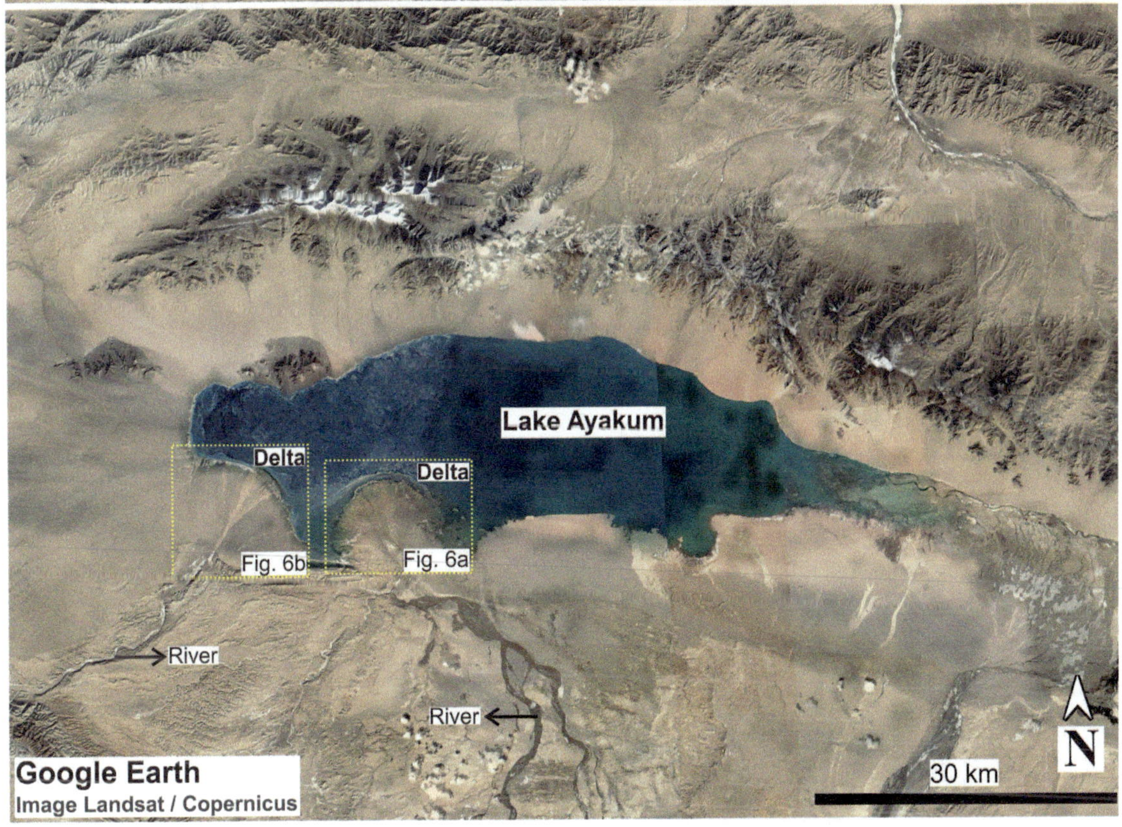

Figure 6.5 Lake Ayakum and river-deltas in Tibetan Plateau, China. Source: Image acquired from Google Earth on 21 February 2021. Actual date of acquisition of the image by the satellite system is 01 April 2006. Image Landsat/Copernicus.

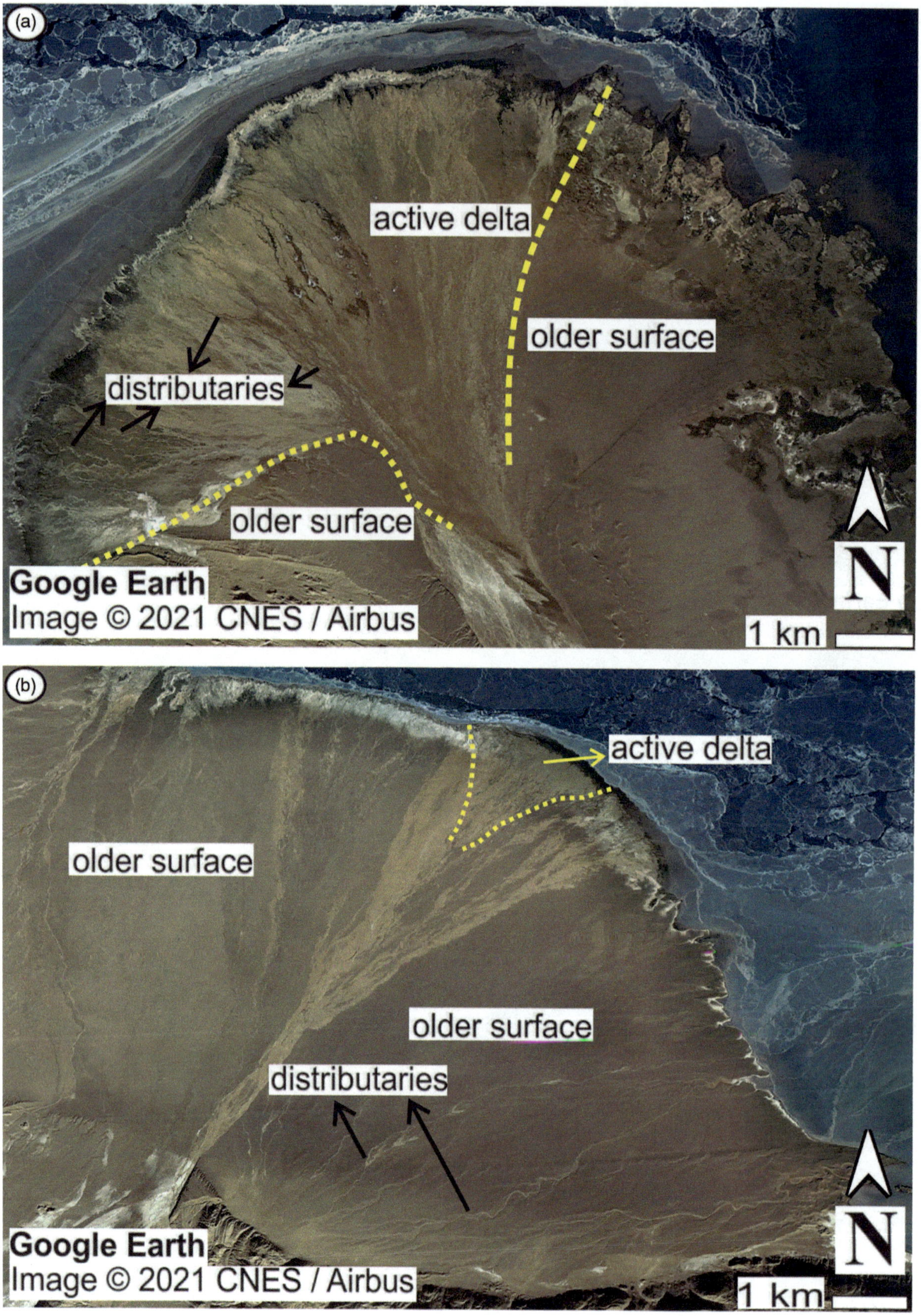

Figure 6.6 (a) and (b) Close-up photos of river-deltas associated with Lake Ayakum in Tibetan Plateau, China. The details (active delta and older surface) and demarcation of delta are made based on NASA (2021). Source: Image (© CNES/Airbus) acquired from Google Earth on 15 September 2021.

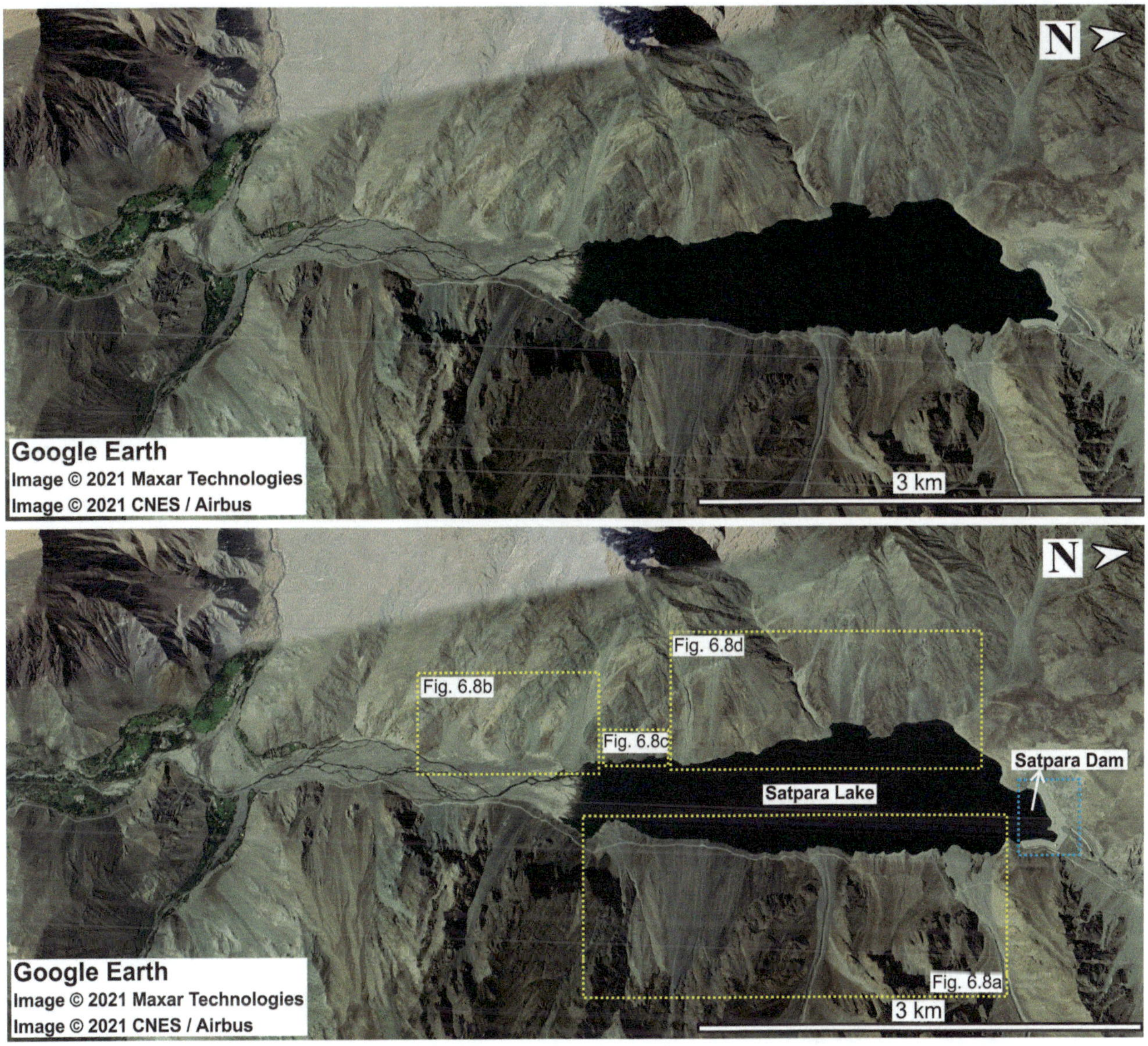

Figure 6.7 Satpara Lake, Satpara Dam, and alluvial fans in Skardu, Gilgit-Baltistan, Pakistan. The alluvial fans are enclosed in the yellow dashed-line boxes. Image acquired from Google Earth on 21 February 2021. Actual date of acquisition of the image by the satellite system is 01 September 2020. Source: Image © 2021 CNES/Airbus and © 2021 Maxar Technologies.

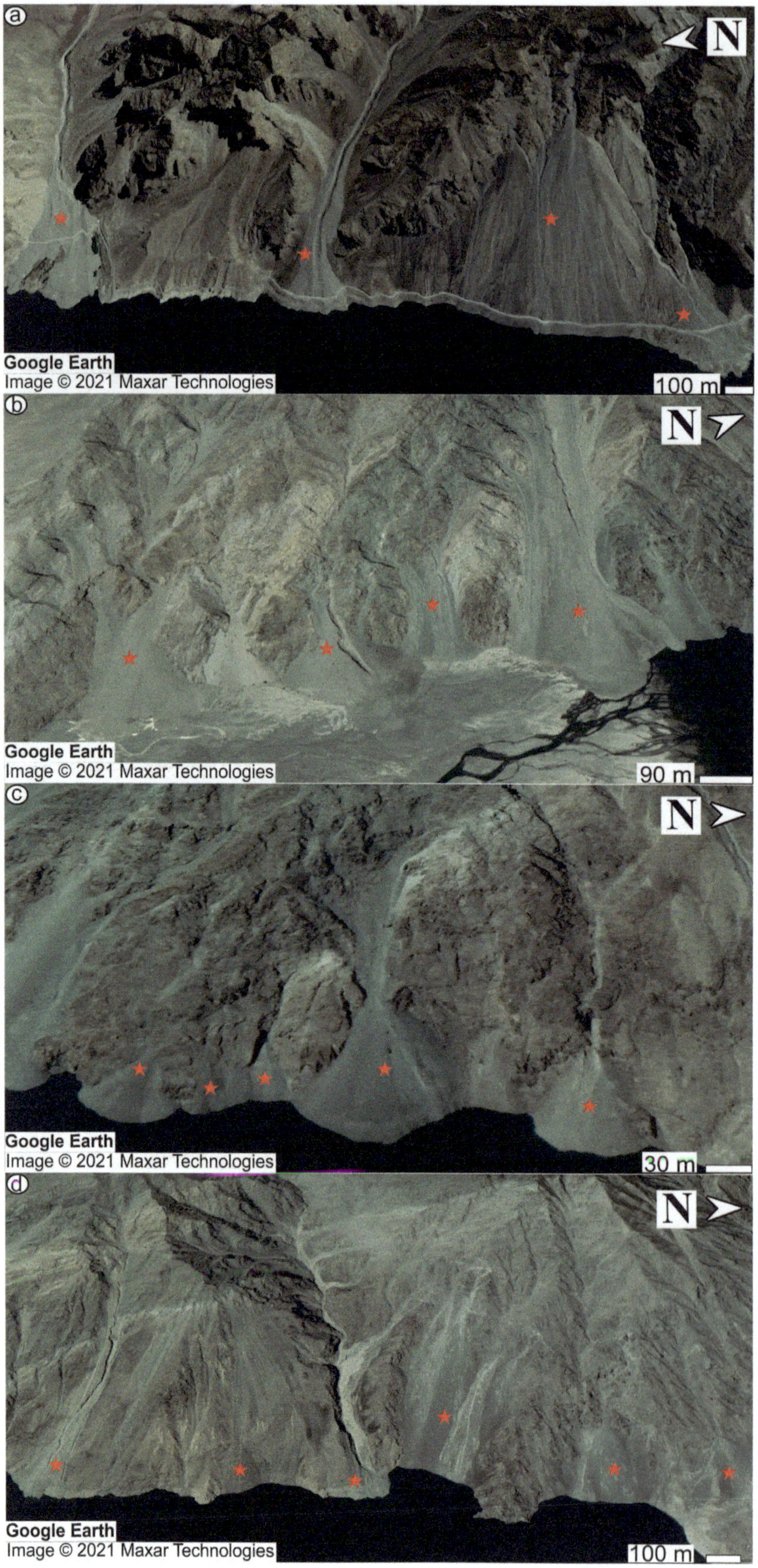

Figure 6.8 Alluvial fans in the vicinity of Satpara Lake, Pakistan. (a) Alluvial fans in east-southeast part of the lake. (b), (c) and (d) Alluvial fans to the west of Satpara Lake. (a) and (c) illustrate well-developed fan shaped alluvial fans. The alluvial fans are marked by red colored stars. Source: Image © 2021 Maxar Technologies.

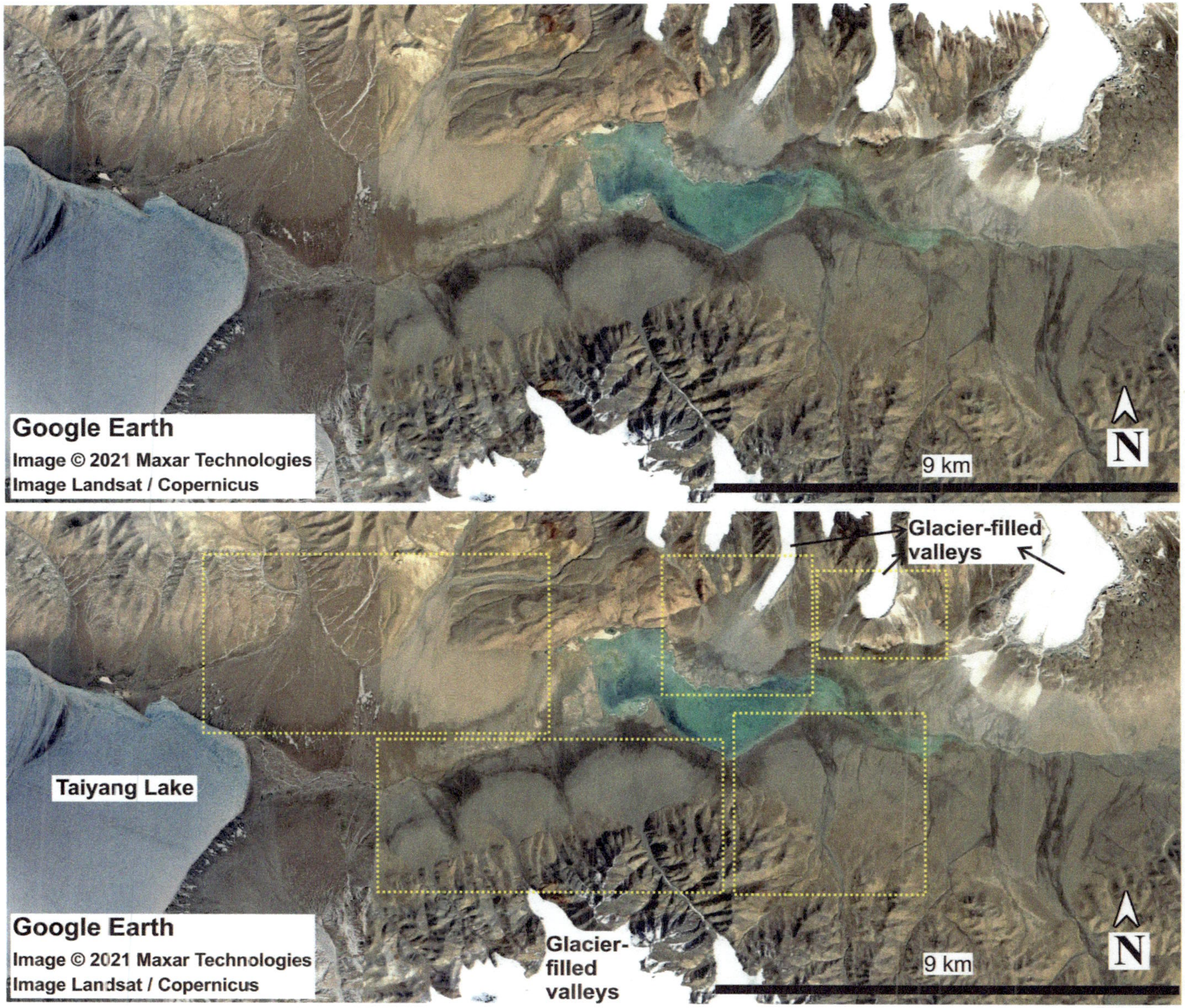

Figure 6.9 Alluvial fans in China. The well-developed alluvial fans are enclosed in yellow dashed-line boxes. Image acquired from Google Earth on 21 February 2021. Actual date of acquisition of the image by the satellite system is 25 March 2015. Source: Image Landsat/Copernicus and © 2021 Maxar Technologies.

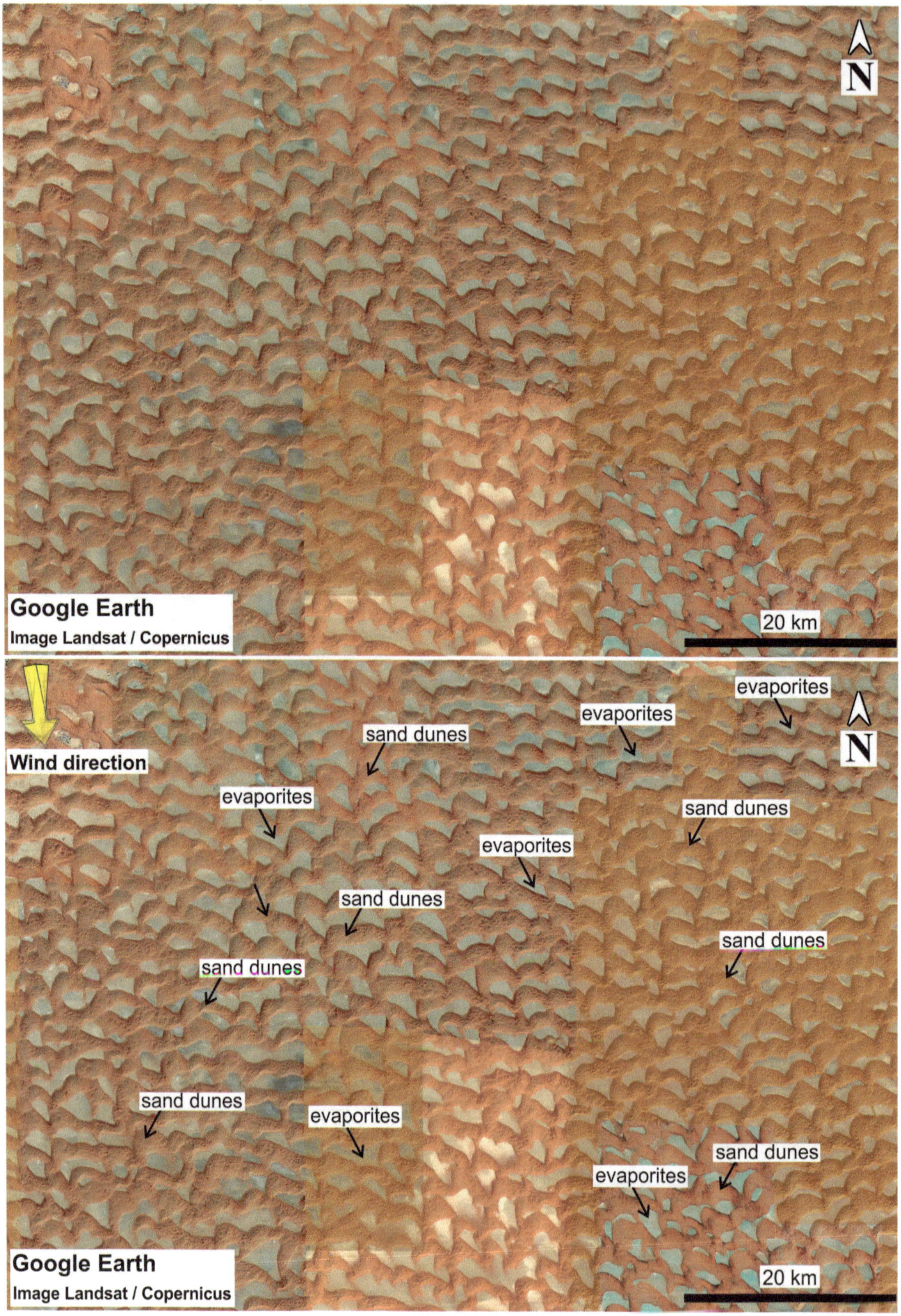

Figure 6.10 Sand dunes and scattered sabkha and evaporites in Rub al-Khali, southern Arabian Peninsula. Source: Image Landsat/Copernicus acquired from Google Earth on 20 February 2021. Actual date of acquisition of the image by the satellite system is 27 February 2017.

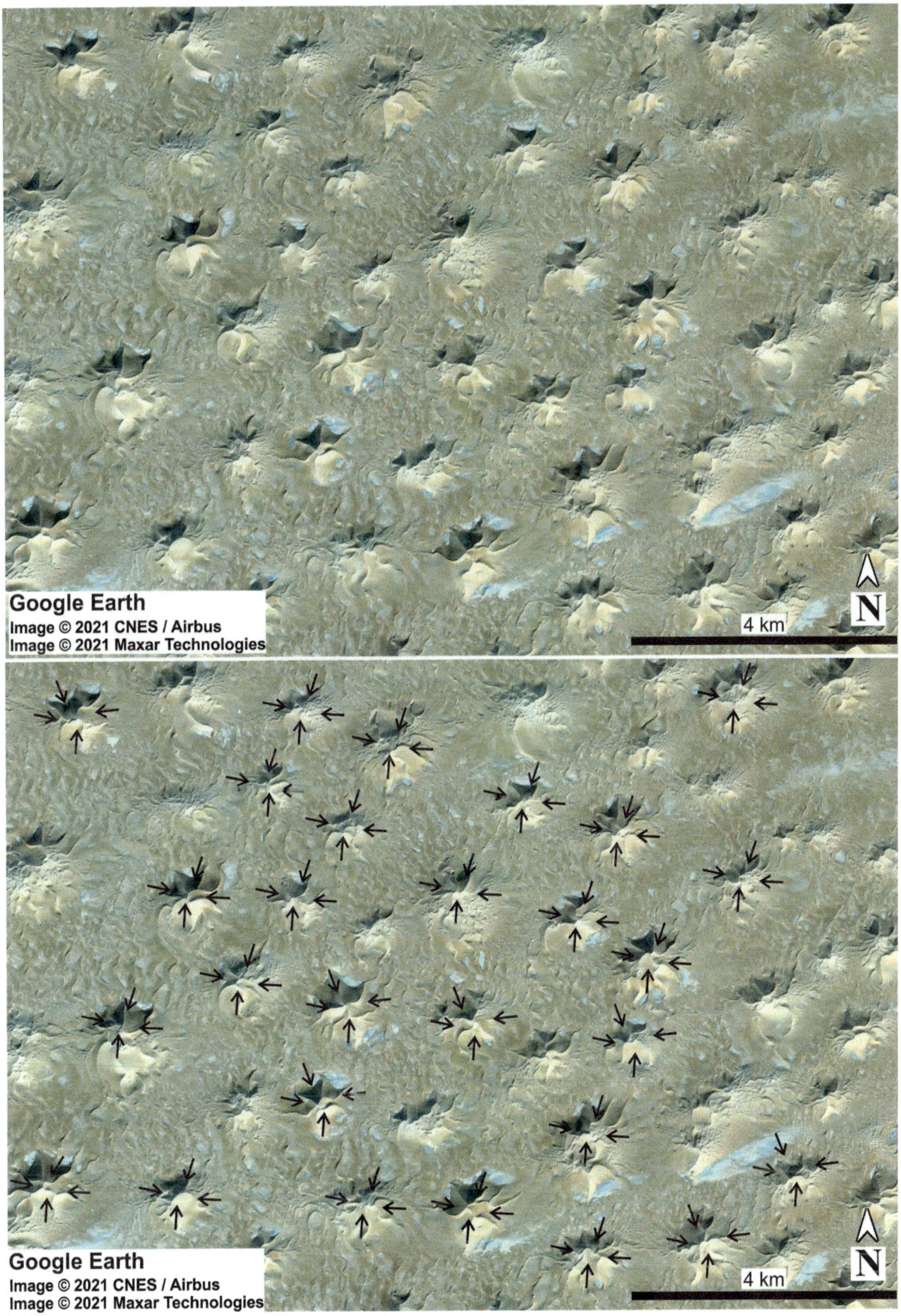

Figure 6.11 Star sand dunes in Algeria. The black arrows indicates multidirectional wind region. Source: Image acquired from Google Earth on 21 February 2021. Actual date of acquisition of the image by the satellite system is 15 September 2013. Image © 2021 CNES/Airbus and Maxar Technologies.

Figure 6.12 Musa Bay, Iran. The white arrows show water movement from the Persian Gulf toward Musa Bay. Source: Image Landsat/Copernicus acquired from Google Earth on 21 February 2021. Actual date of acquisition of the image by the satellite system is 05 February 2019. Data SIO, NOAA, U.S. Navy, NGA, GEBCO.

Acknowledgements

Soumyajit Mukherjee invited for submission of article. Handled by Achyuta Ayan Misra. Wiley Blackwell Acquisition Editors (present: Frank Weinreich, past: Andrew Harrison), Stacey Woods (Handling Editor), and proofreading team. Summarized in Misra and Mukherjee (2022).

References

Charlton, R. (2008). *Fundamentals of Fluvial Geomorphology*, 1e. Routledge.

Dasgupta, S. and Mukherjee, S. (2017). Brittle shear tectonics in a narrow continental rift: asymmetric non-volcanic Barmer basin (Rajasthan, India). *The Journal of Geology* 125: 561–591.

Dasgupta, S. and Mukherjee, S. (2019). Remote sensing in lineament identification: examples from western India. In: *Problems and Solutions in Structural Geology and Tectonics*, Developments in Structural Geology and Tectonics Book Series, vol. 5. Series Editor: Mukherjee, S. (ed. A. Billi and A. Fagereng), 205–221. Elsevier. ISSN: 2542-9000. ISBN: 9780128140482.

Dasgupta, S., Biswas, M., Mukherjee, S., and Chatterjee, R. (2022). Depositional system, morphological signatures, tectonics and sedimentation pattern along the transform margin – Palar-Pennar basin, Indian east coast. *Journal of Petroleum Science and Engineering* 211: 110155. https://doi.org/10.1016/j.petrol.2022.110155.

ESA (European Space Agency). (2017). Musa Bay, Iran. http://www.esa.int/spaceinimages/Images/2017/02/Musa_Bay_Iran.

Fishar, M.R. (2018). Nile Delta (Egypt). In: *The Wetland Book* (ed. C. Finlayson, G. Milton, R. Prentice and N. Davidson), 1251–1260. Dordrecht: Springer https://doi.org/10.1007/978-94-007-4001-3_216.

Glennie, K.W. (1970). *Desert Sedimentary Environments*. Amsterdam: Elsevier.

Gogoi, M.P., Gogoi, B., and Mukherjee, S. (2022). Tectonic instability of the petroliferous upper Assam valley (NE India): a geomorphic approach. *Journal of Earth System Science* 131: 18.

Kaplay, R.D., Md, B., Mukherjee, S., and Kumar, T.V. (2017). Morphotectonic expression of geological structures in eastern part of south east Deccan volcanic province (around Nanded, Maharashtra, India). In: *Tectonics of the Deccan Large Igneous Province*, vol. 445 (ed. S. Mukherjee, A.A. Misra, G. Calvès and M. Nemčok), 317–335. Geological Society, London, Special Publications.

Kumar, A. and Srivastava, P. (2018). Landscape of the Indus River. In: *The Indian Rivers* (ed. D.S. Singh). Springer Hydrogeology/Springer Nature Singapore Pte Ltd. https://doi.org/10.1007/978-981-10-2984-4_4.

Lancaster, N. (1989a). Star dunes. *Progress in Physical Geography* 13: 67.

Lancaster, N. (1989b). The dynamics of star dunes: an example from the gran Desertio, Mexico. *Sedimentology* 36: 273–289.

Mackay, J.R. (1958). The Anderson River map-area, N.W.T. Ottawa, Ontario, Dept. Mines and Technical Surveys, Geographical Branch, Memoir 5, 137 p.

Mackay, J.R. (1981). Dating the Horton River breakthrough, district of Mackenzie. In: Current Research Part B, Geological Survey of Canada, Paper 81-1B: 129–132.

Mackay, J.R. and Slaymaker, O. (1989). The Horton River breakthrough and resulting geomorphic changes in a permafrost environment, Western Arctic coast, Canada. *Geografiska Annaler* 71 A (3-4): 171–184.

Misra, A.A. and Mukherjee, S. (2022). Introduction to "atlas of structural geological and geomorphological interpretation of remote sensing images". In: *Atlas of Structural Geological and Geomorphological Interpretation of Remote Sensing Images* (ed. A.A. Misra and S. Mukherjee). Wiley Blackwell. ISBN: 9781119813354.

Misra, A.A., Bhattacharya, G., Mukherjee, S., and Bose, N. (2014). Near N-S paleo-extension in the western Deccan region in India: does it link strike-slip tectonics with India-Seychelles rifting? *International Journal of Earth Sciences* 103: 1645–1680.

NASA. (2021). River deltas and Lake Ayakum in China (Tibet). https://www.nasa.gov/mission_pages/station/multimedia/gallery/iss027e016922.html).

Rhoads, N. (2020). Channel planform – controls on development and change. In: *River Dynamics: Geomorphology to Support Management*, 186–196. Cambridge University Press.

Scheffers, A.M., May, S.M., and Kelletat, D.H. (2015). *Landforms of the World with Google Earth – Understanding our Environment*, 238. Springer.

Searle, M. (2019). Chapter # 16: Rub al-Khali (empty quarter). In: *Geology of the Oman Mountains, Eastern Arabia*, 429–430. Cham: GeoGuide/Springer Nature.

Stanley, J. (2019). Egypt's Nile Delta in late 4000 years BP: altered flood levels and sedimentation, with archaeological implications. *Journal of Coastal Research* 35 (5): 1036–1050.

Stanley, D.J. and Warne, A.G. (1993). Nile Delta: recent geological evolution and human impact. *Science* 260 (5108): 628–634. https://doi.org/10.1126/science.260.5108.628.

Tooth, S. (2013). Google Earth™ in geomorphology: reenchanting, revolutionising, or just another resource? In: *Treatise on Geomorphology. Vol. 14, Methods in Geomorphology* (ed. A.D. Switzer and D. Kennedy), 53–64. San Diego, CA: Academic Press.

Tooth, S. (2015). Google Earth as a resource. *Geography* 100 (1): 51–56.

Yeknami, S.R., Soltanpour, M., and Ranji, Z. (2016). Analysis of nearshore local effects on tidal behaviour at imam Khomeini port and Musa bay. *Coastal Engineering Proceedings* 1–9.

Zhang, W., Qu, J., Dong, Z. et al. (2000). The air flow field and dynamic processes of pyramid dunes. *Journal of Arid Environments* 45: 357–368.

7

The Lateritic Badlands of Garbeta (West Bengal, India)

Priyank Pravin Patel[*,1], *Sayoni Mondal*[1], *and Rajarshi Dasgupta*[2]

[1] *Department of Geography, Presidency University, Kolkata, West Bengal, India*
[2] *Department of Geography, East Calcutta Girls' College, Kolkata, West Bengal, India*

7.1 Introduction

Remote sensing studies are most helpful for interpreting the geomorphology and structural geology of terrains (e.g. Misra et al. 2014; Kaplay et al. 2017; Dasgupta and Mukherjee 2017, 2019; Dasgupta et al. 2022; Gogoi et al. 2022).

Badlands are zones marked by sustained soil loss via gully erosion. In this process, substantial sediment volumes are entrained and transported by accelerated and concentrated runoff through incised narrow channels (Poesen et al. 2003), with this being one of the major avenues of landscape degradation worldwide. Extensive badland tracts and gully channels appear as a dissected land surface, with a large number of rills and minor channels cutting into the exposed rocks or stratified sediments and usually being devoid of marked vegetation cover. Seen from above, these rills and gullies appear as finger-like projections extending into the landscape.

Most gullies and badlands develop in sedimentary lithologies and only a small percentage are cut into metamorphic substrates (Castillo and Gomez 2016). Laterite is quite an unusual substrate for badland formation (Patel et al. 2021) and is confined mostly to the tropics. In eastern India, however, it abounds, and has been the subject of many previous investigations (Bandyopadhyay 1987, 1988; Das and Bandyopadhyay 1995; Sen et al. 2004; Ghosh and Bhattacharya 2012; Shit et al. 2015). Most of these studies have been conducted within the Rarh region, located in the western and southwestern parts of West Bengal (Bagchi and Mukherjee 1983), which represents a transitional zone between the Ganga-Brahmaputra delta to the east and the Chotanagpur plateau fringe to the west. Geologically, the Rarh region may be considered as the western part of the stable shelf region of the Bengal basin and it has experienced several phases of marine transgression and regression since the Miocene (Alam et al. 2003; Ghosh and Guchhait 2020). A number of basin margin fault systems traverse the region, and the lateritic zones occur as uplifted blocks between these faults (Ghosh and Guchhait 2020).

7.2 Regional Setting of the Gangani Tract

The Gangani region near Garbeta (also spelled Garhbeta) (22°51′47″ N, 87°21′13″ E) in the Paschim Medinipur district, West Bengal, is a classic example of badlands that have developed on the local friable lateritic surface of the Rarh region (Figure 7.1). Its name translates to the *Land of Fire*, derived from the standout orange-red-brown shades of the exposed and dissected compacted alluvium and rocks that contain hydrated iron oxides. This location is often referred to as *The Grand Canyon of Bengal* (Mandal and Chakrabarty 2021) due to the deep ravines that have developed along the right bank of the Silabati River, and this spot has also recently been inducted as a notable Indian geomorphosite (Bandyopadhyay 2017).

The area is part of the lateritic uplands situated in the northern and western parts of the Paschim Medinipur district (Patel et al. 2020a). The compacted older alluvium forming the dissected surface has a lateritic (ferricrete)

*Corresponding Author: priyank.geog@presiuniv.ac.in

Atlas of Structural Geological and Geomorphological Interpretation of Remote Sensing Images, First Edition.
Edited by Achyuta Ayan Misra and Soumyajit Mukherjee.

Companion website: www.wiley.com/go/misra/RemoteSensingImages

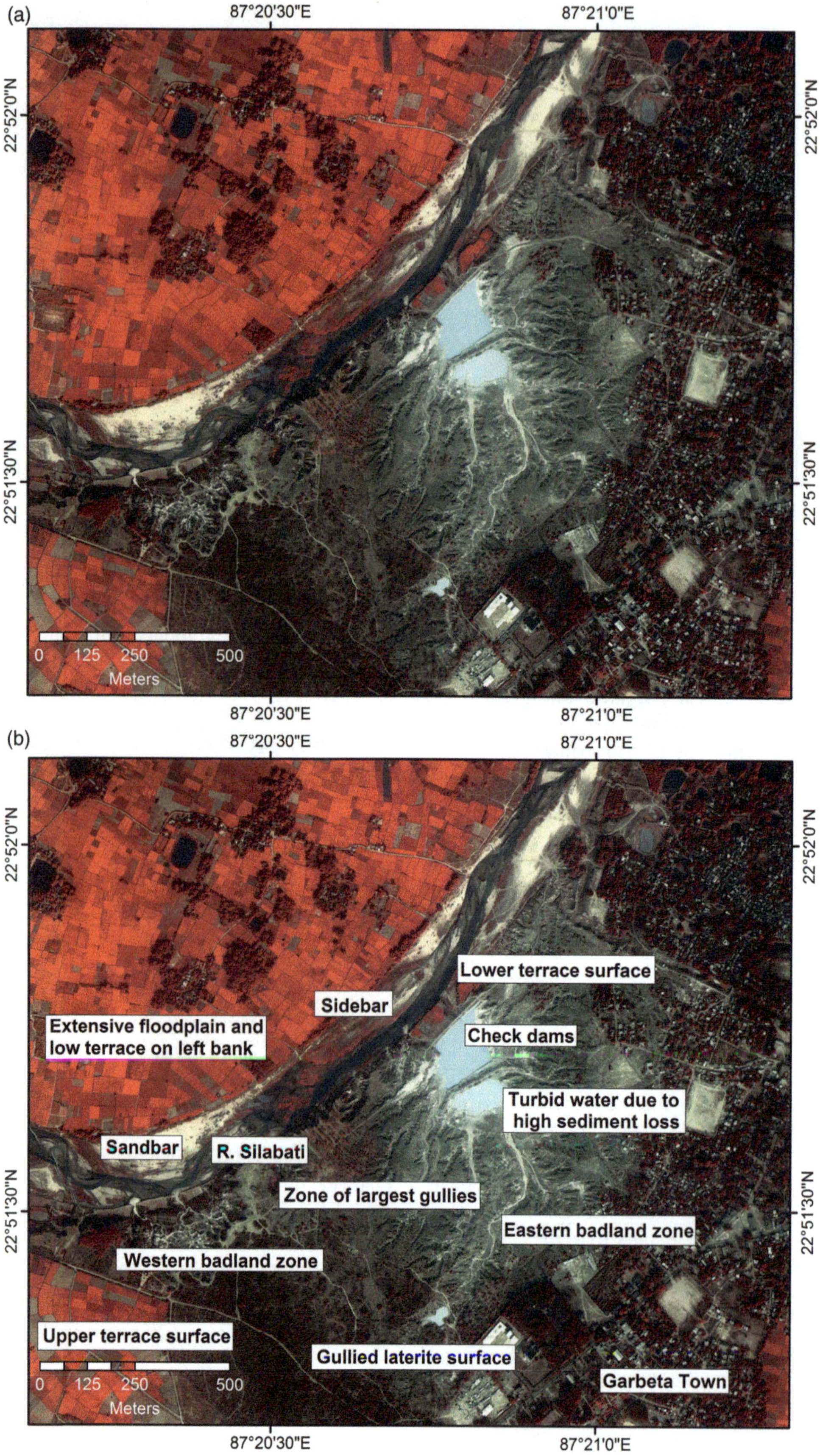

Figure 7.1 High-resolution multispectral image (Worldview-2 4B multispectral bundle; spatial resolution: 1.84 m; image date: 5 January 2011) of the Gangani tract showing the two main gully fields and the other principal geomorphic features in (a) uninterpreted, and (b) interpreted images. The band combination for both the images is that done for a Standard FCC (i.e. NIR, Red and Green in RGB).

mantle and is supposedly of Late Pleistocene age, over underlying Middle Pleistocene mudstone and sandstone strata (Dey et al. 2009) that are often exposed below the eroded upper portions. This lithological composition is also well-reflected in the substrate resistance, which changes rapidly over short distances due to the juxtaposition of these varied lithologies (Patel et al. 2020c, 2021). Receiving 1400 mm of rainfall annually (Patel and Mondal 2019), the Garbeta badlands may be classified as being of arid/semi-arid type (*sensu* Gallart et al. 2013). Consequently, the vegetation cover here is low and the surface is often bare or covered with pisolitic laterite nodules (Sarkar et al. 2020). The ferricrete thickness is between 6 and 25 m (Bandyopadhyay 1988). The laterites here are mottled and dark reddish-brown, being vermicular to scoriaceous vermicular and often have cavities (Singh et al. 1998). They are usually traversed by irregularly spaced vertical joints that can create substantial fissures (Ghosh and Bhattacharya 2012). Ghosh and Guchhait (2019) consider the laterites of Garbeta to be secondary or ex-situ laterites, formed by fluvial transportation of weathered materials from the primary laterites developed on Gondwana and Rajmahal Trap basalts in the western part of the Rarh region and their subsequent re-cementation with gravels, pebbles, ferricrete nodules, and even petrified wood (Haque and Ghosh 2019).

7.3 Badland Formation within Laterites at Gangani

While the Gangani tract's eastern portion has the older and wider gullies (Figure 7.2) that are more pronounced in length and extend further upstream into the lateritic surface (Figure 7.3), the western portion has smaller gully basins that have higher relief (Figure 7.4) and more entrenched channels (Patel et al. 2020b, 2021). The River Silabati flows beside this badland region and undercuts the high lateritic scarp, causing repeated riverbank failures that have gradually eroded the lower river terraces on either side and also enabled the ongoing gully incision (Patel and Mondal 2019). Marked landscape changes have occurred in the area over the last two decades, resulting from continual headward erosion by gullies along with anthropogenic activities like check dam construction for water harvesting and the walking trails developed and progressively widened by tourists attracted to the scenic beauty of this spot (Figure 7.5).

Pipes and earth pillars have developed along the gully walls, with these being hollowed out to form small cave-like forms in some locations, alongside local mounds and other minor erosional and depositional features (Haque and Ghosh 2019). Even from very high-resolution satellite images or digital elevation models and vertical top-down looking UAV surveys, such features are less apparent. However, these can be succinctly mapped and measured from oblique ground-based close range photogrammetry using the structure-from-motion (SfM) technique (Patel et al. 2020c). The 3D models developed from such surveys clearly locate rill and gully heads and reveal the alternate more eroded and relatively resistant strata, along with piping channels and micro-slope forms (Figure 7.6). These soil pipes and earth pillars form as a result of processes such as basal sapping, percolation, and weathering and are washed away by pluvial erosion and seepage (to download/access a textured 3D model of these features, which has been prepared using the SfM technique and can be zoomed in or rotated, please see the online repository and the related instructions for its use in Appendix A). Thus, they often collapse and link up with gully heads, thereby extending the incised channels further, both laterally and longitudinally (Patel et al. 2021). Other prominent micro-features here include numerous overhangs and undercuts. While the overhanging slopes are the result of basal erosion, the undercuts usually result from erosion during the rainy season when the otherwise dry gully channels have some flow. These overhanging parts also have hollows formed by the combined processes of solution, seepage, and weathering of the slope faces (Patel et al. 2020c). A number of field photographs are provided to help visualize the abovementioned different morphological attributes of the Gangani Badlands, which are a standout example of extensive gullying on a lateritic terrain (Figures 7.7–7.9). The particular fieldwork during which these photographs were taken (along with those for creating the SfM-based 3D model in Figure 7.6) was conducted in September 2017. The locations of the features and the areal extents depicted in Figures 7.6–7.9 within the Gangani Badlands are shown in Appendix B.

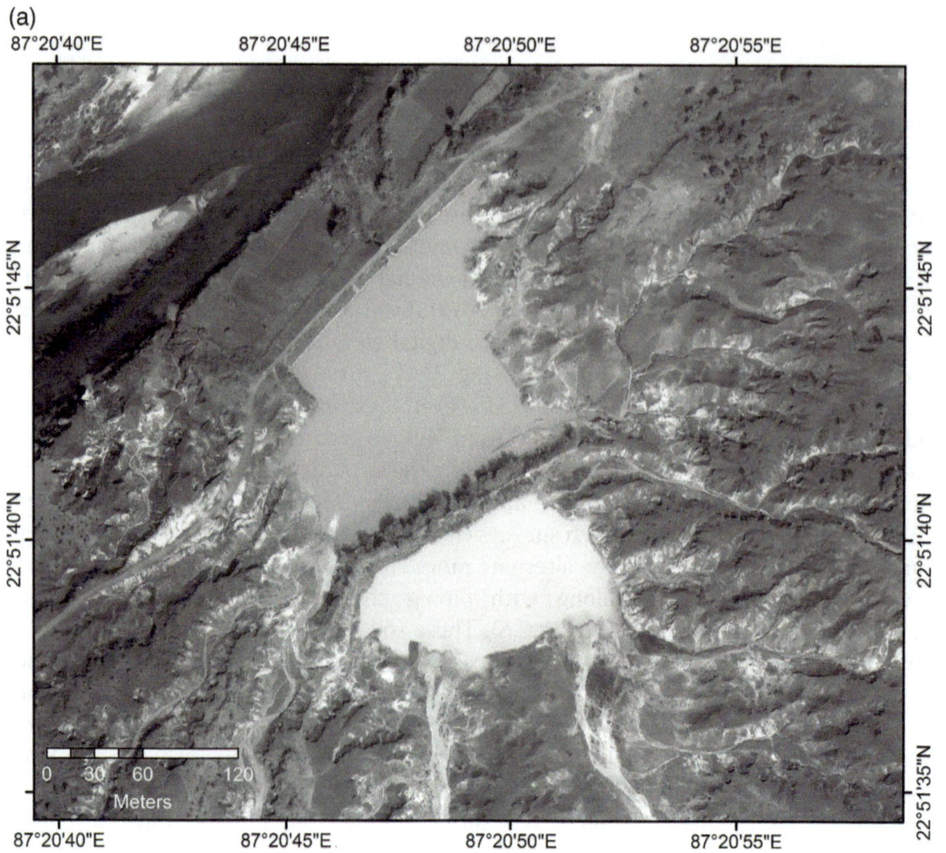

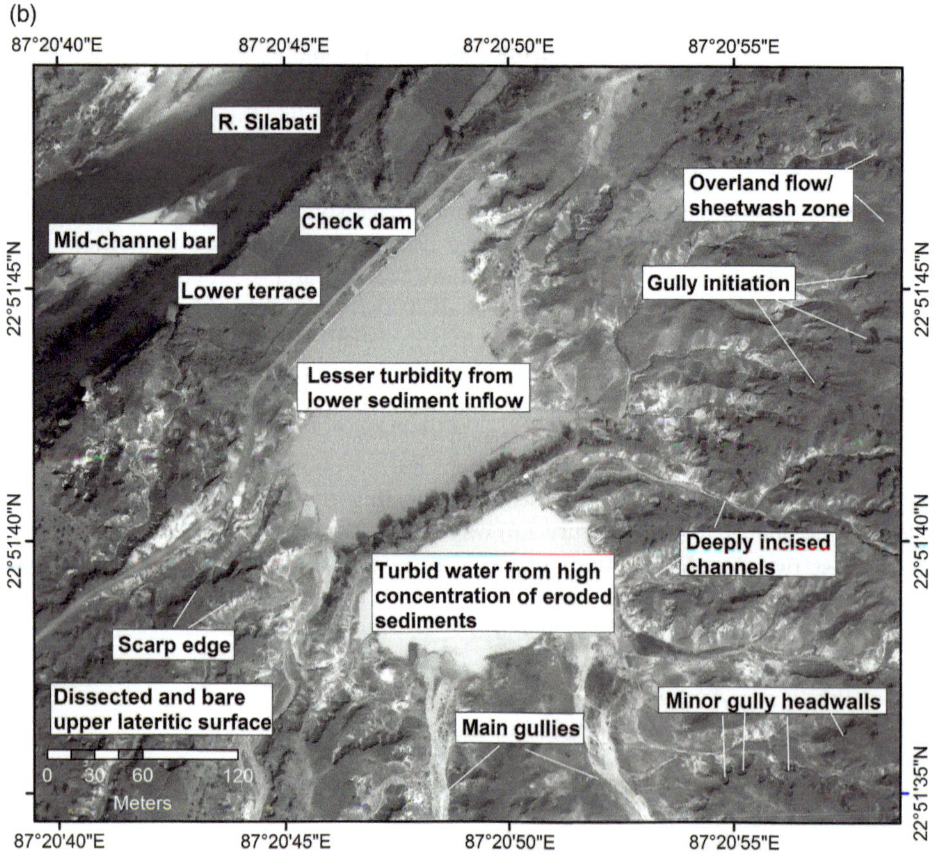

Figure 7.2 Very high-resolution panchromatic image (Worldview-3 PAN; spatial resolution: 0.31 m; image date: 12 November 2014) of the Gangani's eastern portion, showing the main gullies developed herein in (a) uninterpreted, and (b) interpreted images. These gullies are longer and wider than their western counterparts. Two check dams have been constructed for local water harvesting. The marked turbidity in the collected runoff attests to the high sediment volumes eroded from the gully basins.

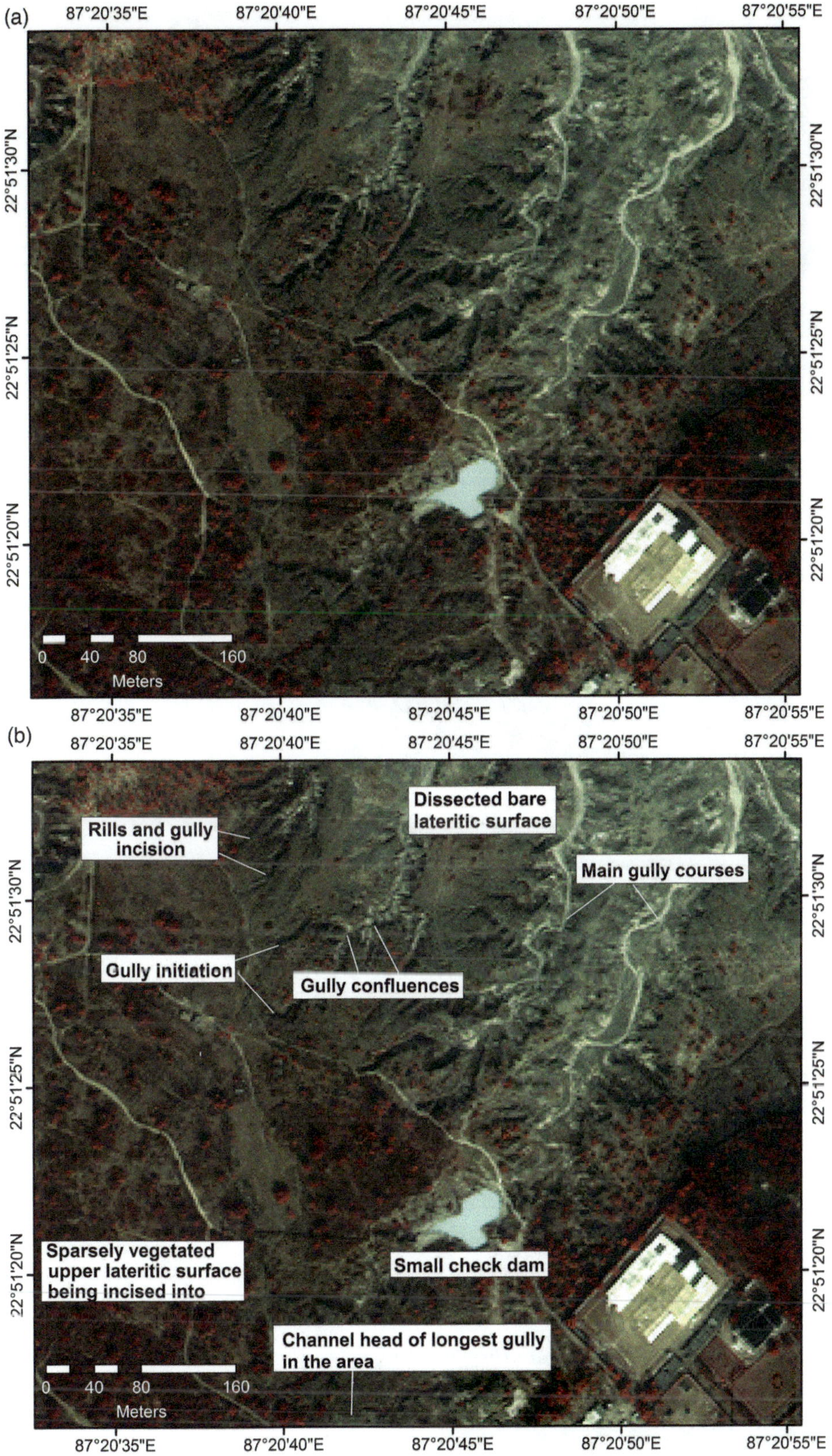

Figure 7.3 Gully initiation zones of the longest gullies in Gangani that traverse across its eastern portion as discerned from the Worldview-2 4B multispectral image bundle (spatial resolution: 1.84 m; Image date: 5 January 2011) from the (a) uninterpreted, and (b) interpreted images. The minor elevation drops in the sparsely vegetated or bare upper surface are clearly distinguishable, while the rills that emanate from the channel initiation points beyond the overland flow zones gradually coalesce and grow into larger gullies as these progressively incise downstream. The band combination for both the images is that done for a Standard FCC (i.e. NIR, Red and Green in RGB).

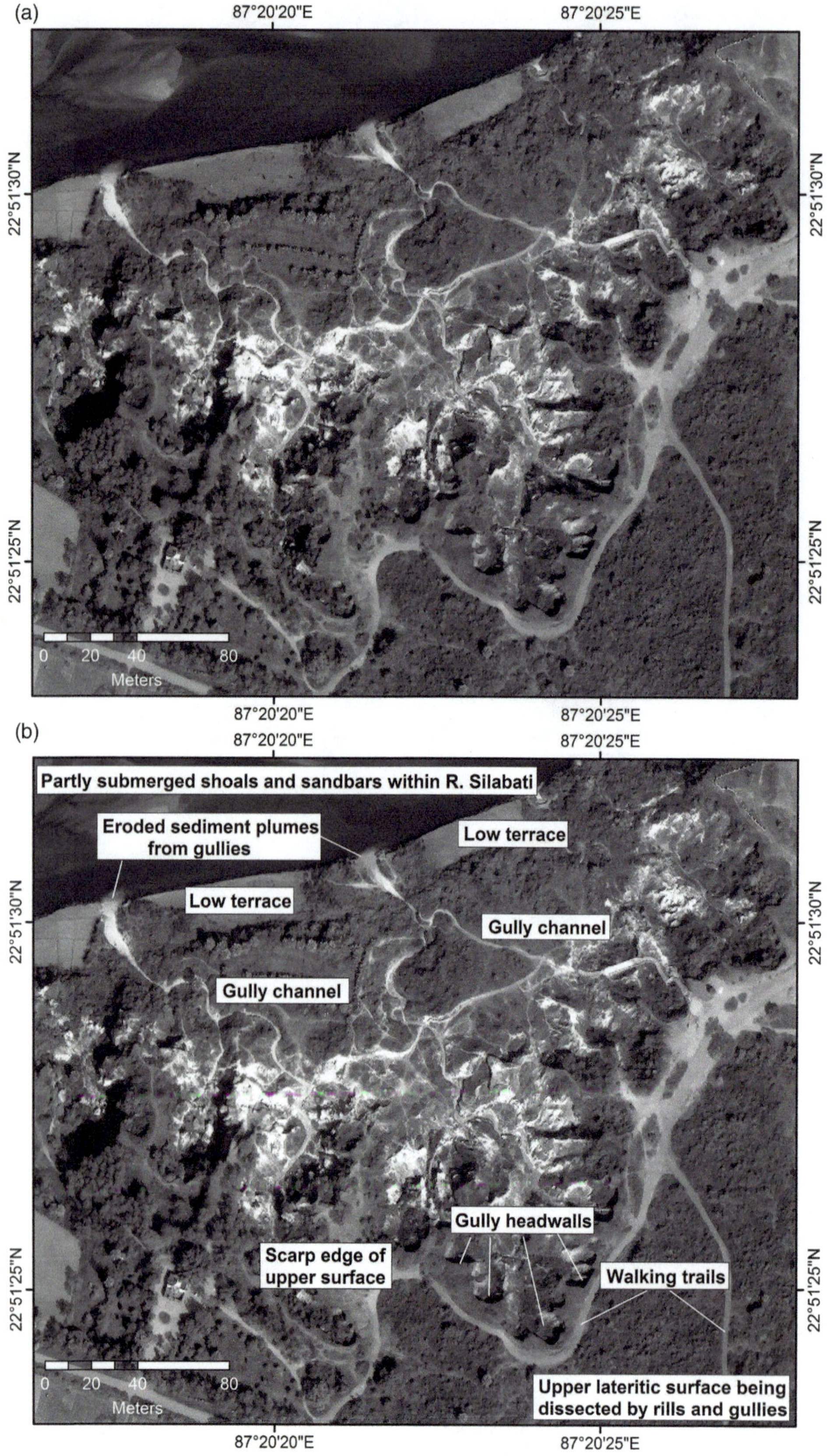

Figure 7.4 Worldview-3 PAN image of the western portion of the Gangani badland tract showing the upper and lower surfaces, gully headwalls and scarp edge and main gullies that debouch into the River Silabati in (a) uninterpreted, and (b) interpreted images.

Figure 7.5 Historical (a) and recent (b) Google Earth images show the main landscape changes that have occurred in Gangani over the last two decades.

Figure 7.6 Generated 3D model of a gully wall using close range photogrammetry and SfM techniques showing piping, hollowing, and minor weathering features. The GCPs for the survey were collected during a Total Station survey conducted in September 2017 and a Nikon CoolPix camera was used to obtain the required overlapping photographs. Note: See the PDF file of the same figure provided in the online repository for the interactive 3D model, Appendix A, for this model's description and how to open and operate it, and Appendix-B to discern where this spot is situated within the Gangani Badlands.

(a)

(b)

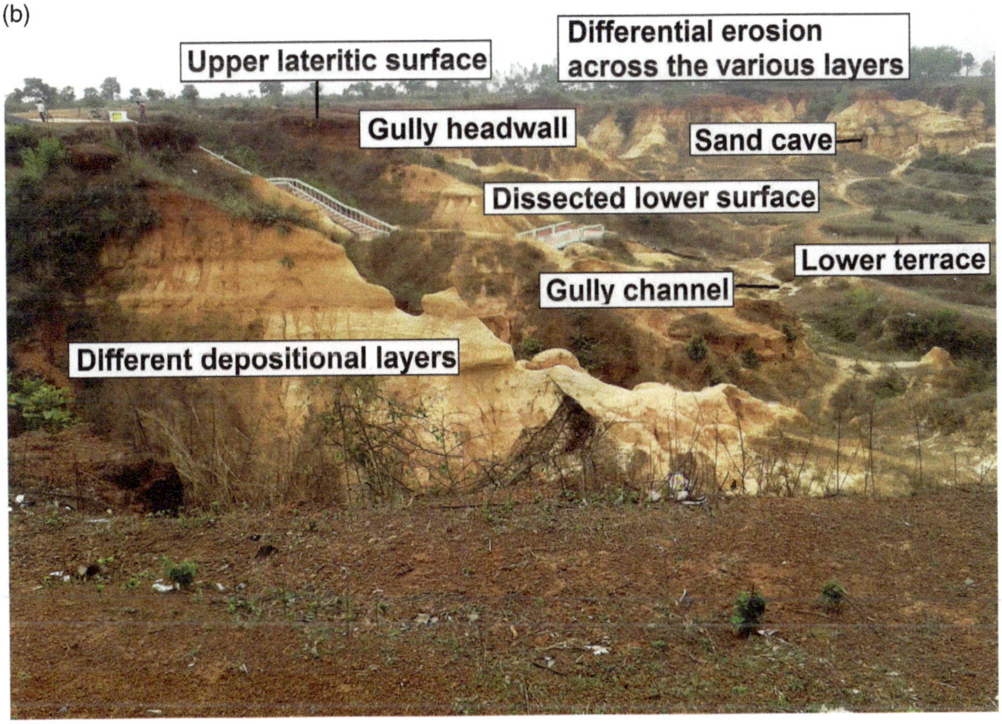

Figure 7.7 View across the western section of the Gangani badland tract with discerned morphological features in (a) uninterpreted, and (b) interpreted images. Note: See Appendix B, where this location has been marked in a Google Earth screenshot of the area.

(a)

(b)

Figure 7.8 Incised main gullies and the differentially eroded/weathered escarpment separating the upper and lower surfaces at Gangani, as identified from (a) uninterpreted, and (b) interpreted images. Note: See Appendix B, where this location has been marked in a Google Earth screenshot of the area.

Figure 7.9 View upstream from within one of the main gully channels in the western section of the Gangani badlands, showing the gully head and sidewall features in (a) uninterpreted, and (b) interpreted images. Note: See Appendix B, where this location has been marked in a Google Earth screenshot of the area.

Acknowledgements

Soumyajit Mukherjee invited to submit an article in this book. Achyuta Ayan Misra handled this chapter. Wiley Blackwell Acquisition Editors (present: Frank Weinreich, past: Andrew Harrison), Stacey Woods (Handling Editor), and proofreading team. Summarized in Misra and Mukherjee (2022).

Funding

Funding for the study was obtained from UGC FRPS BSR Start-Up Grant (F.30-78/2014(BSR) dated 22 January 2015). The high resolution satellite images were obtained courtesy of the DigitalGlobe Foundation via an Imagery Grant (3 August 2015). Both grants were awarded to PP Patel.

References

Alam, M., Alam, M.M., Curray, J.R. et al. (2003). An overview of the sedimentary geology of the Bengal Basin in relation to the regional tectonic framework and basin-fill history. *Sedimentary Geology* 155: 179–208.

Bagchi, K. and Mukherjee, K.K.N. (1983). *Diagnostic Survey of Rarh Bengal (Part II)*. Calcutta: University of Calcutta.

Bandyopadhyay, S. (1987). Man-initiated gullying and slope formation in a lateritic terrain at Santiniketan, West Bengal. *Geographical Review of India* 49: 21–26.

Bandyopadhyay, S. (1988). Drainage evolution in a badland terrain at Gangani in Medinipur District, West Bengal. *Geographical Review of India* 50: 10–20.

Bandyopadhyay, S. (2017). The Garhbeta badlands. In: *Atlas of Geomorphosites in India* (ed. V.S. Kale), 37–39. Allahabad, India: Indian Institute of Geomorphologists.

Castillo, C. and Gomez, J.A. (2016). A century of gully erosion research: urgency, complexity and study approach. *Earth Science Reviews* 160: 300–319.

Das, K. and Bandyopadhyay, S. (1995). Badland development over lateritic duricrusts. In: *Indian Geomorphology*, vol. 1 (ed. S.R. Jog), 31–41. New Delhi, India: Rawat Publications.

Dasgupta, S. and Mukherjee, S. (2017). Brittle shear tectonics in a narrow continental rift: asymmetric non-volcanic Barmer basin (Rajasthan, India). *The Journal of Geology* 125: 561–591.

Dasgupta, S. and Mukherjee, S. (2019). Remote sensing in lineament identification: examples from western India. In: *Problems and Solutions in Structural Geology and Tectonics*, Developments in Structural Geology and Tectonics Book Series, vol. 5. Series Editor: Mukherjee, S. (ed. A. Billi and A. Fagereng), 205–221. Elsevier. ISSN: 2542-9000. ISBN: 9780128140482.

Dasgupta, S., Biswas, M., Mukherjee, S., and Chatterjee, R. (2022). Depositional system, morphological signatures, tectonics and sedimentation pattern along the transform margin- Palar-Pennar basin, Indian east coast. *Journal of Petroleum Science & Engineering* 211: 110155. https://doi.org/10.1016/j.petrol.2022.110155.

Dey, S., Ghosh, S., Debbarma, C., and Sarkar, P. (2009). Some regional indicators of the tertiary-quaternary geodynamics in the paleo-coastal part of the Bengal basin (India). *Russian Geology and Geophysics* 50: 884–894. https://doi.org/10.1016/j.rgg.2008.10.007.

Gallart, F., Marignani, M., Perez-Gallego, N., and Maccherini, S. (2013). Thirty years of studies on badlands, from physical to vegetational approaches. A succinct review. *Catena* 106: 4–11.

Ghosh, S. and Bhattacharya, K. (2012). Multivariate erosion risk assessment of lateritic badlands of Birbhum (West Bengal, India): a case study. *Journal of Earth System Science* 121: 1441–1454. https://doi.org/10.1007/s12040-012-0243-1.

Ghosh, S. and Guchhait, S.K. (2019). Modes of formation, Palaeogene to early quaternary Palaeogenesis and geochronology of laterites in Rajmahal basalt traps and Rarh Bengal of lower Ganga Basin. In: *Quaternary Geomorphology in India* (ed. B.C. Das, S. Ghosh and A. Islam), 25–60. Singapore: Springer.

Ghosh, S. and Guchhait, S.K. (2020). *Laterites of the Bengal Basin- Characterization, Geochronology and Evolution*. Cham, Switzerland: Springer Nature.

Gogoi, M.P., Gogoi, B., and Mukherjee, S. (2022). Tectonic instability of the petroliferous upper Assam valley (NE India): a geomorphic approach. *Journal of Earth System Science* 131: 18.

Haque, S.M. and Ghosh, S. (2019). Microstructural evidence of palaeo-coastal landform from westernmost fringe of lower ganga-Brahmaputra delta. In: *Quaternary Geomorphology in India: Case Studies from the Lower Ganga Basin* (ed. B.C. Das, S. Ghosh and A. Islam), 61–78. Cham, Switzerland: Springer International https://doi.org/10.1007/978-3-319-90427-6_3.

Kaplay, R.D., Md, B., Mukherjee, S., and Kumar, T.V. (2017). Morphotectonic expression of geological structures in eastern part of south East Deccan volcanic province (around Nanded, Maharashtra, India). In: *Tectonics of the Deccan Large Igneous Province*, vol. 445 (ed. S. Mukherjee, A.A. Misra, G. Calvès and M. Nemčok), 317–335. Geological Society, London, Special Publications.

Mandal, R. and Chakrabarty, P. (2021). Badlands of Gangani, West Bengal, India: an assessment on account of geotourism development. *International Journal of Geoheritage and Parks*. 9 (2): 147–156. https://doi.org/10.1016/j.ijgeop.2021.02.001.

Misra, A.A. and Mukherjee, S. (2022). Introduction to "atlas of structural geological and geomorphological interpretation of remote sensing images". In: *Atlas of Structural Geological and Geomorphological Interpretation of Remote Sensing Images* (ed. A.A. Misra and S. Mukherjee). Wiley Blackwell. ISBN: 9781119813354.

Misra, A.A., Bhattacharya, G., Mukherjee, S., and Bose, N. (2014). Near N-S paleo-extension in the western Deccan region in India: does it link strike-slip tectonics with India-Seychelles rifting? *International Journal of Earth Sciences* 103: 1645–1680.

Patel, P.P. and Mondal, S. (2019). Terrain – Landuse relation in Garbeta-I block, Paschim Medinipur District, West Bengal. In: *Importance and Utilities of GIS* (ed. S. Mukherjee), 82–101. Burdwan, India: Avenel Press.

Patel, P.P., Mondal, S., and Prasad, R. (2020a). Modifications of the geomorphic diversity by anthropogenic interventions in the Silabati River basin. In: *Anthropogeomorphology of Bhagirathi-Hooghly River System in India* (ed. B.C. Das, S. Ghosh, A. Islam and S. Roy), 331–356. Routledge https://www.taylorfrancis.com/chapters/modifications-geomorphic-diversity-anthropogenic-interventions-silabati-river-basin-priyank-pravin-patel-sayoni-mondal-rishikesh-prasad/e/10.1201/9781003032373-10.

Patel, P.P., Mondal, S., Dasgupta, R. (2020b). Morphometric and channel erosivity analysis of lateritic gully catchments using high resolution DTM and repeat survey Structure-from-Motion datasets. *Proceedings of the Geomorphometry 2020 Conference* (pp. 251–253). Consiglio Nazionale delle Ricerche, Perugia, Italy. http://www.irpi.cnr.it/conference-files/geomorphometry-2020/67.pdf.

Patel, P.P., Dasgupta, R., and Mondal, S. (2020c). Using ground-based photogrammetry for fine-scale gully morphology studies: some examples. In: *Gully erosion Studies from India and Surrounding Regions* (ed. P.K. Shit, H.R. Pourghasemi and G.S. Bhunia), 207–220.

Cham, Switzerland: Springer https://doi.org/10.1007/978-3-030-23243-6_12.

Patel, P.P., Dasgupta, R., Chanda, S., and Mondal, S. (2021). An Investigation into Longitudinal Forms of Gullies within The 'Grand Canyon' of Bengal, Eastern India. *Transactions in GIS* https://doi.org/10.1111/tgis.12828.

Poesen, J., Nachtergaele, J., Verstraeten, G., and Valentin, C. (2003). Gully erosion and environmental change: importance and research needs. *Catena* 50: 91–133. https://doi.org/10.1016/S0341-8162(02)00143-1.

Sarkar, T., Mishra, M., and Chatterjee, S. (2020). On detailed field-based observations of laterite and laterization: a study in the Paschim Medinipur lateritic upland of India. *Journal of Sedimentary Environments* 5: 219–245. https://doi.org/10.1007/s43217-020-00017-4.

Sen, J., Sen, S., and Bandyopadhyay, S. (2004). Geomorphological investigation in badlands- a case study of Garhbeta, West Medinipur, India. In: *Geomorphology and Environment* (ed. S. Singh, H.S. Sharma and S.K. De), 204–234. Kolkata, India: ACB Publications.

Shit, P.K., Paira, R., Bhunia, G.S., and Maiti, R.K. (2015). Modelling of potential gully erosion hazard using geo-spatial technology at Garbheta block, West Bengal in India. *Modeling Earth Systems and Environment* 1: 2. https://doi.org/10.1007/s40808-015-0001-x.

Singh, L.P., Parkash, B., and Singhvi, A.K. (1998). Evolution of lower Gangetic plain landforms and soils in West Bengal, India. *Catena* 33: 75–104. https://doi.org/10.1016/S0341-8162(98)00066-6.

Appendix A

Details about 3D model prepared via SfM for image given in Figure 7.6 and how to open/operate it

The 3D model for Figure 7.6 was prepared via the Structure-from-Motion (SfM) technique using multiple photographs of this landscape element. A number of minor geomorphic features can be distinguished, which have formed along the channel walls of one of the main gullies in Gangani (this location has been marked in the Google Earth screenshot included in the Appendix). From the 3D model, piping features are evident, along with the development of a hollow in the left portion of the figure. Weathered and broken-down blocks that have detached from the main gully wall are seen toward the right. The alternate fine depositional beds are distinguishable by their differential colors, while nodules of weathered ferricrete are visible on some of the minor slope segments.

In the 3D pdf file for Figure 7.6 (available in the online repository), do the following to visualize the 3D model:

Open the pdf file in Adobe Reader. From the top right corner of the opened file, select the Options button (in yellow) and click on the option "Trust this document always." Then in the top left corner, click on the Question Mark (?) button that appears after the previous step. The 3D image will now appear. Once it does, use the mouse scroll button (central wheel) to zoom in and zoom out of the 3D model. To rotate the model and move it around, press and hold the left click button on the mouse and then drag it.

Appendix B

Google Earth screenshot to show locations of spots denoted in Figures 7.6–7.9

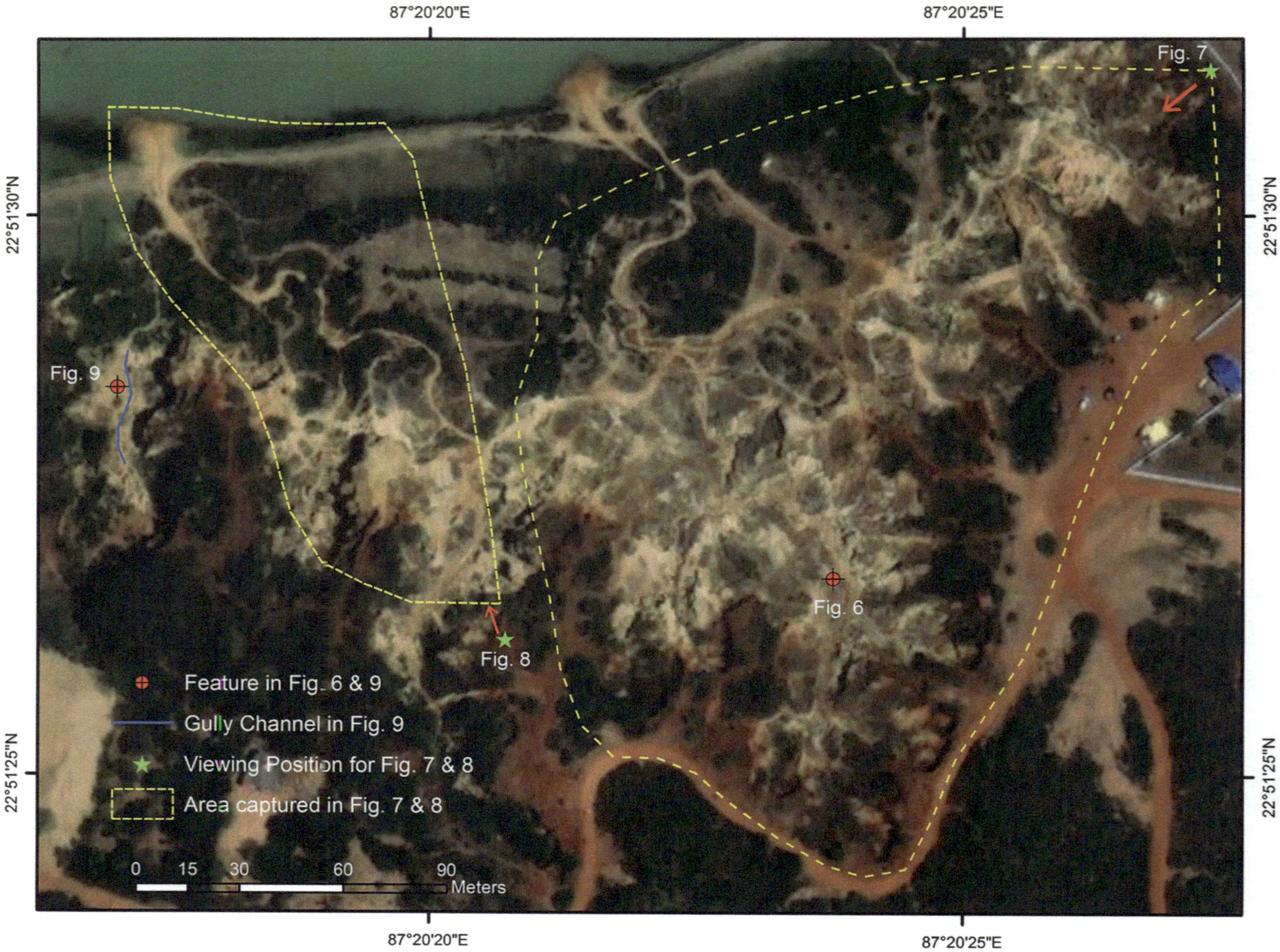

Source: ESRI World Imagery Map Base Layer (primarily, Maxar [DigitalGlobe], ~1 m resolution); Earthstar Geographics 2.5 m resolution.

8

Geomorphology along the West Coast of India, Through Remote Sensing

Chinmay U. Dongare[*,1], Bhushan S. Deota[1], Aditya U. Joshi[2], and Manoj A. Limaye[1]*

[1] Department of Geology, The M.S. University of Baroda, Vadodara, Gujarat, India
[2] Department of Civil Engineering, Manipal Institute of Technology, Manipal Academy of Higher Education, Manipal, Karnataka, India

8.1 Introduction

Remote sensing is an effective tool to decipher the structural geology and geomorphology of terrains (Misra et al. 2014; Dasgupta and Mukherjee 2017, 2019; Kaplay et al. 2017; Dasgupta et al. 2022; Gogoi et al. 2022). The Indian west coast has been of great importance for tectonic and petroleum geological viewpoints (e.g. Misra et al. 2015; Misra and Mukherjee 2017; Misra and Mukherjee 2018; Mukherjee et al. 2020). The coastline of Goa (Figure 8.1), which forms the southern part of Kokan coast originated after the India–Madagascar breakup at ~88 Ma, trends parallel to the west coast fault of India and is characterized by west-facing scarp extending for 1500 km (Sukhtankar 1995; Subrahmanya 1998).

8.2 Geomorphic Characteristics of the Goa Coast

The coast of Goa possesses significant geomorphic features such as headlands, sea cliffs, wave-cut platforms, islands, estuaries and bays, beaches, spits and bars, dunes, and recent broad alluvial plains (Figure 8.1). It is located between N 14° 55′ 0″-15° 45′ 00″ latitude and E 73° 40′ 00″ – 74°10′ 00″ longitude and stretches for about 100 km from the River Terekhol in the north to the River Talpona in the south. The Goa coast is highly indented with a presence of peculiar erosional and depositional landforms. Erosional landforms include headlands (at Baga, Aguada, Dona Paula, and Marmagoa), rocky platforms, cliffs (at Aguada and Marmagoa bays; Figure 8.3), estuaries (at Zuari and Mandovi). and islands (Piqueno, Grande, St. George, and St. Jacinto), indicative of palaeo regional submergence (Fieo 1956; Krishnan 1956; Davies and Clayton 1980; Kale and Rajguru 1983; Ghate 1986; Wagle 1987, 1991; Dholakia 1995; Bloom 1998; Dholakia et al. 2000; Martha et al. 2013). Depositional landforms characterize beaches (at Calangute and Colva), parallel coastal dunes, and spit (at Betul), indicative of emergent coastline. The entire setup of the Goa coast has preserved the imprints of physical changes that have occurred over geologic time in the form of geomorphic features, which are manifestations of combined effects of neotectonics, eustacy, subaerial processes of erosion, and deposition (Fieo 1956; Krishnan 1956; Kale and Rajguru 1983; Ghate 1986; Wagle 1987, 1991; Dholakia 1995; Dholakia et al. 2000; Martha et al. 2013) Figure 8.1.

These erosional and depositional features are mapped on False Color Composite (FCC) (NIR, Red, and Green band) prepared from October 2020 Landsat-8 Operational Land Imager (OLI) data with a resolution of 30 m (https://earthexplorer.usgs.gov) (Figure 8.2). However, the features such as rocky platforms (Figures 8.4a, b, 8.5b, and 8.6a), cliffs (Figures 8.4a, b) and spit (Figures 8.9a, 8.9b; Figure 8.11a) are more appreciable when viewed through high-resolution Google Earth imageries in 3D. Coasts with alternating headlands and half-heart shaped beaches are described as zeta-form beaches (Figure 8.12a, b). Each straight segment of zeta-form beach is rotated from the mean trend of the coast toward the dominant swell direction and offset relative to its neighbors (Yasso 1965; Davies and Clayton 1980; Bloom 1998; Claudino Sales et al. 2018). The synoptic view of Google Earth image reveals the significant development of zeta-form beaches (Figure 8.12a, b) at Bhatkal, Karnataka, India.

*Corresponding Author: chinmayd.avs@gmail.com

Atlas of Structural Geological and Geomorphological Interpretation of Remote Sensing Images, First Edition.
Edited by Achyuta Ayan Misra and Soumyajit Mukherjee.

Companion website: www.wiley.com/go/misra/RemoteSensingImages

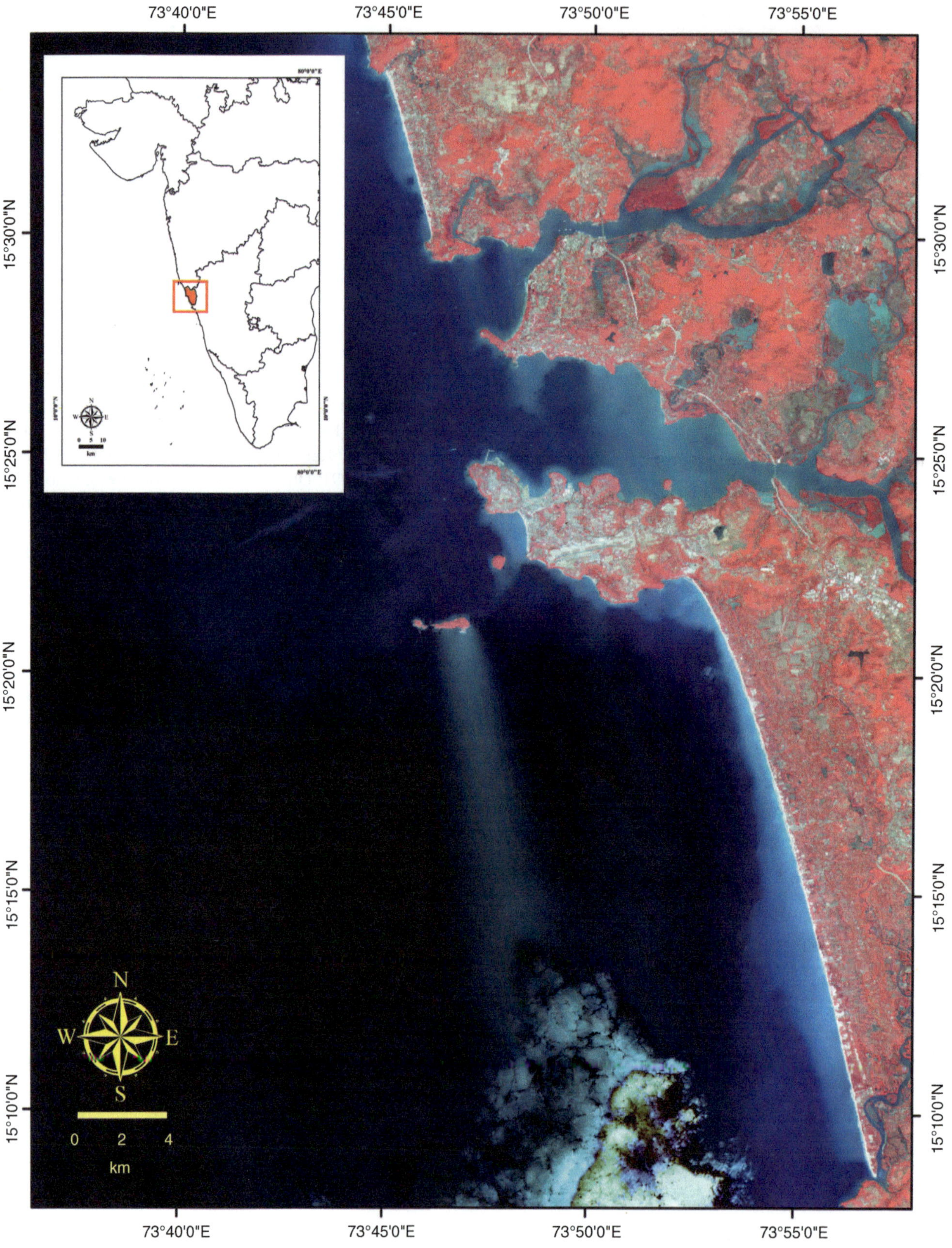

Figure 8.1 Uninterpreted False Color Composite (FCC) (NIR, Red and Green band) of Landsat – 8 OLI/TIRS satellite image between Baga in the North and Betul in the south. Subset displaying a political map of India with the location of Goa state highlighted by red color box and fill.

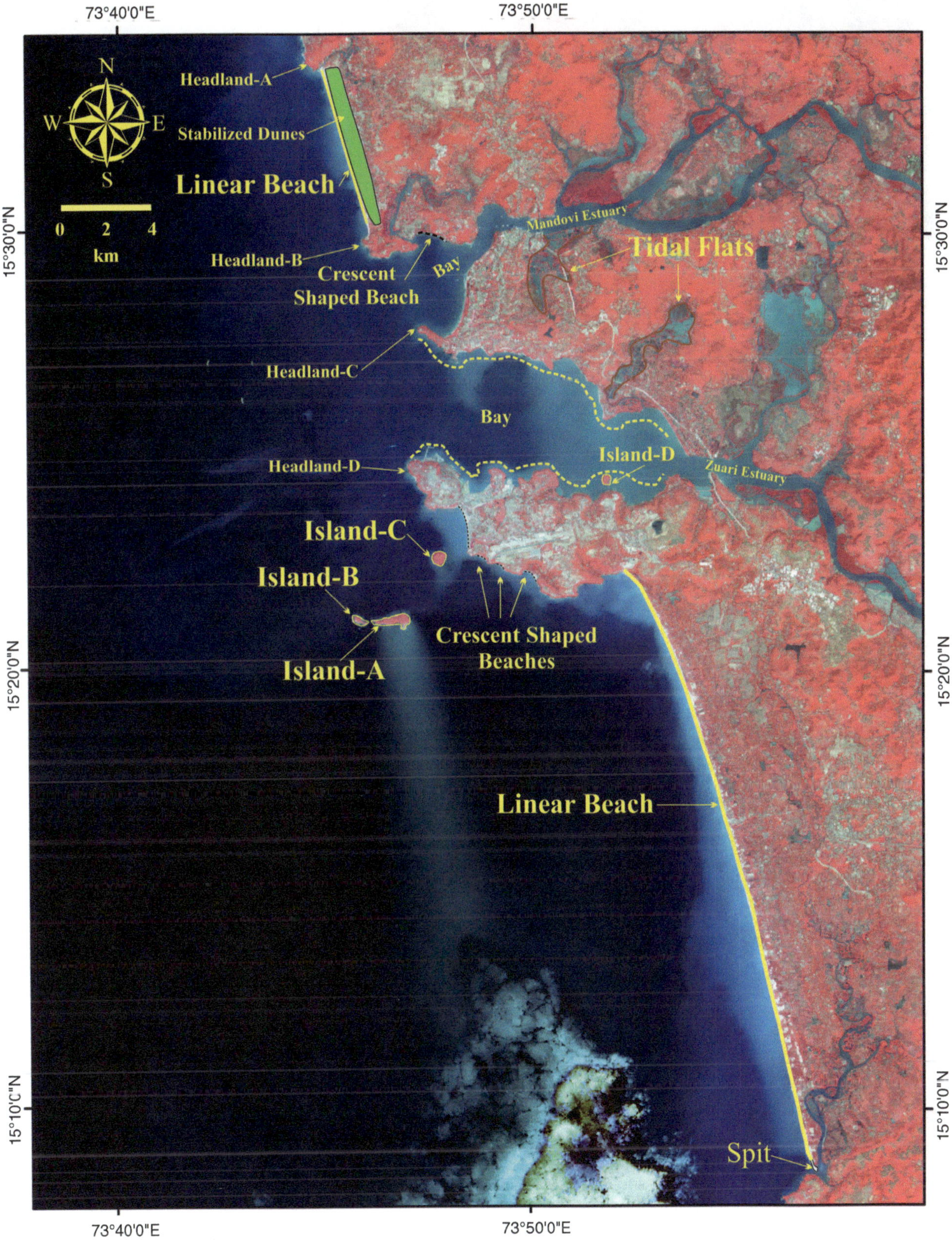

Figure 8.2 Interpreted Landsat 8 OLI/TIRS False Color Composite (FCC) (NIR, Red and Green band) satellite image of indented coastline of Goa displaying geomorphic features. viz. Headland-A = Baga; Headland-B = Aguada; Headland-C = Dona Paula; Headland-D = Marmagoa; Bay = Aguada (North) & Marmagoa (South); Island-A = Grande; Island-B = St. George; Island-C = Pequeno; Island-D = St. Jacinto; Linear Beach = Calangute (North) and Colva (South); Crescent Shape Beachs = At East of Aguada Headland and South of Marmagoa Headland; Spit = At the end of Colva beach near Betul; Stabilized Dunes = Parallel to Calangute Beach; Tidal Flats = Near Mandovi and Zuari Estuary.

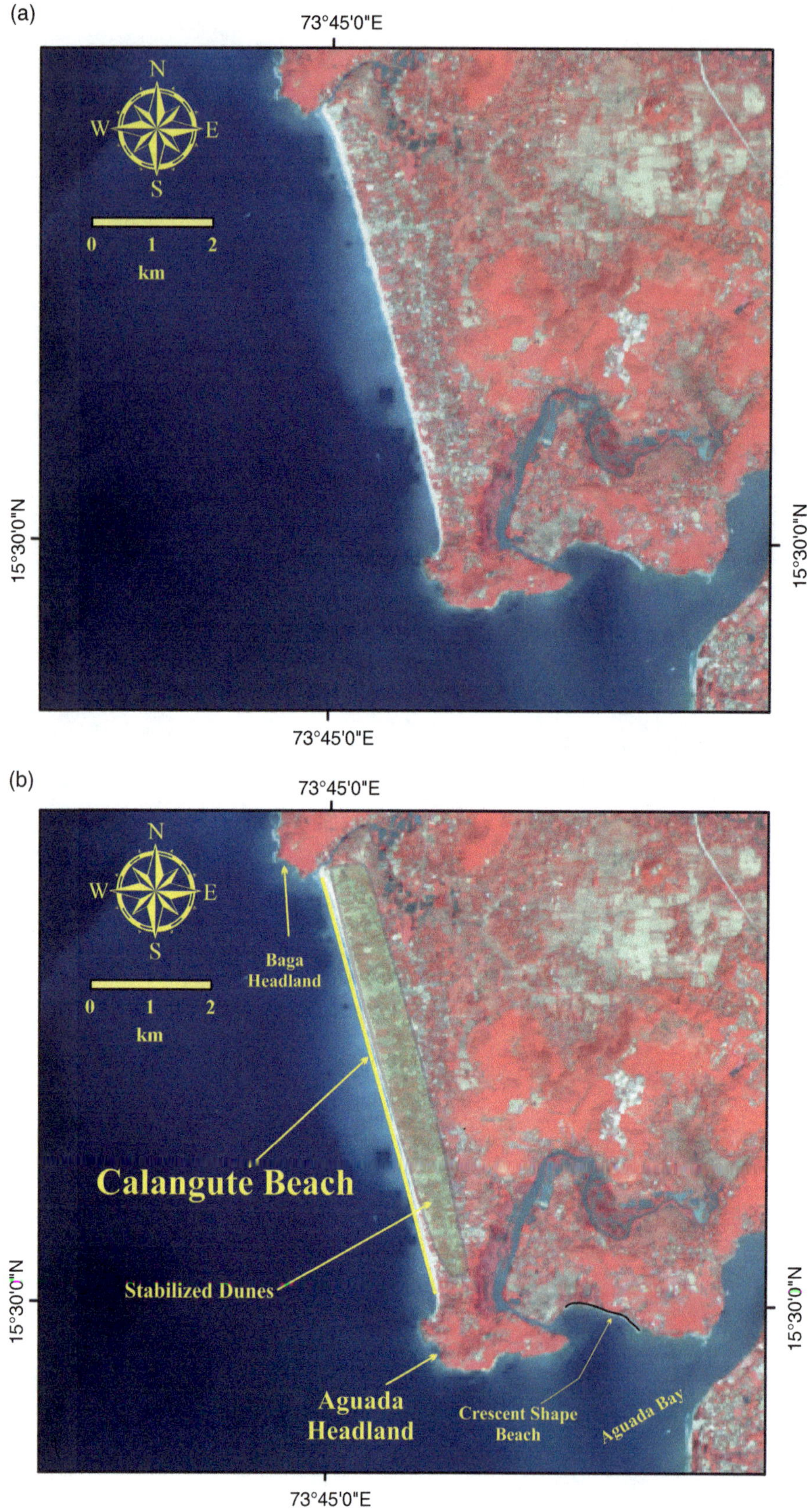

Figure 8.3 In the satellite data, headland exhibits red color and medium to coarse texture due to presence of vegetation, e.g. Baga and Aguada. Funnel-shaped blue color Aguada bay between Aguada and Dona Paula headlands is prominently seen on satellite data. Stabilized dunes behind Calangute beach exhibit mixed red and blue color and medium to mottled texture due to scattered vegetation.

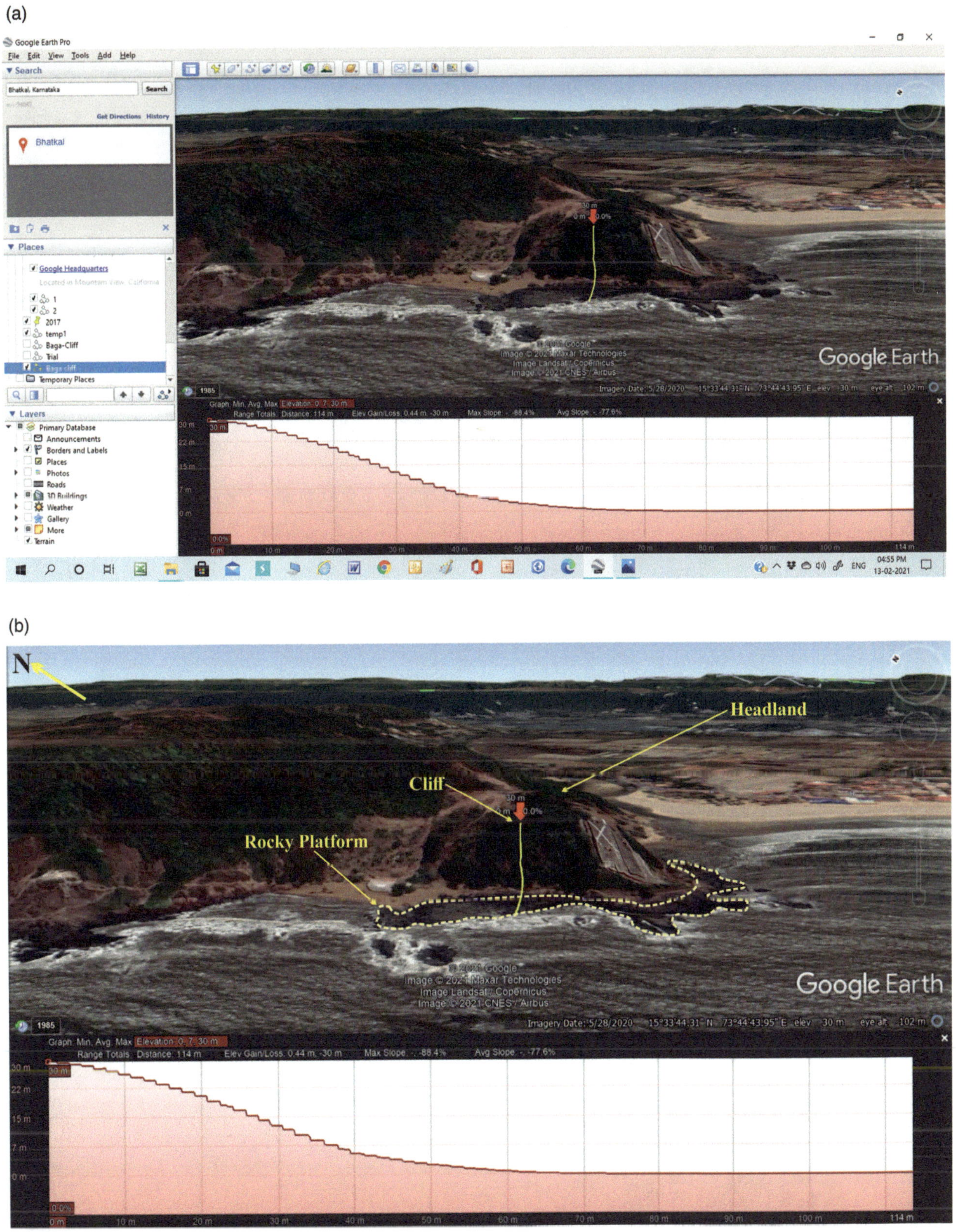

Figure 8.4 (a) Uninterpreted Google Earth 3 – Dimensional view of headland, cliff, and rocky platforms at Baga. (b) Interpreted Google Earth 3 – Dimensional view of headland, cliff, and rocky platforms at Baga.

(a)

Figure 8.5 (a) Photograph showing headland, cliff, and rocky platform at Baga. (b) Uninterpreted Google Earth 3-Dimensional view of headland and rocky platform at Aguada.

(a)

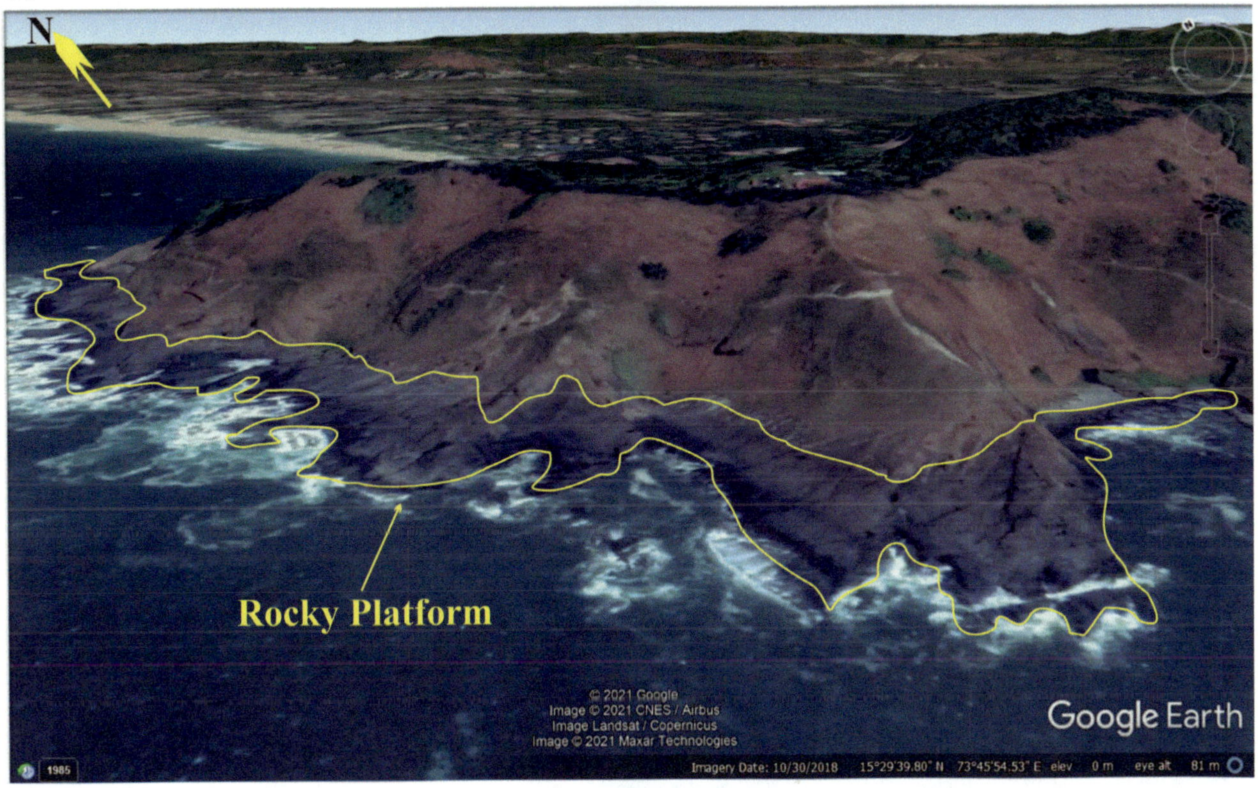

(b)

Figure 8.6 (a) Interpreted Google Earth three-dimensional view of rocky platforms at Aguada. (b) Photograph depicting seaward sloping rocky platforms at Aguada.

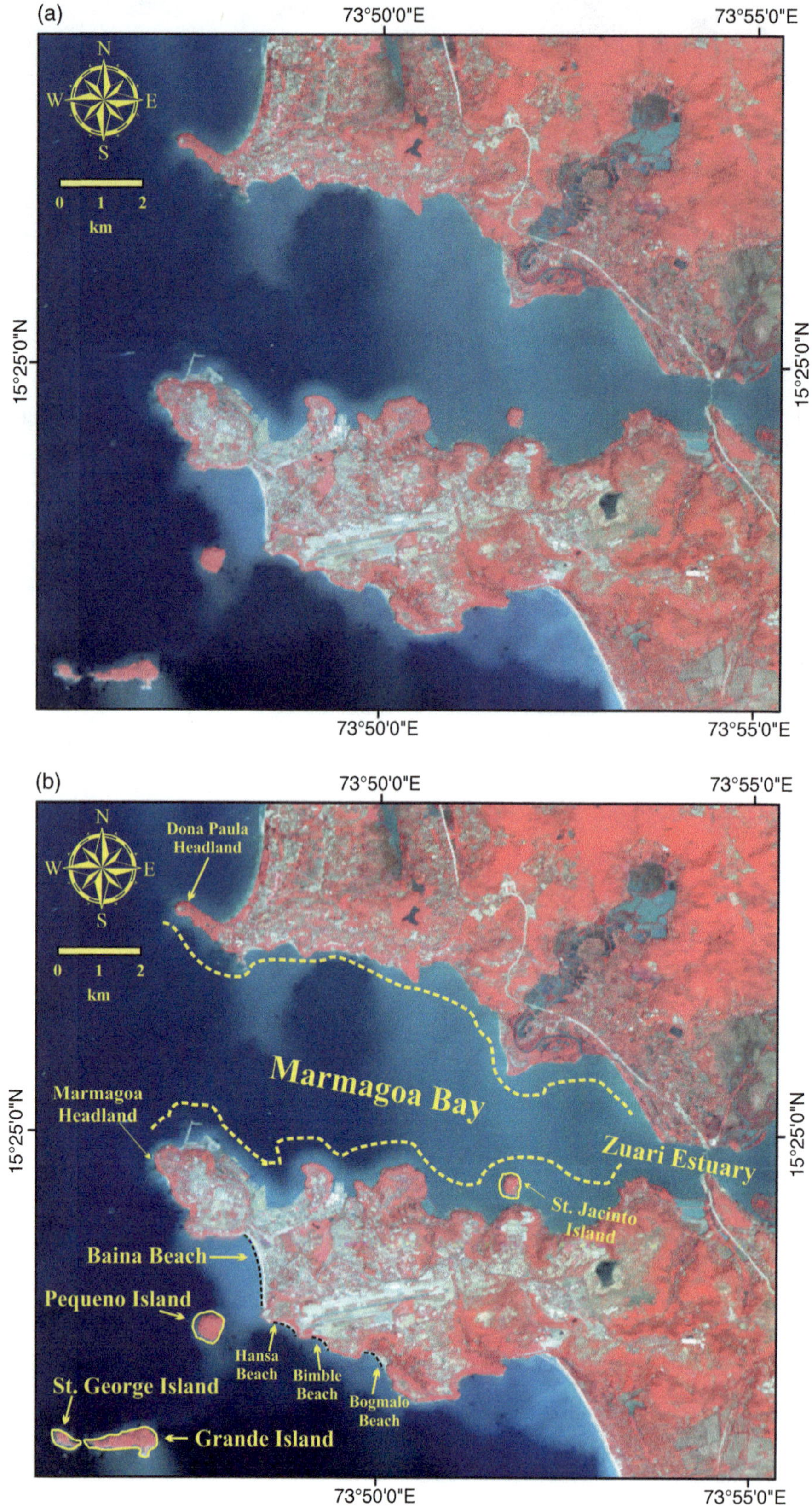

Figure 8.7 (a) Uninterpreted image and (b) interpreted image. Funnel-shaped blue color, fine texture on satellite data marks Marmagoa bay between Dona Paula and Marmagoa headlands. Islands of St. Jacinto, Pequeno, Grande, and St. George are conspicuously seen within and outside the Marmagoa bay on satellite image by their rounded to elliptical shape, moderate texture, and red color due to presence of vegetation.

(a)

(b)

Figure 8.8 (a) Photograph of St. Jacinto Island. (b) Photograph of Grande and St. George Islands captured from Marmagoa headland.

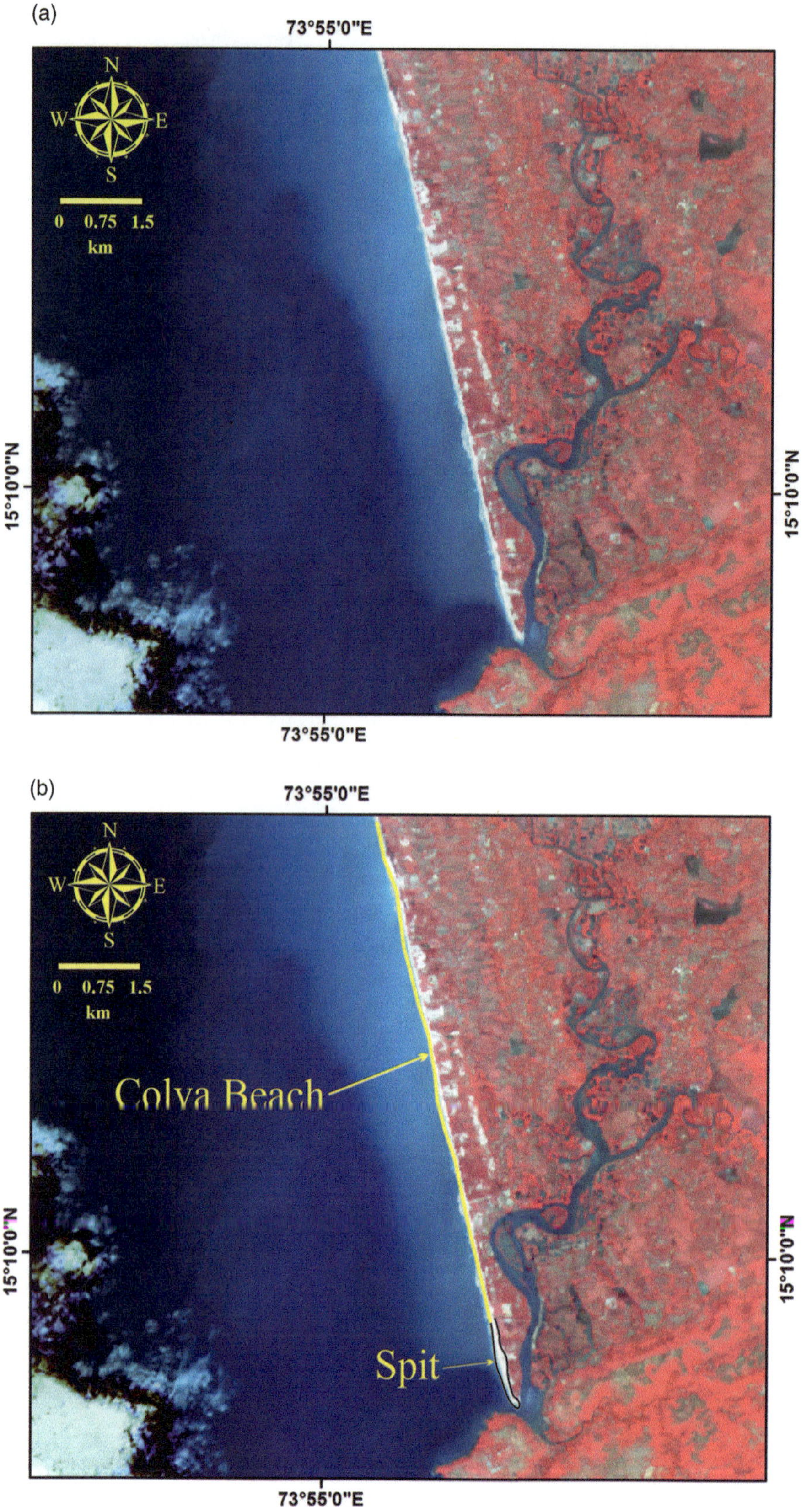

Figure 8.9 Linear beach at Calanguate (Figure 8.2) and Colva appear as white color and fine texture on the satellite data. Image (a) is uninterpreted and (b) is interpreted. Southward extension of Colva beach with white color and medium to fine texture projecting into the water body is marked as Betul spit. It is more clearly appreciated when seen through the three-dimensional view of Google Earth (Figure 8.11a).

(a)

(b)

Figure 8.10 (a and b) Photograph of straight linear beach at Calangute and Colva, respectively.

(a)

(b)

Figure 8.11 (a) Three-dimensional Google Earth view of anchor-shaped spit of Betul. (b) Photograph revealing crescent-shaped beach to the east of Aguada (Figure 8.2b).

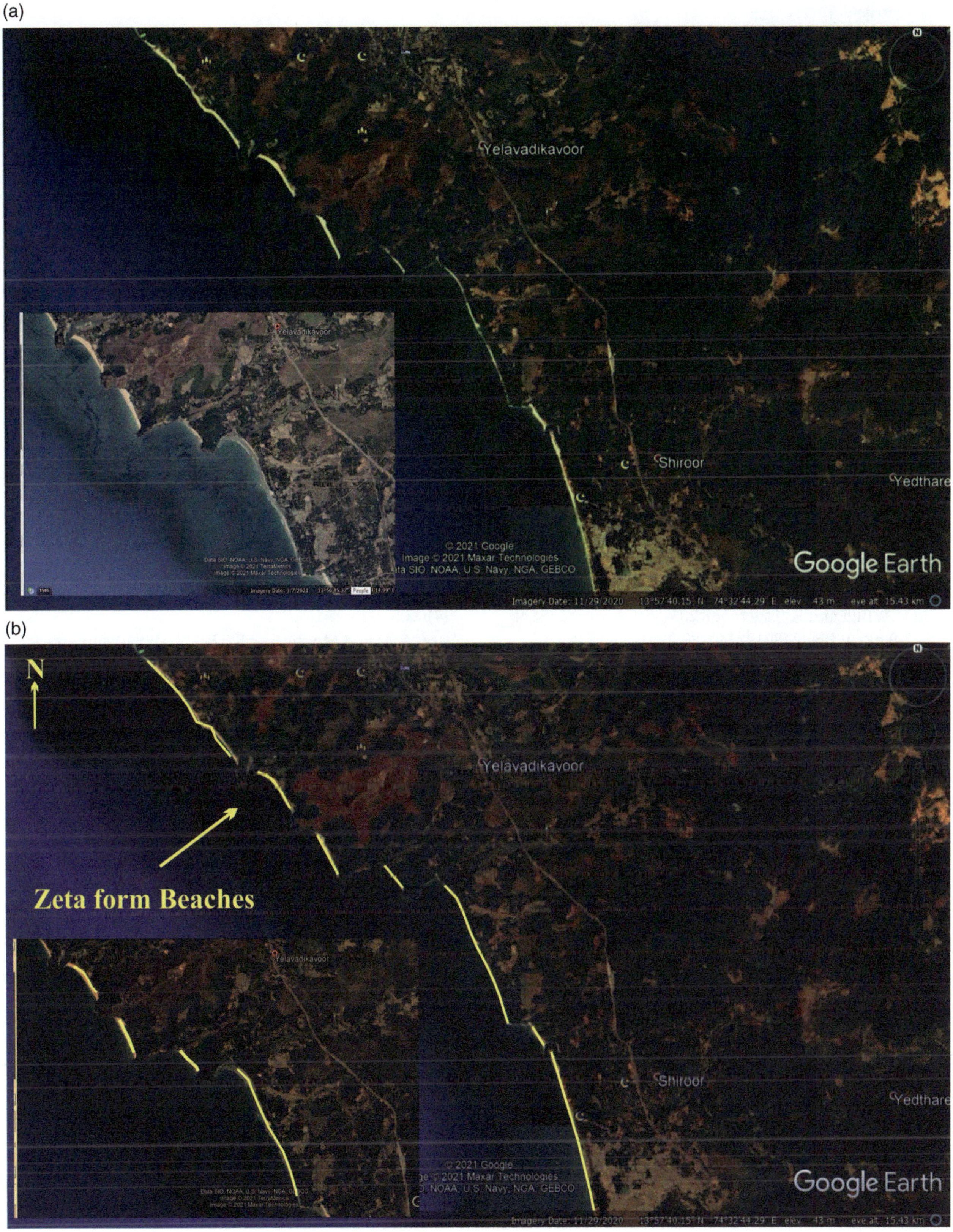

Figure 8.12 (a) Uninterpreted synoptic view of Google Earth image of Zeta form beaches at Bhatkal, Karnataka. (b) Interpreted Google Earth image representing alternate headlands and half-heart shaped (Zeta form) beaches, Bhatkal, Karnataka.

Acknowledgements

Soumyajit Mukherjee invited to submit an article in the edited book. Achyuta Ayan Misra handled this article. Wiley Blackwell Acquisition Editors (present: Frank Weinreich, past: Andrew Harrison), Stacey Woods (Handling Editor) and proofreading team. Summarized in Misra and Mukherjee (2022).

References

Bloom, A.L. (1998). *Geomorphology: A Systematic Analysis of Late Cenozoic Landforms* (No. 551.41 B5.)., 436–437. Upper Saddle River, New Jersey: Prentice Hall.

Claudino-Sales, V., Wang, P., and Carvalho, A.M. (2018). Interactions between various headlands, beaches, and dunes along the coast of Ceará state, Northeast Brazil. *Journal of Coastal Research* 34 (2): 413–428.

Dasgupta, S. and Mukherjee, S. (2017). Brittle shear tectonics in a narrow continental rift: asymmetric non-volcanic Barmer basin (Rajasthan, India). *The Journal of Geology* 125: 561–591.

Dasgupta, S. and Mukherjee, S. (2019). Remote sensing in lineament identification: examples from western India. In: *Problems and Solutions in Structural Geology and Tectonics*, Developments in Structural Geology and Tectonics Book Series, vol. 5. Series Editor: Mukherjee, S (ed. A. Billi and A. Fagereng), 205–221. Elsevier. ISSN: 2542-9000. ISBN: 9780128140482.

Dasgupta, S., Biswas, M., Mukherjee, S., and Chatterjee, R. (2022). Depositional system, morphological signatures, tectonics and sedimentation pattern along the transform margin-Palar-Pennar basin, Indian east coast. *Journal of Petroleum Science & Engineering.* 211: 110155. https://doi.org/10.1016/j.petrol.2022.110155.

Davies, J.L. and Clayton, K.M. (1980). *Geographical Variation in Coastal Development*, vol. 1522168. London: Longman.

Dholakia, A. (1995). Studies on the Quaternary landforms of Goa coast. Unpublished thesis. Department of Geology, The M.S University of Baroda

Dholakia, A., Deota, B., Desai, N. (2000). Manifestation of Holocene sea-level changes along Goa coast. *Proceedings of the International Quarterly Seminar on INQUA shoreline Indian Sub-Commission*. pp. 70–73

Fieo, M. (1956). Problems geomorphologicas de Goa Garcia de orta. Rev. Junta Miss Geogr. Invest. UltranNumero especial, 78. p.39

Ghate. S.N. (1986). Karal – An intertidal formation of Konkan, its petrology and problems of dating. *Proceedings of the National Workshop of Carbonates P.R.L.*, Ahmedabad (Abs) pp. 41–42.

Gogoi, M.P., Gogoi, B., and Mukherjee, S. (2022). Tectonic instability of the petroliferous upper Assam valley (NE India): a geomorphic approach. *Journal of Earth System Science* 131: 18.

Kale, V.S. and Rajguru, S.N. (1983). Mid – Holocene fossil wood from Colva, Goa. *Current Science* 52: 778.

Kaplay, R.D., Md, B., Mukherjee, S., and Kumar, T.V. (2017). Morphotectonic expression of geological structures in eastern part of south East Deccan volcanic province (around Nanded, Maharashtra, India). In: *Tectonics of the Deccan Large Igneous Province*, vol. 445 (ed. S. Mukherjee, A.A. Misra, G. Calvès and M. Nemčok), 317–335. Geological Society, London, Special Publications.

Krishnan, M.S. (1956). Tectonics of India. Bull. of the Nat. Inst. of Sci. India, 32.

Martha, T.R., Ghosh, D., Kumar, K.V. et al. (2013). Geospatial technologies for national geomorphology and lineament mapping project – a case study of Goa state. *Journal of the Indian Society of Remote Sensing* 41 (4): 905–920.

Misra, A.A. and Mukherjee, S. (2017). Dyke-brittle shear relationships in the Western Deccan strike slip zone around Mumbai (Maharashtra, India). In: *Tectonics of the Deccan Large Igneous Province*, vol. 445 (ed. S. Mukherjee, A.A. Misra, G. Calvès and M. Nemčok), 269–295. Geological Society, London, Special Publications.

Misra, A.A. and Mukherjee, S. (2018). *Atlas of Structural Geological Interpretation from Seismic Images*. Wiley Blackwell. ISBN: 978-1-119-15832-5.

Misra, A.A. and Mukherjee, S. (2022). Introduction to "atlas of structural geological and geomorphological interpretation of remote sensing images". In: *Atlas of Structural Geological and Geomorphological Interpretation of Remote Sensing Images* (ed. A.A. Misra and S. Mukherjee). Wiley Blackwell. ISBN: 9781119813354.

Misra, A.A., Bhattacharya, G., Mukherjee, S., and Bose, N. (2014). Near N-S paleo-extension in the western Deccan region in India: does it link strike-slip tectonics with India-Seychelles rifting? *International Journal of Earth Sciences* 103: 1645–1680.

Misra, A.A., Sinha, N., and Mukherjee, S. (2015). Repeat ridge jumps and microcontinent separation: insights from NE Arabian Sea. *Marine and Petroleum Geology* 59: 406–428.

Mukherjee, S., Dole, G., Yatheesh, V., and Kale, V. (2020). Tectonics of the deccan trap: Focus on Indian geoscientists' contribution in the last four years. *Glimpse of Geoscience Researches in India. Proceedings of the Indian National Science Academy* 86: 237–244. https://doi.org/10.16943/ptinsa/2020/49816.

Subrahmanya, K.R. (1998). Tectono-magmatic evolution of the west coast of India. *Gondwana research* 1 (3–4): 319–327.

Sukhtankar, R.K. (1995). An evolutionary model based on geomorphologic and tectonic characteristics of the Maharashtra coast, India. *Quaternary International* 26: 131–137.

Wagle, B.G. (1987). Geomorphology and evolution of coastal and offshore areas of Maharashtra and Goa, India. Unpubl. Ph.D. Thesis, Bombay University.

Wagle, B.G. (1991). Shores of the central west coast of India-a case study using remote sensing data. *Mahasagar* 24 (1): 13–20.

Yasso, W.E. (1965). Plan geometry of headland-bay beaches. *The Journal of Geology* 73 (5): 702–714.

9

Fluvial Geomorphology in a Part of the Spiti River Basin, Himachal Pradesh, India

*Achyuta Ayan Misra**

Reliance Industries Ltd., Reliance Corporate Park, Mumbai, Maharashtra, India

9.1 General Geology

Remote sensing has been a powerful tool to interpret the structural geology and geomorphology of terrains (e.g. Misra et al. 2014; Kaplay et al. 2017; Dasgupta and Mukherjee 2017, 2019; Dasgupta et al. 2022; Gogoi et al. 2022). The Spiti River, a tributary of the Sutlej River, flows through the rain-shadow zone of the picturesque valleys of the Lahaul-Spiti district of Himachal Pradesh (Binita et al. 2009; Gill 2010; Srivastava et al. 2013). It starts from a glacier in the Kunzum Range with the mouth of the glacier at (Lat: 32°32′34″N; Long: 77°37′ E). The Spiti River constitutes a series of minor snow-fed tributaries from the glaciers on the northern slopes of the Great Himalayan Ranges. It flows through the Spiti Valley in a general southeasterly direction and drains into the Sutlej at Namgia. Rivers draining the famous Pin Valley drainage area are also a part of the Spiti river system (Srivastava et al. 2013). The general geomorphology of the Spiti River basin is complex (Phartiyal et al. 2009b; Kujur 2013). The topography of the terrain is immature (Phartiyal and Kothyari 2012) and consists of deep valleys with steep edges, high mountain ranges, and a number of lakes (Binita et al. 2009; Phartiyal et al. 2009a). The entire area is a "cold desert" and receives scant rainfall. The rivers are all glacier-fed during the summers when the glaciers melt. The area is underlain by rocks of Neoproterozoic, Paleozoic, Mesozoic, and Cenozoic eras (Bhargava 2008). The lithostratigraphic layers generally trend NNE–SSW and NE–SW with steep slopes, and can be accessed along the river valleys.

9.2 Image Interpretation

The Spiti valley is characterized by glacio-fluvial geomorphology. However, this work depicts only the various fluvial geomorphic features along the Spiti River on high-resolution images provided by Google Earth (Figures 9.1–9.16). These images are high-resolution (typically <1 m) data products offered by Maxar Technologies and/or CNES Airbus. The red, green, and blue EM bands are merged into red, green, and blue channels in image processing, generating "True Color Composites" that generate naturally colored images. The imagery dates are different for each image and are mentioned in each image. The satellite image interpretation is supported with ground truth photographs from the field area.

*Corresponding Author: achyutaayan@gmail.com

Atlas of Structural Geological and Geomorphological Interpretation of Remote Sensing Images, First Edition.
Edited by Achyuta Ayan Misra and Soumyajit Mukherjee.

Companion website: www.wiley.com/go/misra/RemoteSensingImages

Figure 9.1 Hill-shaded topographic map of the Spiti River basin showing the geomorphology of the area. High mountain ranges appear in the entire area, traversed by deep gorges cut by swift-flowing rivers in their youth phases. River Beas and River Spiti are the major rivers cutting across this region. There are numerous glaciers that accumulate snow and feed these rivers. In this work, only fluvial geomorphology is depicted, so the glacial geomorphic features are out of scope here. Source: Image courtesy of Google maps in "Terrain view" mode. Inset: India map with the study area marked in red. Red dots: positions of ground truth locations.

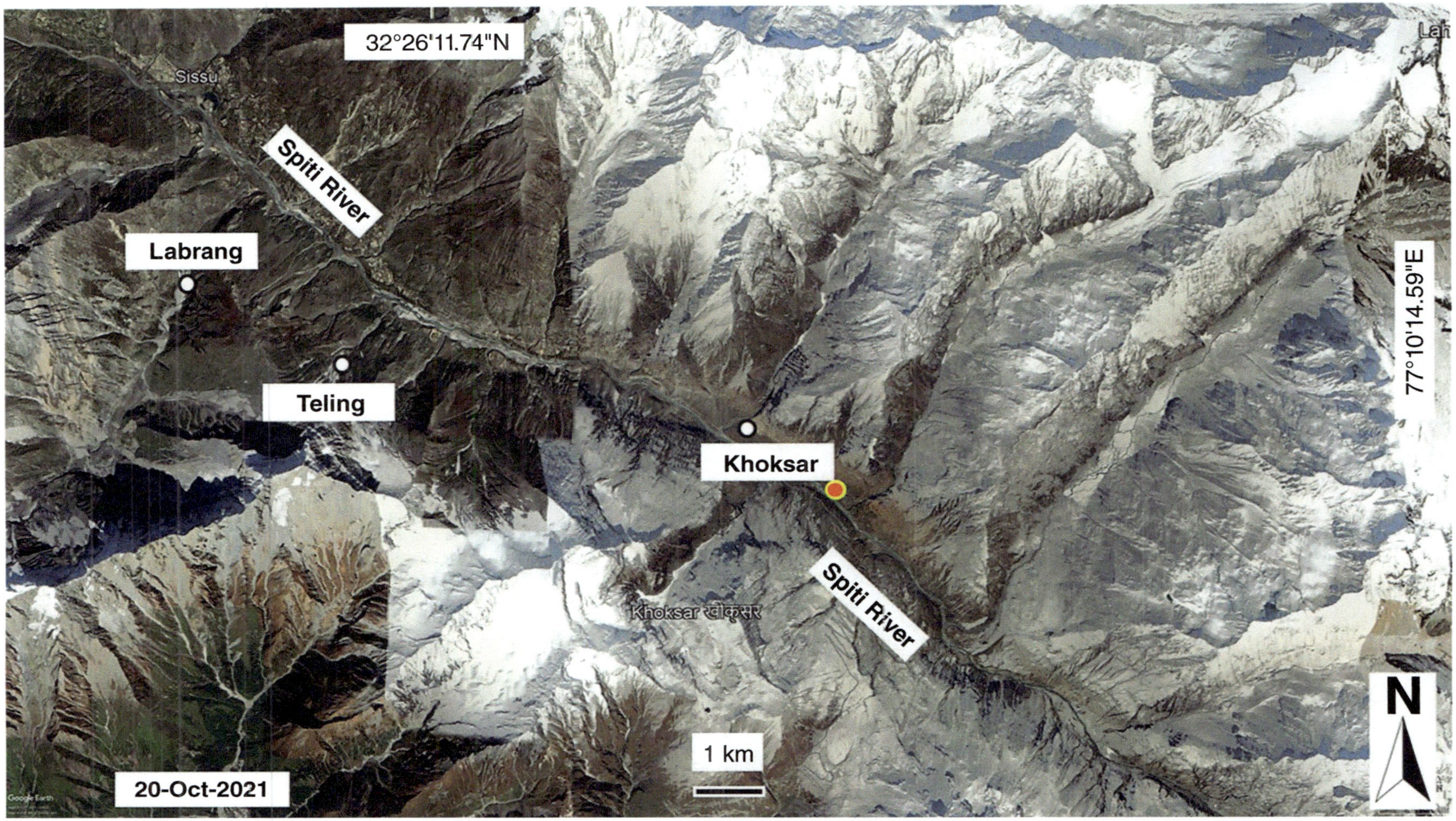

Figure 9.2 Satellite image showing the Spiti River valley flowing from NW to SE, and the high mountain ranges flanking either side with steep valleys. There are a number of glacier-fed tributaries feeding water into the Spiti River. All these tributaries are short and drain either from NE or SW into the trunk channel of the Spiti River. Red dot: ground truth location in Figure 9.3. Source: Image courtesy of Google Earth.

Figure 9.3 Field photograph showing the deep gorge formed by the Spiti River flanked by the steep edges of the valley. Location of photograph marked as a red dot in Figure 9.2.

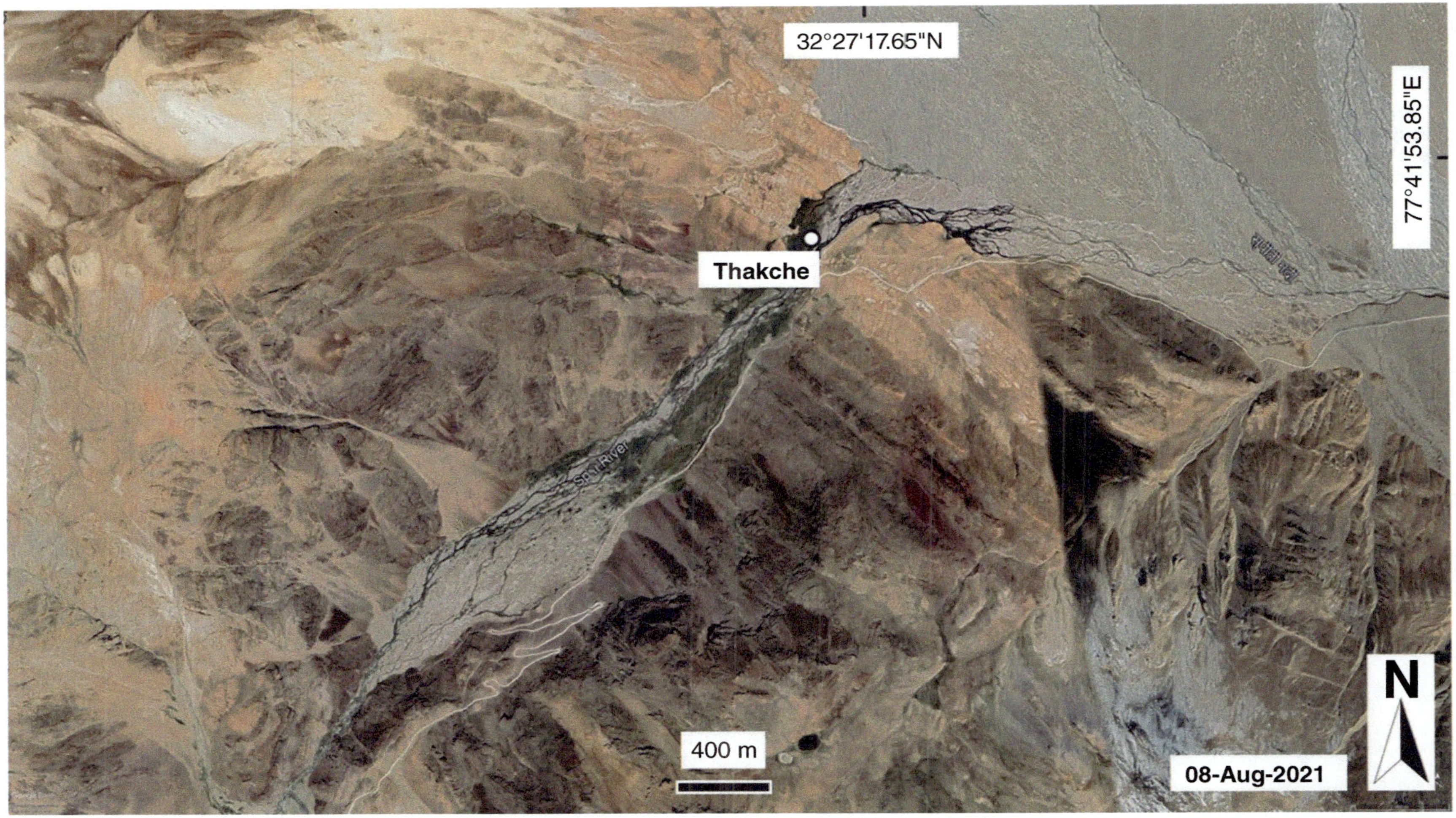

Figure 9.4 Clean satellite image of Thakche area showing the Spiti River bed and a tributary of the Spiti River, called ThakcheNalla. Source: Image courtesy of Google Earth.

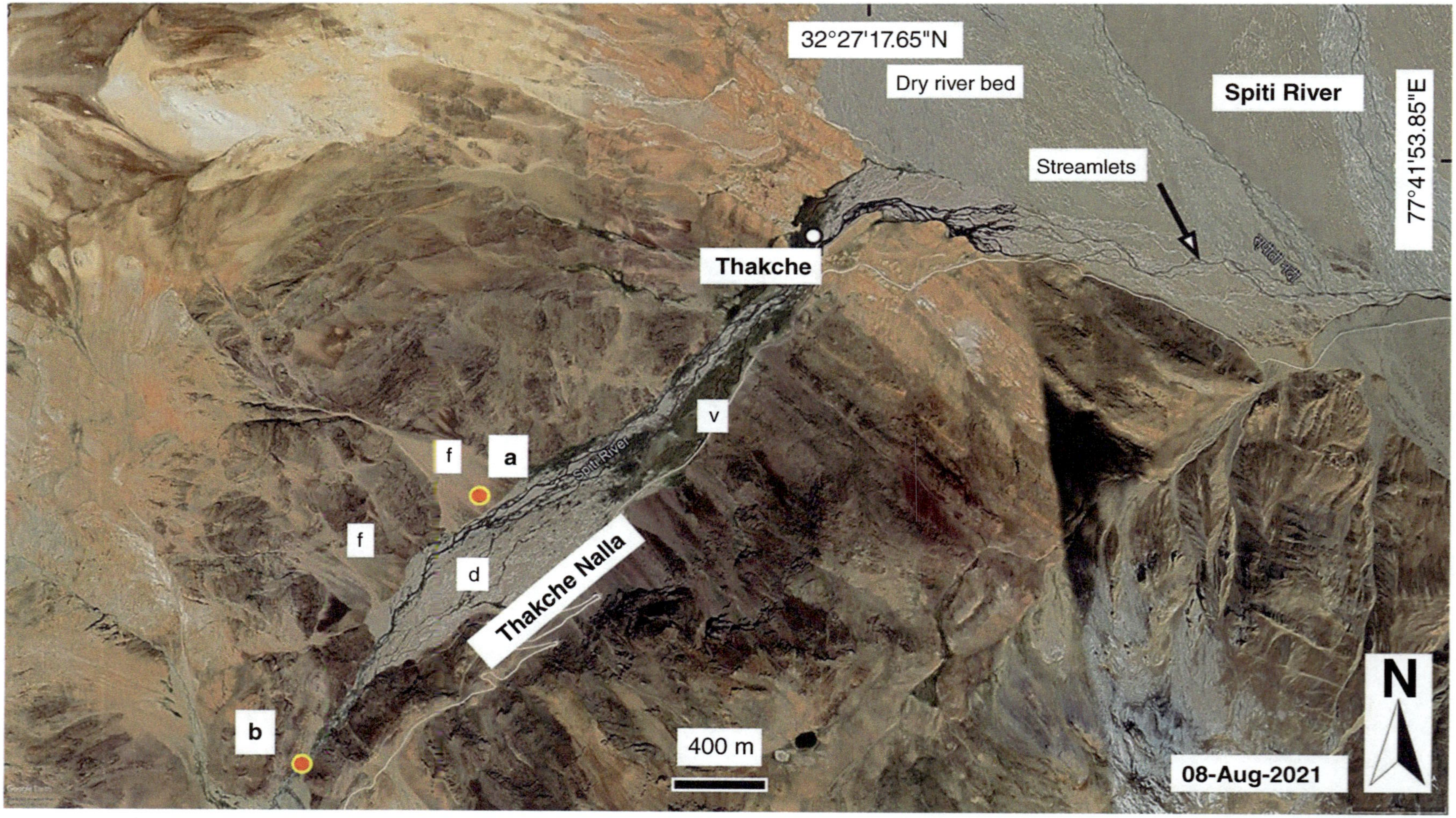

Figure 9.5 Annotated version of satellite image in Figure 9.4. The expansive dry riverbed of the Spiti River is visible in the upper right of the image. The riverbed has large boulders, indicating the ephemeral high-energy streams that bring in sediments in the system. The image also shows thin streamlets that carry low-volume streams from the tributaries into the main channel. The tributary of the Spiti River called ThakcheNalla is also a seasonal channel, which had a broad unvegetated dry riverbed (marked "d") and a vegetated riverbed (marked "v"). The vegetated area has finer sediments that allow for the growth of vegetation on the riverbed. There are a numerous colluvial fans (marked "f") formed on the edges of the ThakcheNalla, developed by the activity of the short streams bringing sediments during seasonal floods. The ThakcheNalla is a narrow stream in the SW and develops into a broad stream as it reaches the Spiti River, possibly because of the change of slope along the course of the stream. Red dots: ground truth location in Figure 9.6. Source: Image courtesy of Google Earth.

Figure 9.6 Field photographs for ground truth for the location shown in Figure 9.5. (a) Colluvial fans forming at the valley floor; (b) narrow channel of the ThakcheNalla at the upper part of the stream.

Figure 9.7 Clean satellite image of Thakche area showing the Spiti River bed. Source: Image courtesy of Google Earth.

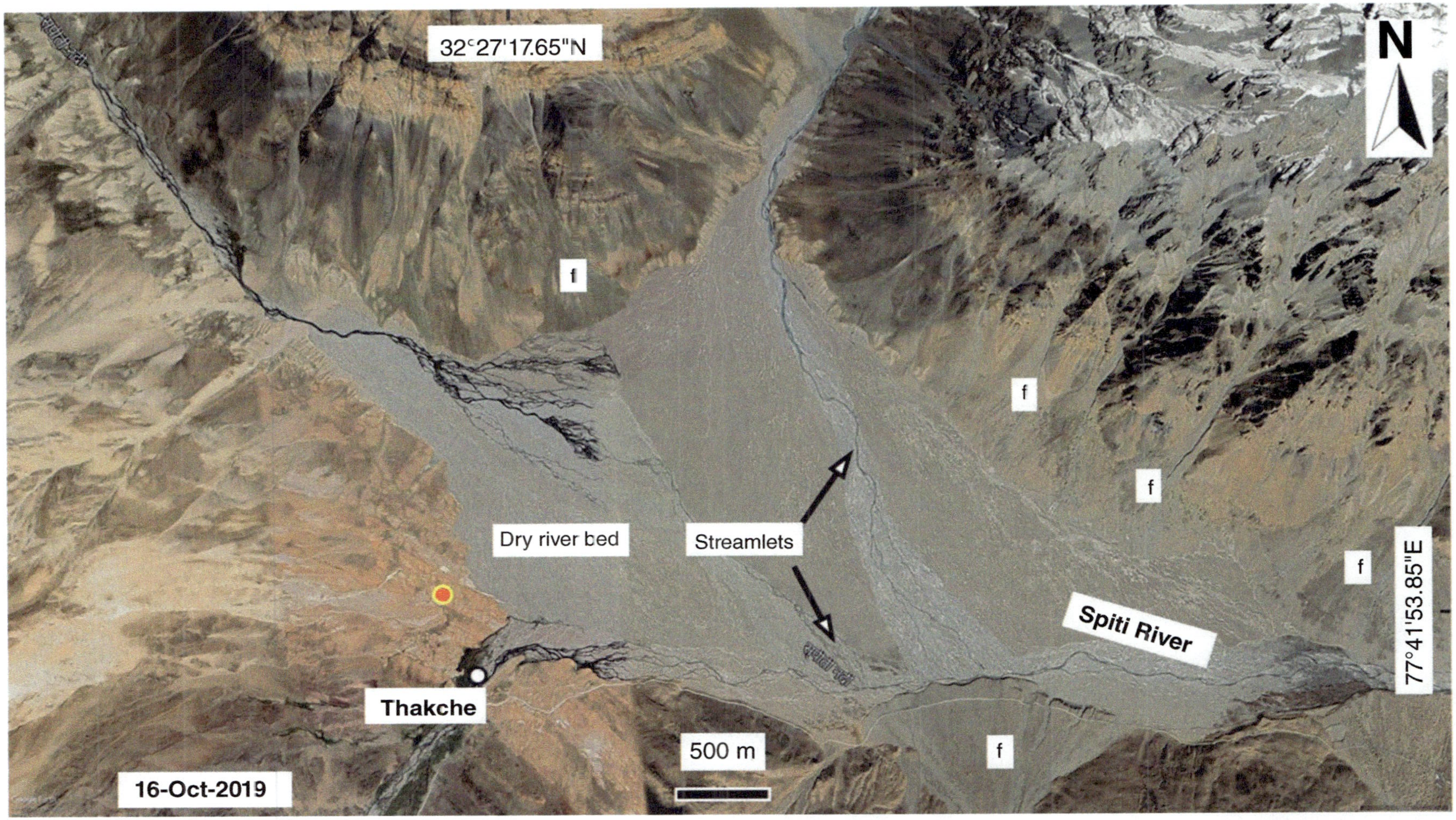

Figure 9.8 Annotated version of satellite image in Figure 9.7. The vast unvegetated dry riverbed of the Spiti River is visible in most of the image. There are numerous colluvial fans (marked "f") in the image. The riverbed deposits have different colors viz. mauve, light gray, ochre, which indicate the various layers of rocks eroded at the provenance during each flooding event. Such differences can only be picked up on a satellite image. Red dot: location from where the ground truth image (Figure 9.9) was taken. Source: Image courtesy of Google Earth.

Figure 9.9 Field photograph looking toward E from the location marked in Figure 9.8. Note the broad riverbed and the colluvial fans on the base of the hills in the distance.

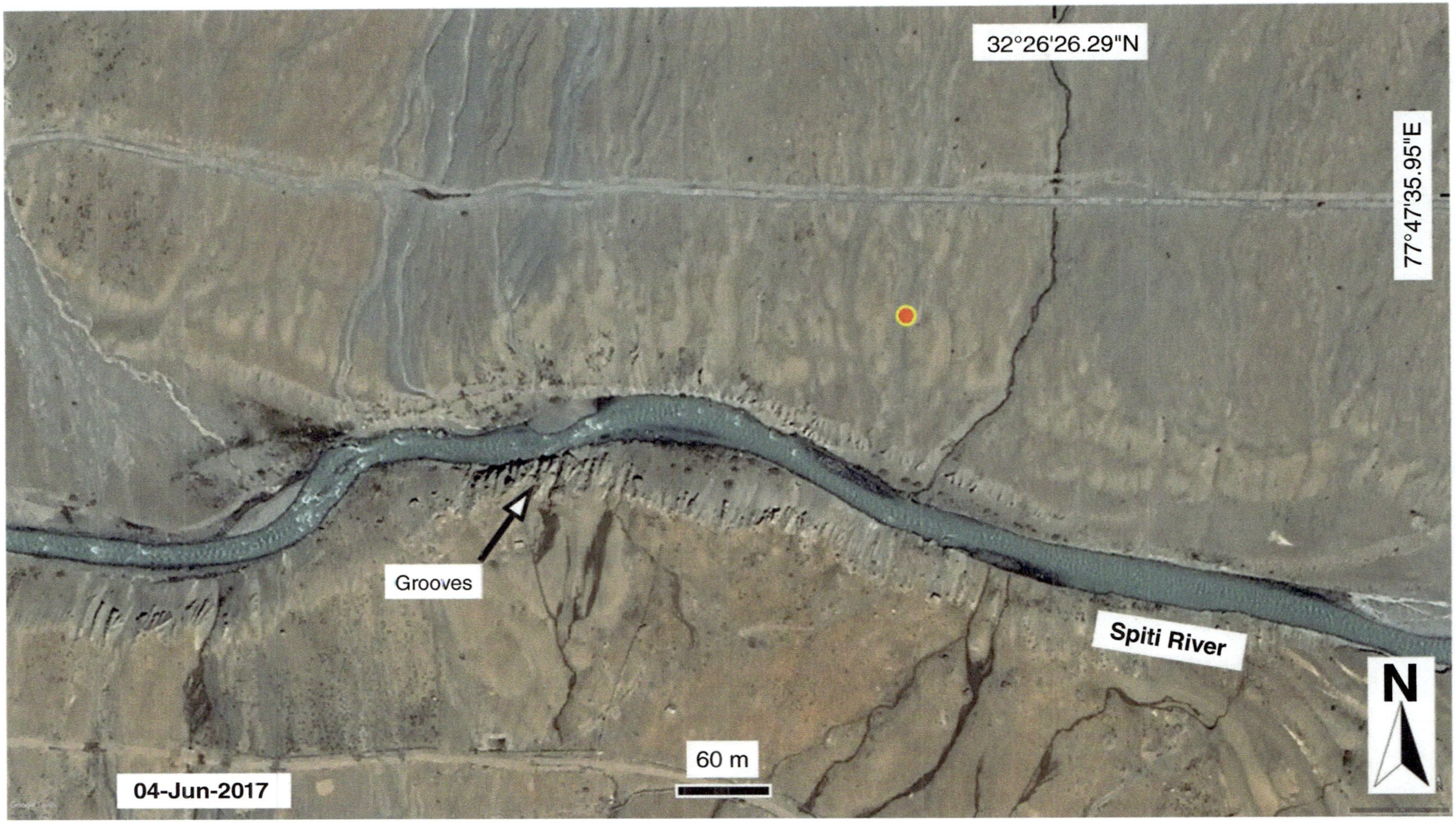

Figure 9.10 Annotated satellite image showing the Spiti River and grooves on the bank. Red dot: location from where the ground truth image (Figure 9.11) was taken. Source: Image courtesy of Google Earth.

Figure 9.11 Field photograph looking toward S from the location marked in Figure 9.10. Note the grooves on the banks, which form because of differential erosion of the banks due to wind and river action. The dark colored horizontal level is the maximum flooding level during the previous season.

Figure 9.12 Clean satellite image of Key Monastery area showing the steep mountain cliffs to the NE and Spiti River to the SW. Source: Image courtesy of Google Earth.

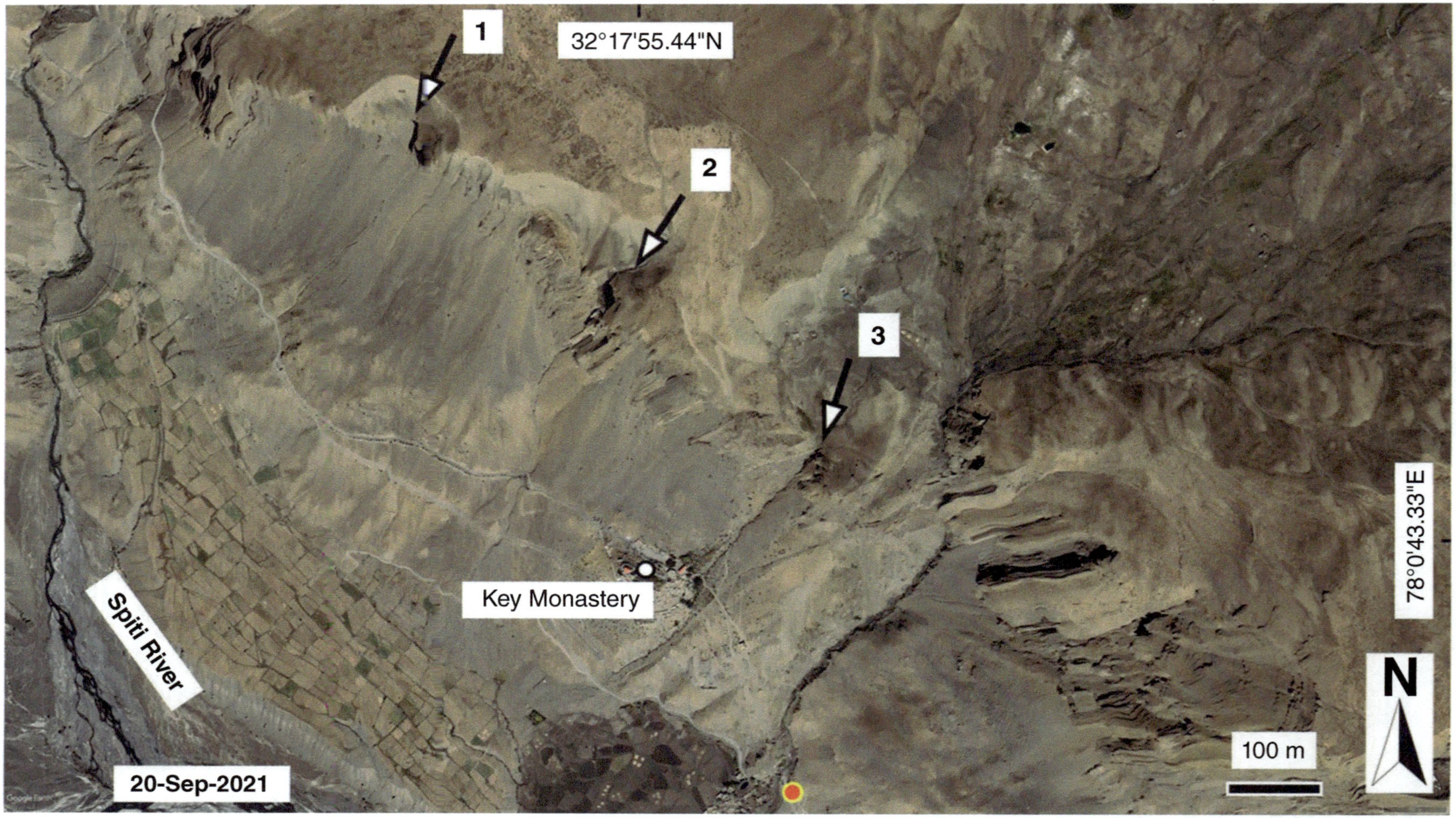

Figure 9.13 Annotated version of satellite image in Figure 9.12. The Key Monastery and the Spiti River are marked. 1, 2, and 3 are three hanging valleys, which were formed due to glacial action in the past, however they currently are fluvial river valleys, bringing in sediments in ephemeral (1) to perineal (2 and 3) flow along their streams. Red dot: location from where the ground truth image (Figure 9.14) was taken. Source: Image courtesy of Google Earth.

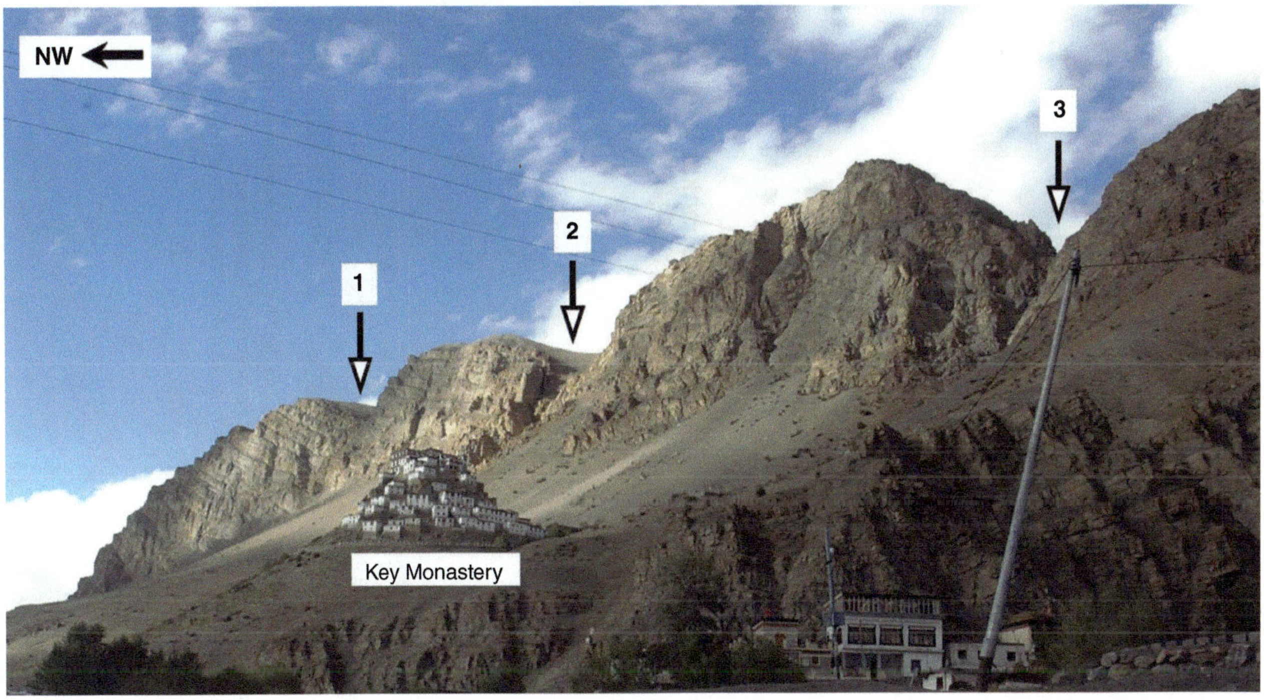

Figure 9.14 Field photograph looking toward S from the location marked in Figure 9.13. 1, 2, and 3 are the corresponding hanging valleys in Figure 9.13.

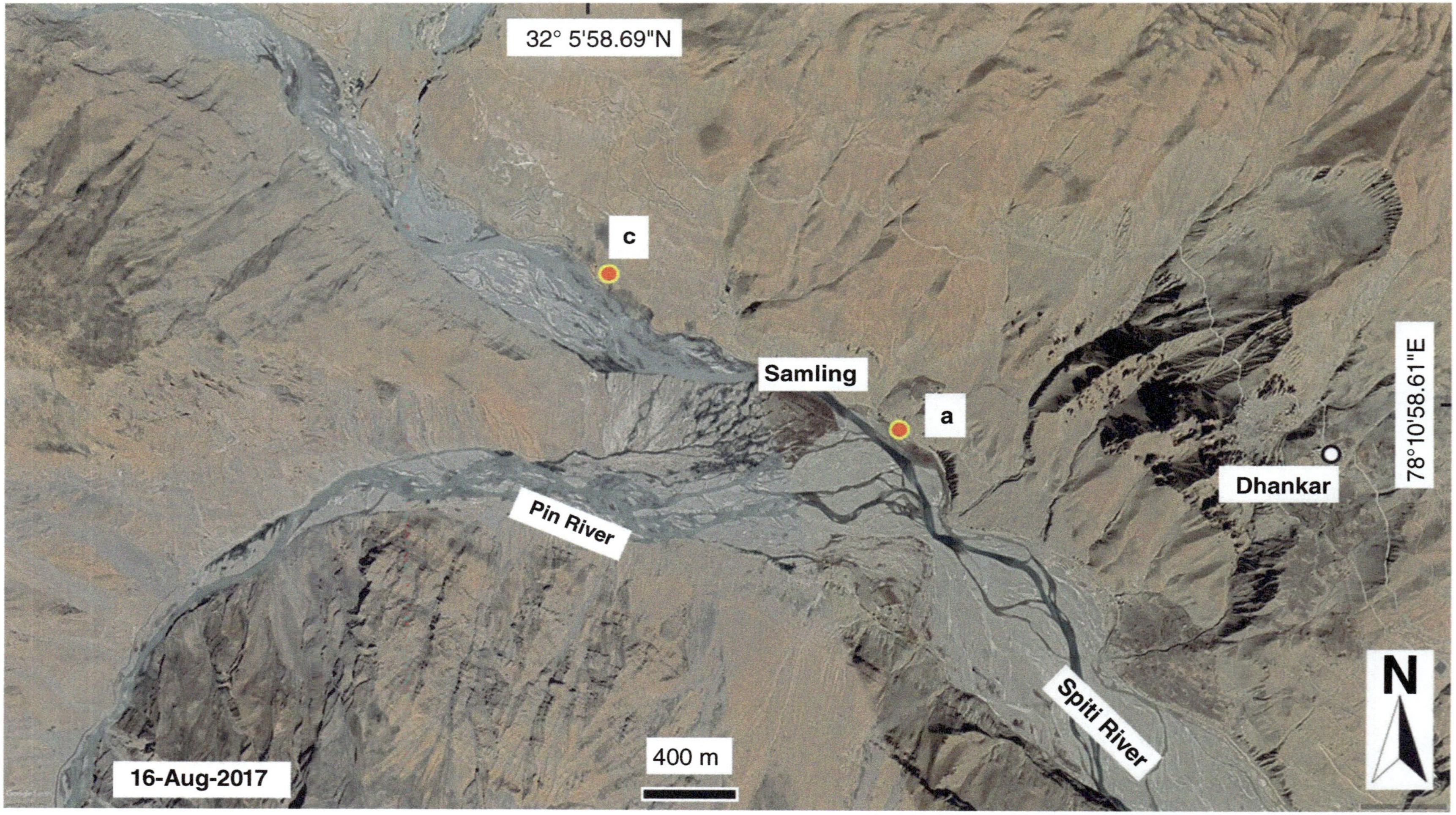

Figure 9.15 Annotated satellite image showing the Spiti River and presence of large, braided trunk channels (similar to Phartiyal et al. 2018). The Pin River, another important river in the Spiti Valley, also joins the Spiti River forming a braided channel system. The braided system in the youth stage of a river is formed because of very high energy sedimentation due to slight local decrease of the slope. The braided bars can be either vegetated or non-vegetated depending on the fine fraction in the sediment. Red dots: locations from where the ground truth image (Figure 9.16) was taken.

Figure 9.16 Field photographs for supporting ground truth. (a) Corresponding to location "a" marked in Figure 9.15, looking toward SW at the confluence of the Pin and Spiti rivers. (b) Braided part of the Pin River near Mud Village (c) Corresponding to location "c" marked in Figure 9.15, looking toward NW at the upper slope of the Spiti River. Note the highly vegetated braided bars of the trunk channel.

Acknowledgements

Soumyajit Mukherjee (IIT Bombay) handled this article and provided minor review comments. Mery Biswas is thanked for her review. Reliance Industries Ltd. is thanked for funding the fieldwork. Wiley Blackwell Acquisition Editors (present: Frank Weinreich, past: Andrew Harrison), Stacey Woods (Handling Editor), and proofreading team. Summarized in Misra and Mukherjee (2022).

References

Bhargava, O.N. (2008). An updated introduction to the Spiti geology. *Journal of the Palaeontological Society of India* 53 (2): 113–128.

Binita, P., Anupam, S., Pradeep, S., and Yogesh, R. (2009). Chronology of relict lake deposits in the Spiti River, NW trans Himalaya: implications to late Pleistocene-Holocene climate-tectonic perturbations. *Geomorphology* 108 (3/4): 264–272.

Dasgupta, S. and Mukherjee, S. (2017). Brittle shear tectonics in a narrow continental rift: asymmetric non-volcanic Barmer basin (Rajasthan, India). *The Journal of Geology* 125: 561–591.

Dasgupta, S. and Mukherjee, S. (2019). Remote sensing in lineament identification: examples from western India. In: *Problems and Solutions in Structural Geology and Tectonics*, Developments in Structural Geology and Tectonics Book Series, vol. 5. Series Editor: Mukherjee, S (ed. A. Billi and A. Fagereng), 205–221. Elsevier. ISSN: 2542-9000. ISBN: 9780128140482.

Dasgupta, S., Biswas, M., Mukherjee, S., and Chatterjee, R. (2022). Depositional system, morphological signatures, tectonics and sedimentation pattern along the transform margin- Palar-Pennar basin, Indian east coast. *Journal of*

Petroleum Science & Engineering 211: 110155. https://doi.org/10.1016/j.petrol.2022.110155.

Gill, M.S. (2010). *Himalayan Wonderland: Travels in Lahaul and Spiti*. New Delhi: Penguin Books India.

Gogoi, M.P., Gogoi, B., and Mukherjee, S. (2022). Tectonic instability of the petroliferous upper Assam valley (NE India): a geomorphic approach. *Journal of Earth System Science* 131: 18. https://doi.org/10.1007/s12040-021-01752-6.

Kaplay, R.D., Md, B., Mukherjee, S., and Kumar, T.V. (2017). Morphotectonic expression of geological structures in eastern part of south East Deccan volcanic province (around Nanded, Maharashtra, India). In: *Tectonics of the Deccan Large Igneous Province*, vol. 445 (ed. S. Mukherjee, A.A. Misra, G. Calvès and M. Nemčok), 317–335. Geological Society, London, Special Publications.

Kujur, A., 2013. Groundwater information booklet-Lahaul and Spiti District, Himachal Pradesh. Central Ground Water Board, Ministry of Water Resources, Government of India Publication. http://cgwb.gov.in/District_Profile/HP/Lahul%20Spiti.pdf.

Misra, A.A. and Mukherjee, S. (2022). Introduction to "atlas of structural geological and geomorphological interpretation of remote sensing images". In: *Atlas of Structural Geological and Geomorphological Interpretation of Remote Sensing Images* (ed. A.A. Misra and S. Mukherjee). Wiley Blackwell. ISBN: 9781119813354.

Misra, A.A., Bhattacharya, G., Mukherjee, S., and Bose, N. (2014). Near N-S paleo-extension in the western Deccan region in India: does it link strike-slip tectonics with India-Seychelles rifting? *International Journal of Earth Sciences* 103: 1645–1680.

Phartiyal, B. and Kothyari, G.C. (2012). Impact of neotectonics on drainage network evolution reconstructed from morphometric indices: case study from NW Indian Himalaya. *Zeitschrift für Geomorphologie* 56 (1): 121.

Phartiyal, B., Sharma, A., Srivastava, P., and Ray, Y. (2009a). Chronology of relict lake deposits in the Spiti River, NW trans Himalaya: implications to late Pleistocene–Holocene climate-tectonic perturbations. *Geomorphology* 108 (3–4): 264–272.

Phartiyal, B., Srivastava, P., and Sharma, A. (2009b). Tectono-climatic signatures during late quaternary period from upper Spiti Valley, NW Himalaya, India. *Himalayan Geology* 30 (2): 10–17.

Phartiyal, B., Singh, R., and Nag, D. (2018). Trans-and Tethyan Himalayan Rivers: in reference to Ladakh and Lahaul-Spiti, NW Himalaya. In: *The Indian Rivers*, 367–382. Singapore: Springer.

Srivastava, P., Ray, Y., Phartiyal, B., and Sharma, A. (2013). Late Pleistocene-Holocene morphosedimentary architecture, Spiti River, arid higher Himalaya. *International Journal of Earth Sciences* 102 (7): 1967–1984.

Section C

Structural Geology

10

Deformation Bands Mapped in the Miocene Sandstone-Dominated Outcrops, Sengkurong, Brunei Darussalam, SE Asia

Syaakiirroh Sahari[1], Dk Aaisyah[1], Amirul Shahbuddin[2], and Afroz A. Shah[*,1]*

[1] *Department Geosciences, Universiti Brunei Darussalam, Bandar Seri Begawan, Brunei*
[2] *Department of Petroleum Engineering, Politeknik Brunei, Bandar Seri Begawan, Brunei*

10.1 Mapping of Deformation Bands

To decipher structural geology and geomorphology of terrains, application of remote sensing is ubiquitous (e.g. Misra et al. 2014; Kaplay et al. 2017; Dasgupta and Mukherjee 2017, 2019; Dasgupta et al. 2022; Gogoi et al. 2022). Deformation bands are some of the most distinguished features of the sandstone-dominated Miocene rock outcrops in Brunei Darussalam (this study), which is part of Borneo Island (Figure 10.1). These structures are formed in response to the strain localization processes in highly porous rocks and are routinely observed in various types of sandstones, limestones, siltstones, poorly welded volcanic tuffs, and breccias (Schultz and Siddharthan 2005; Fossen et al. 2007; Torabi and Fossen 2009). The geologic outcrops in Brunei Muara are dominated by the shallow marine sedimentary sequence with interbedded sandstone and shale lithologies. These rocks are asymmetrically folded, which is shown by the well-developed synclines and narrow anticlines (Figure 10.1). The Berakas syncline and Jerudong anticline are the structures that we have mapped in detail (Figure 10.1). The outcrop locations shown are located near the Jerudong anticline where steeply dipping beds, and we have mapped a number of faults in the area (Figure 10.2). Broadly, the faults are observed to have laterally displaced the marker beds, which are mapped either as sinistral faults (labeled F1 and F3), or dextral strike-slip faults (F2) when viewed from the top. In this ATLAS, we have only been focused on one main fault in the Sengkurong outcrop, which is F1 (Figure 10.3). Aerial images (Figures 10.2, 10.3, and 10.6) were taken using the DJI Mavic 2 Pro while the field photographs (Figures 10.4 and 10.5) were taken using the Fujifilm XT20. The Sengkurong outcrop is part of Miri Formation, where sandstone dominates and interbedded layers are a common occurrence. The fault has clearly displaced a number of lithological layers, and since sandstone dominates, a number of textbook examples of deformation bands can be observed. Some of the examples are shown in Figures 10.3–10.6. Fossen et al. (2007) have shown various types of deformation bands, and we have observed two such types at the Sengkurong outcrop, which are classified either as disaggregation (Figure 10.4–10.6) or phyllosilicate bands (Figures 10.4 and 10.6). Disaggregation bands are distinguished by characteristic features involving rolling, boundary sliding, and breaking of bonding cements in grains along the brittle shear zone (Fossen et al. 2007). The phyllosilicate bands are easily identified because of the typical mm-scale offset of various bands and foliations that are usually distinguished by variations in color, fabric, and a higher content of platy minerals (>10–15%), which causes the grains to slide and not fracture (Fossen et al. 2007). Some phyllosilicate bands can develop into clay smears, as observed in Figures 10.4–10.6, which is usually attributed to the host rock lithology that contains 40% clay content (Fisher and Knipe 2001).

*Corresponding Author: afroz.shah@ubd.edu.bn

Atlas of Structural Geological and Geomorphological Interpretation of Remote Sensing Images, First Edition.
Edited by Achyuta Ayan Misra and Soumyajit Mukherjee.

Companion website: www.wiley.com/go/misra/RemoteSensingImages

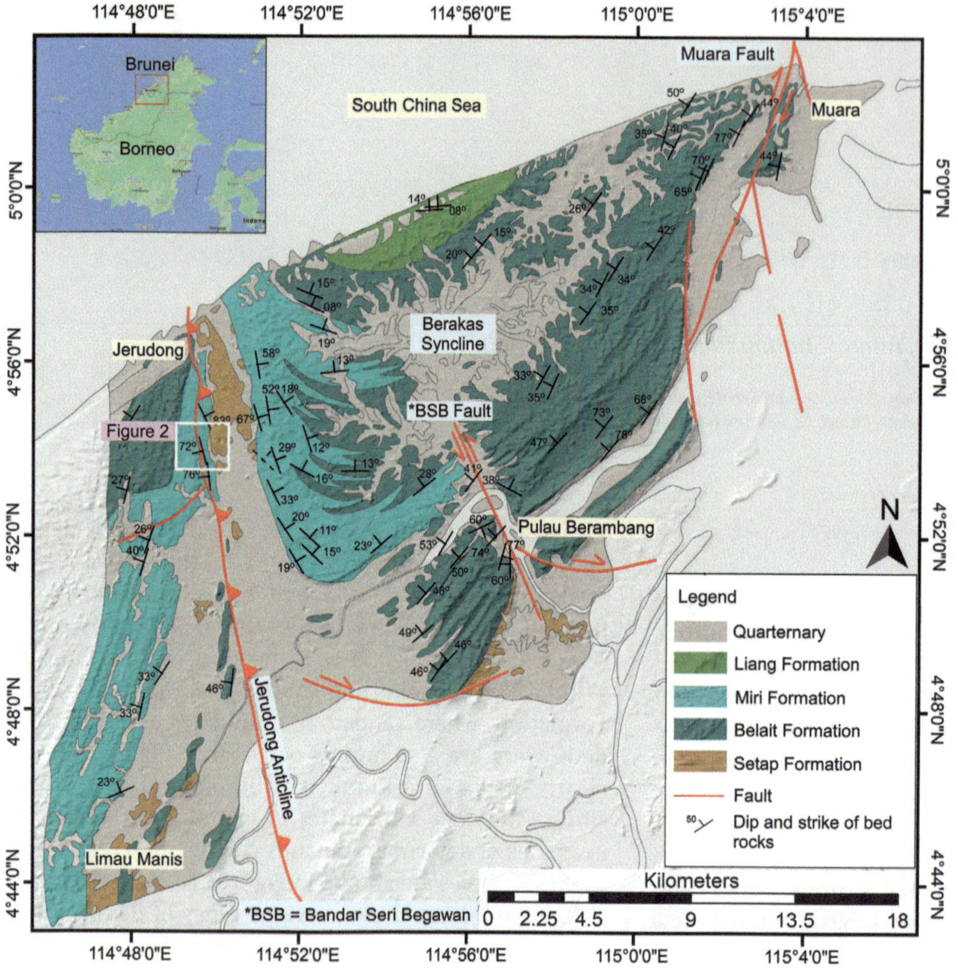

Figure 10.1 The geological and structural map of Brunei Muara modified after Wilford (1960) and Morley et al. (2003), overlaid on the 30 m shuttle radar topography. The location map is shown on the left upper corner. The outcrop structural information is based on our own work (Jamalullail et al. 2021). The location of the study area labeled Figure 10.2 is located in Brunei Muara district, Brunei (4.900998, 114.830440).

(a)

Figure 10.2 (a) The map-view aerial image of the Sengkurong outcrop captured 100 m above sea level using the DJI Mavic 2 Pro. With its 20MP 1″ CMOS sensor, the drone was able to take high-resolution images, which shows the exposed shallow marine faulted rock sequence of Miri Formation of the site. (b) The faults traced and numbered as F1, F2, and F3 have laterally displaced the steeply dipping lithological beds. The F1 and F3 show evidence for sinistral displacement when viewed in map view, and with F2 is mapped as a dextral strike-slip fault. The sedimentary beds dip 63° toward SW.

(b)

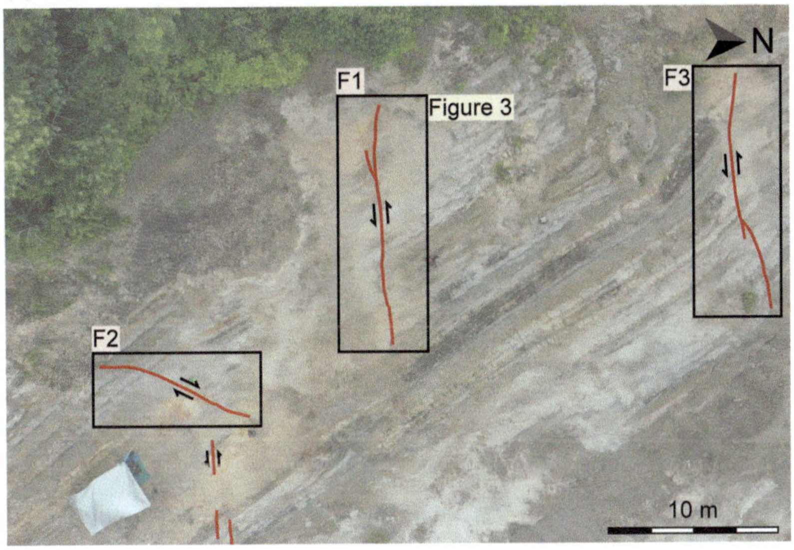

Figure 10.2 (*Continued*)

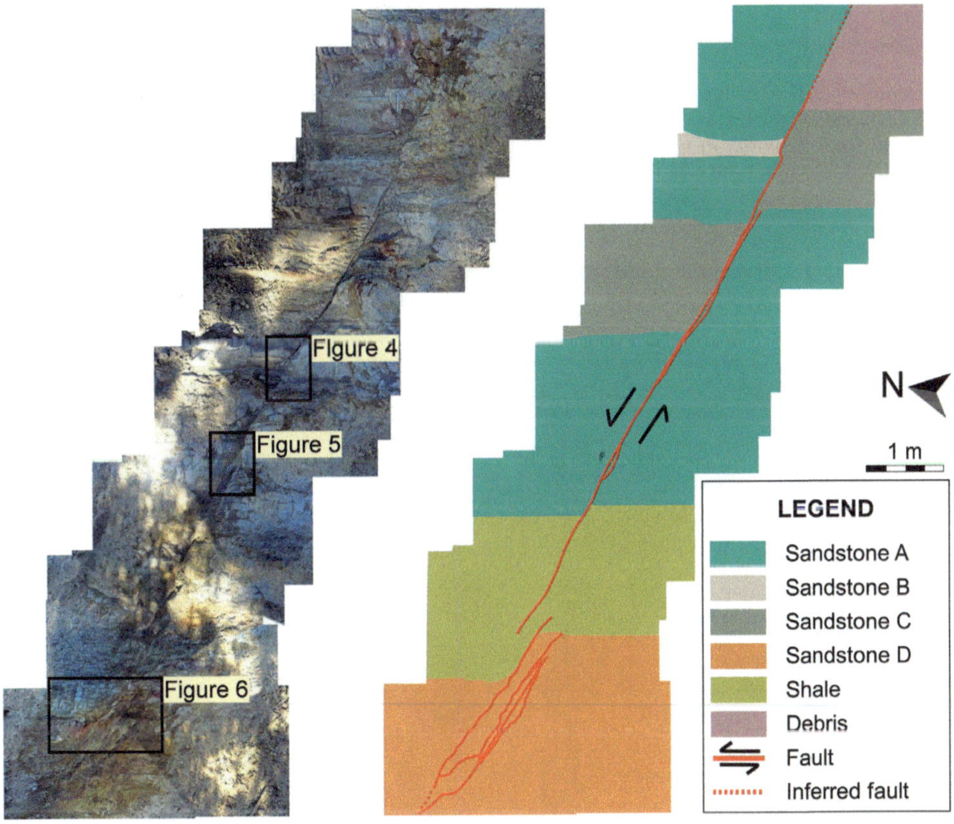

Figure 10.3 Uninterpreted (left) and interpreted (right) drone image of F1 taken 60 m above sea level. Flight time took less than 15 minutes where images were taken at a constant height along the fault. The images were merged using Adobe Photoshop to create a single image showing the 11 m of the sinistral strike-slip fault that dips 83° toward the south. Black rectangle: Figures 10.4–10.6.

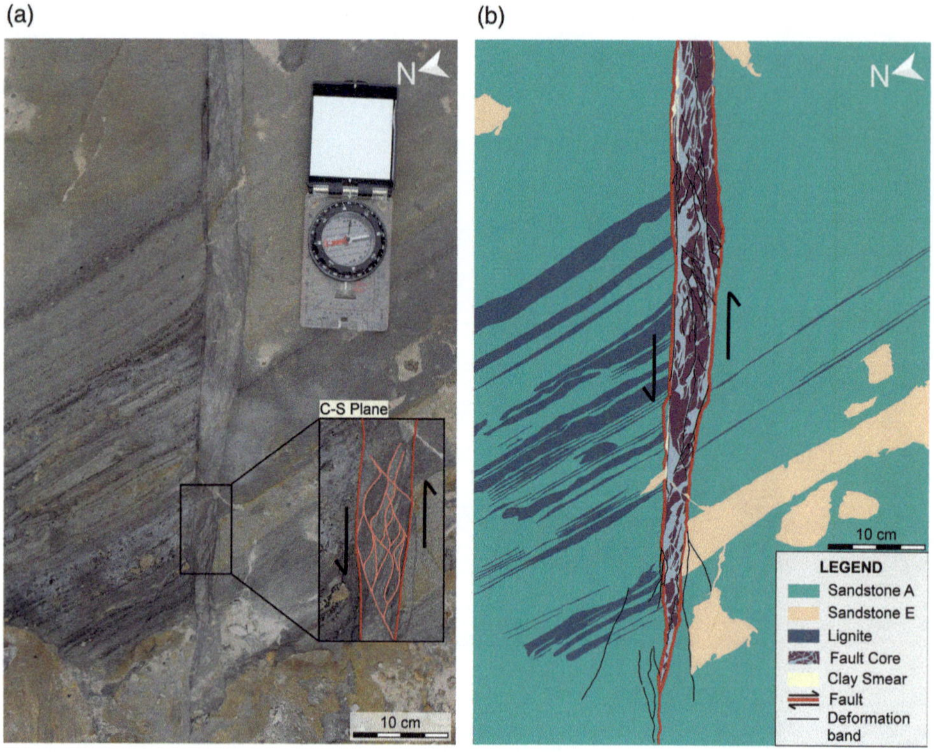

Figure 10.4 Uninterpreted (a) and interpreted (b) field image taken using the Fujifilm XT20 handheld mirrorless camera. With its 24.3MP 23.5 mm × 15.6 mm X-Trans CMOS III sensor, the camera was able to capture the details of the complex deformation bands. The feature here shows both the disaggregation and phyllosilicate band formation with clearly developed P-Y plane, according to Passchier and Trouw (2005). This feature clearly shows a left lateral strike-slip displacement.

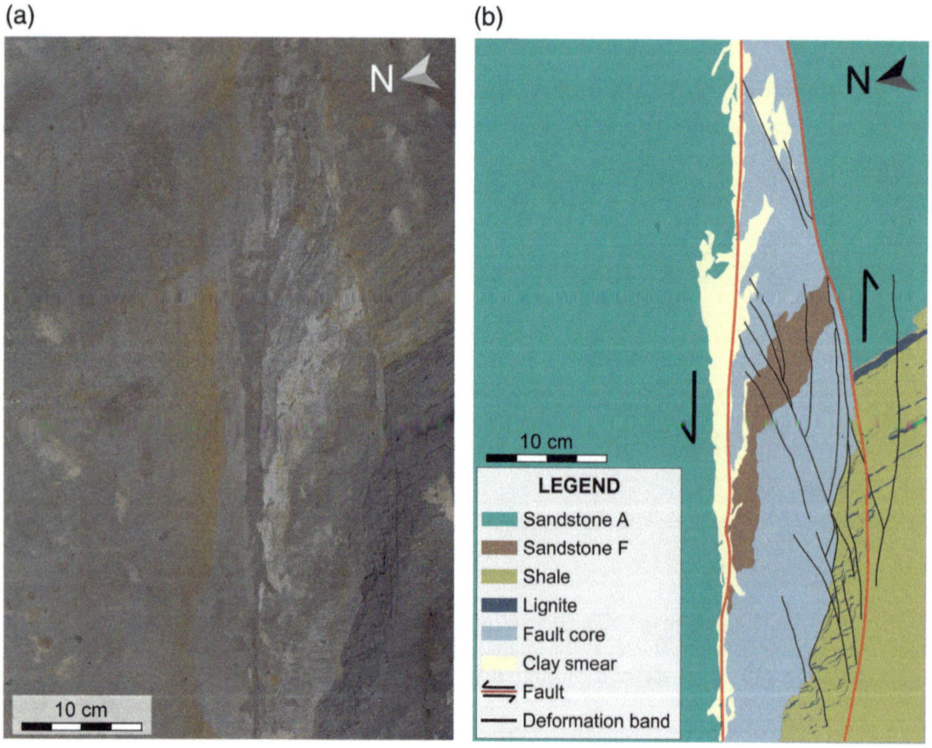

Figure 10.5 Uninterpreted (a) and interpreted (b) field image taken using the handheld camera. The detail of the image shows a compressional deformation band formation zone with textbook examples of disaggregation bands and clay smears. From the high resolution (6000 × 4000 pixels), the details of the deformation bands were easily traced and illustrated.

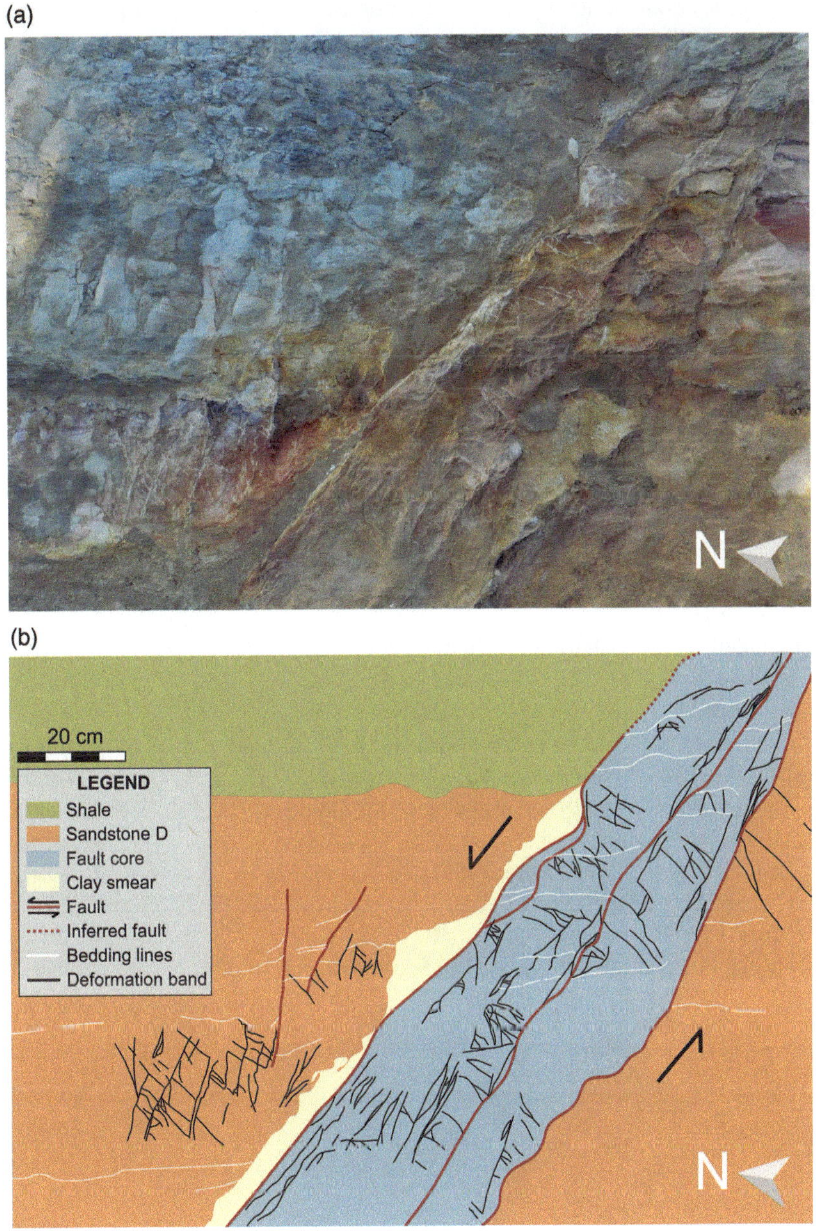

Figure 10.6 Uninterpreted (a) and interpreted (b) aerial image taken 60 m above sea level whose disaggregation and phyllosilicate band present intensely in fault core of Sandstone D. By using the drone, we captured an image of the fault core, which was not possible using the handheld camera due to its large aerial extent. The image clearly shows that the deformation bands are more brittle, sheared, and fractured compared to other types of sandstone along the fault.

Acknowledgements

Soumyajit Mukherjee invited to submit an article in the edited book. Achyuta Ayan Misra handled this article. Wiley Blackwell Acquisition Editors (present: Frank Weinreich, past: Andrew Harrison), Stacey Woods (Handling Editor), and proofreading team. Summarized in Misra and Mukherjee (2022).

References

Dasgupta, S. and Mukherjee, S. (2017). Brittle shear tectonics in a narrow continental rift: asymmetric non-volcanic Barmer basin (Rajasthan, India). *The Journal of Geology* 125: 561–591.

Dasgupta, S. and Mukherjee, S. (2019). Remote sensing in lineament identification: examples from western India. In: *Problems and Solutions in Structural Geology and Tectonics*, Developments in Structural Geology and

Tectonics Book Series, vol. 5. Series Editor: Mukherjee, S (ed. A. Billi and A. Fagereng), 205–221. Elsevier. ISSN: 2542-9000. ISBN: 9780128140482.

Dasgupta, S., Biswas, M., Mukherjee, S., and Chatterjee, R. (2022). Depositional system, morphological signatures, tectonics and sedimentation pattern along the transform margin- Palar-Pennar basin, Indian east coast. *Journal of Petroleum Science & Engineering* 211: 110155. https://doi.org/10.1016/j.petrol.2022.110155.

Fisher, Q.J. and Knipe, R.J. (2001). The permeability of faults within siliciclastic petroleum reservoirs of the North Sea and Norwegian continental shelf. *Marine and Petroleum Geology* 18: 1063–1081.

Fossen, H., Schultz, R.A., Shipton, Z.K., and Mair, K. (2007). Deformation bands in a sandstone: a review. *Journal of the Geological Society, London* 164: 1–15.

Gogoi, M.P., Gogoi, B., and Mukherjee, S. (2022). Tectonic instability of the petroliferous upper Assam valley (NE India): a geomorphic approach. *Journal of Earth System Science* 131: 18.

Jamalullail, S.N.R., Sahari, S., Shah, A.A., and Batmanathan, N. (2021). Preliminary analysis of landslide hazard in Brunei Darussalam, SE Asia. *Environmental Earth Sciences* 80 (16): 1–16.

Kaplay, R.D., Md, B., Mukherjee, S., and Kumar, T.V. (2017). Morphotectonic expression of geological structures in eastern part of south East Deccan volcanic province (around Nanded, Maharashtra, India). In: *Tectonics of the Deccan Large Igneous Province* (ed. S. Mukherjee, A.A. Misra, G. Calvès and M. Nemčok), 317–335. London, Special Publications 445: Geological Society.

Misra, A.A. and Mukherjee, S. (2022). Introduction to "Atlas of structural geological and geomorphological interpretation of remote sensing images". In: *Atlas of Structural Geological and Geomorphological Interpretation of Remote Sensing Images* (ed. A.A. Misra and S. Mukherjee). Wiley Blackwell. ISBN: 9781119813354.

Misra, A.A., Bhattacharya, G., Mukherjee, S., and Bose, N. (2014). Near N-S paleo-extension in the western Deccan region in India: does it link strike-slip tectonics with India-Seychelles rifting? *International Journal of Earth Sciences* 103: 1645–1680.

Morley, C.K., Back, S., Van, R.P. et al. (2003). Characteristics of repeated, detached, Miocene-Pliocene tectonic inversion events, in a large delta province on an active margin, Brunei Darussalam, Borneo. *Journal of Structural Geology* 25: 1147–1169.

Passchier, C.W. and Trouw, R.A.J. (2005). *Microtectonics: Shear Sense Indicators in the Brittle Regime*, 157. Berlin: Springer.

Schultz, R.A. and Siddharthan, R. (2005). A general framework for the occurrence and faulting of deformation bands in porous granular rocks. *Tectonophysics* 411: 1–18.

Torabi, A. and Fossen, H. (2009). Spatial variation of microstructure and petrophysical properties along deformation bands in reservoir sandstones. *AAPG Bulletin* 93 (7): 919–938.

Wilford, G.E. (1960). Geological Map of Brunei and Adjacent Parts of Sarawak. Sheet 2, 1:125000. Brunei Darussalam, The Surveyor General Brunei

11

Disaggregation Deformation Bands Dominate the Trapping and Sealing Process at the Lion King Fault Zone, Brunei, SE Asia

Dk Aaisyah[1], Syaakiirroh Sahari[1], Afroz A. Shah,[1], Ain Said[2], Ezra Jayasuriya[2], and Prassana[2]*

[1] *Department of Geosciences, Universiti Brunei Darussalam, Bandar Seri Begawan, Brunei*
[2] *Department of Geology, Curtin University, Miri, Sarawak, Malaysia*

11.1 Deformation Bands

Remote sensing helps to decipher structural geology and geomorphology in mega-scale (e.g. Misra et al. 2014; Kaplay et al. 2017; Dasgupta and Mukherjee 2017, 2019; Dasgupta et al. 2022; Gogoi et al. 2022). Brunei Darussalam is part of Borneo Island, which is tectonically surrounded by the plates of India, Australia, and the Pacific (Shah et al. 2018). The geologic map (Figure 11.1) shows the Brunei-Muara district, which is dominantly composed of shallow marine sedimentary rocks of the Miri and Belait Formation that were sourced from the fold and

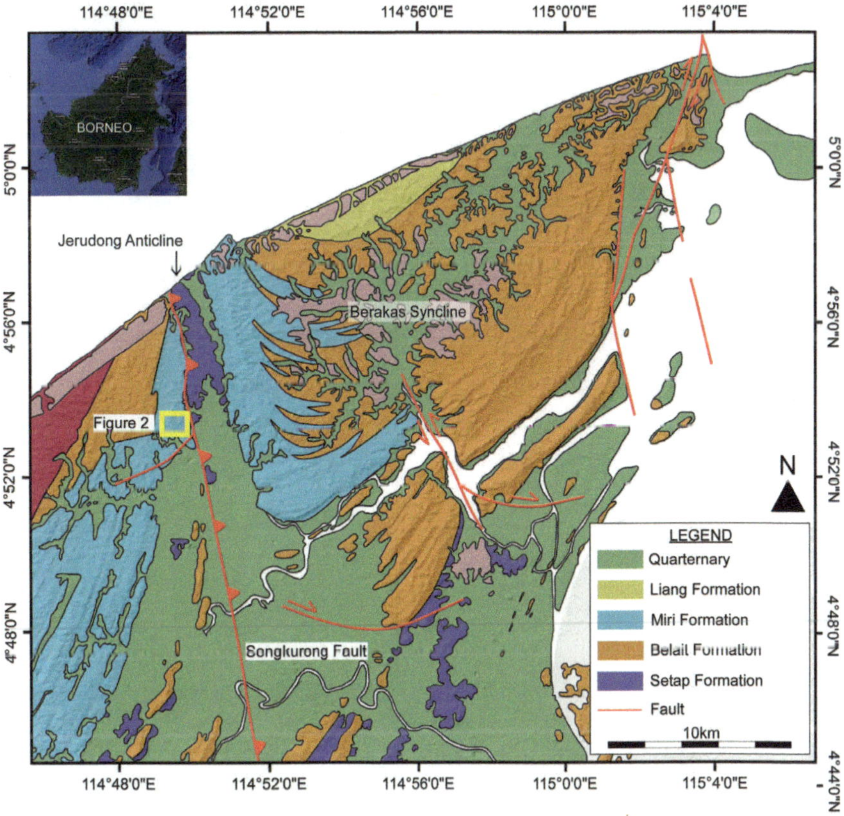

Figure 11.1 Geologic map of Brunei-Muara district, modified after Morley et al. (2003) and Aaisyah et al. (2020a, 2020b), showing that the area is mostly made up of alternating sandstones and shales of the Belait and Miri Formation. Two major structural features – Berakas syncline and Jerudong anticline – are shown. The selected site shown here is located on the western flank of the Jerudong anticline, which is cut by a thrust fault known as the Sengkurong fault.

*Corresponding Author: afroz.shah@ubd.edu.bn

Atlas of Structural Geological and Geomorphological Interpretation of Remote Sensing Images, First Edition.
Edited by Achyuta Ayan Misra and Soumyajit Mukherjee.

Companion website: www.wiley.com/go/misra/RemoteSensingImages

thrust belt in the east and have Setap shale at the base (Morley et al. 2003). The rocks are asymmetrically folded with broadly developed synclines and narrow anticlines, which are structurally controlled by faults at depth (e.g. Sandal 1996; Morley et al. 1998, 2003; Aaisyah et al. 2020a, 2020b). We investigated one of the best geological outcrops in Brunei, locally known as the "Lion King" outcrop, which shows textbook examples of faults, deformation bands, and shale smears. The map view exposure is quite large, which allows drone-based mapping with ease in the otherwise canopy-covered tropical regions of Borneo. The aerial images (Figures 11.2 and 11.3) were captured using the DJI Mavic 2 Pro with close-up images (Figures 11.4–11.7) taken using the Nikon D7000, a handheld camera. Images captured by both pieces of equipment show that the subvertical sedimentary beds are crisscrossed by faults and the length of a major fault zone can be continuously mapped for 60 m along the strike, which makes the site a perfect place to map the fault in detail (Figures 11.2–11.7). In this chapter we show some examples of the deformation bands, shale smears, compressional bends, and deformation lens (Figures 11.4–11.7) that we have mapped to understand the trapping and sealing potential of the outcrops. These structures are well developed; therefore, we have used high-resolution images to capture the kilometer to millimeter resolution scale that illustrates the structure in much more detail.

Figure 11.2 The drone image of the "Lion King" outcrop located along the western limb of the Jerudong Anticline. The aerial image was taken by the DJI Mavic 2 Pro that is equipped with a 20MP 1" CMOS sensor. Even at a height of more than 100 m, the cropped image of the outcrop clearly shows that the site is sandstone dominant with steeply west dipping (77°) beds that are pierced by a steeply south dipping (~ 70°) fault. This site is one of a few outcrops in Brunei Darussalam where textbook examples of brittle deformational structures can be observed. The drone image shown above captures the entire extent of the outcrop, where 60 m strike of the main sinistral strike-slip fault can be seen.

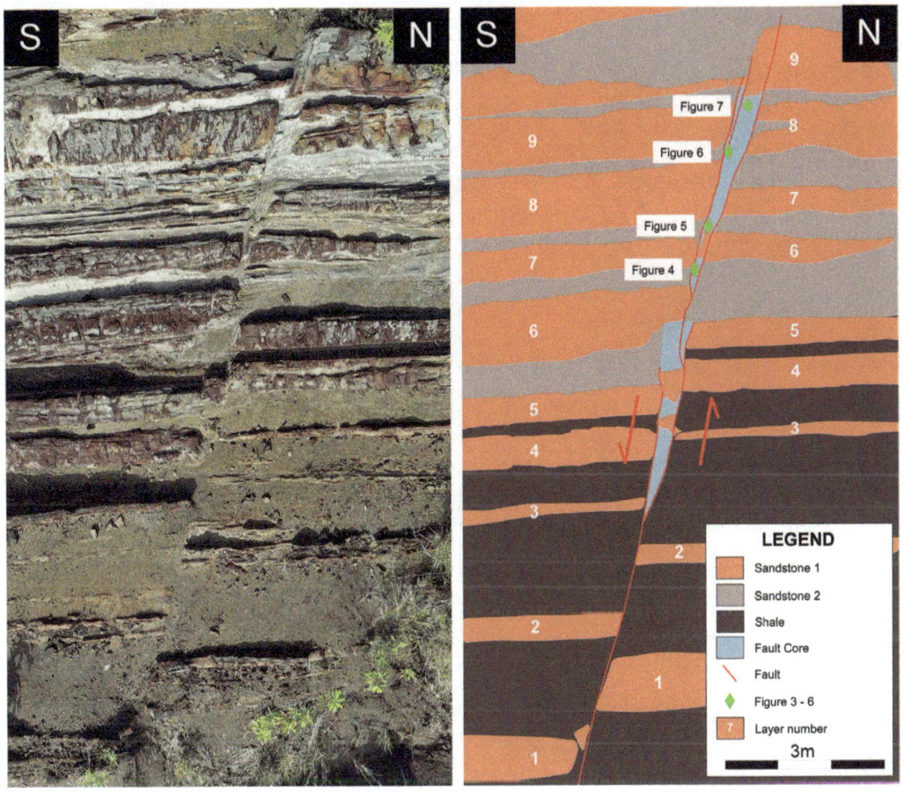

Figure 11.3 Uninterpreted (left) and interpreted (right) close-up drone image where the drone was flown lower to show the lithology and the fault zone along 10 m strike length of the main E–W trending fault. In the high-resolution (5472 × 3648 pixels) image, it is clear that the displacement varies along the strike of the fault, which is due to changes in lithology, fault interaction, branching, and deformation bands. The decreasing displacement toward the west shows that the deformation is mainly compensated by the formation of deformation bands, which are shown in Figures 11.4–11.7. Location is shown in Figure 11.2.

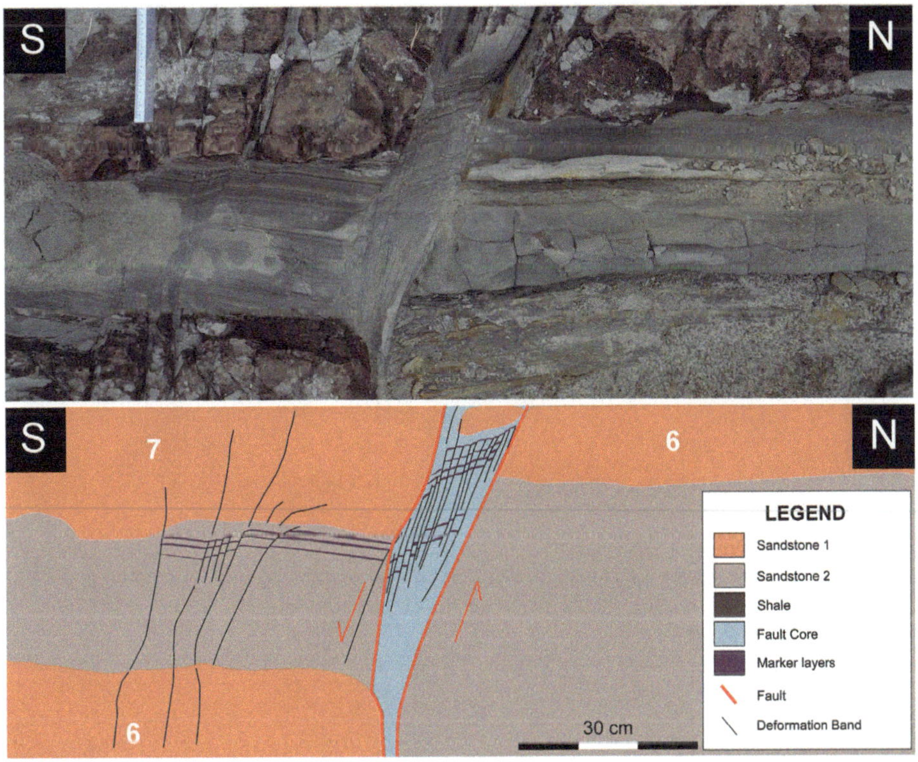

Figure 11.4 Uninterpreted (top) and interpreted (bottom) field images taken using the Nikon D7000, which has a 16.2MP 23.6 mm × 15.6 mm CMOS sensor. Field photos were taken at a constant height along the fault to show the deformation zone. A compressional deformation bend with variable slip across the fault zone. The outcrop map view shows thin clay layers are offset along the fault, which is greater at the fault margins than within the core. The deflection of the sedimentary laminations within the fault zone is consistent with the left-lateral slip of the main fault. Location in Figure 11.3.

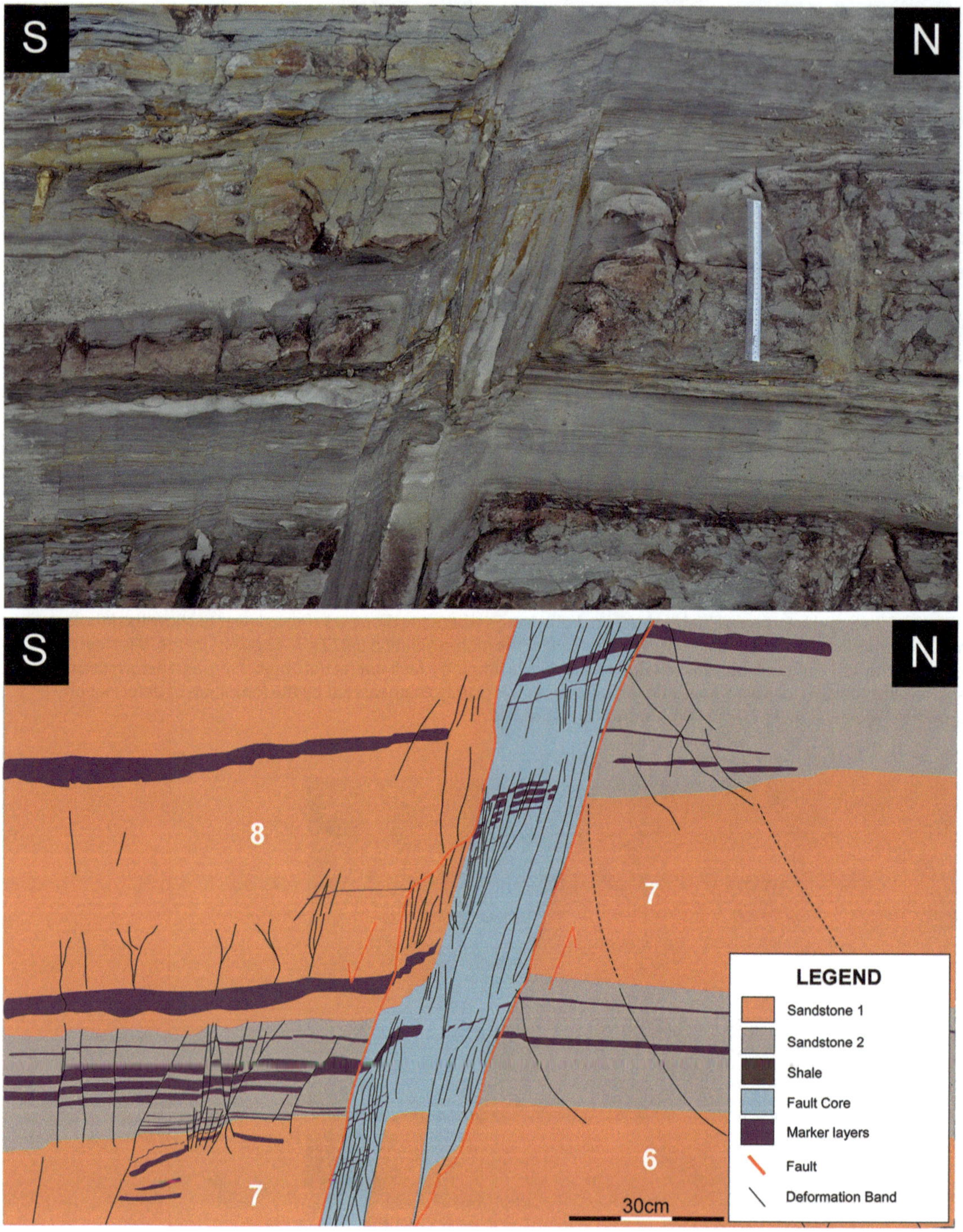

Figure 11.5 Uninterpreted (top) and interpreted (bottom) field image showing the formation of disaggregation and phyllosilicate bands within the thin layers of laminated sand and clay. The displacement/slip of marker beds also reflect the left-lateral strike-slip displacement of the main fault, which has also caused the reduction of the pore spaces and permeability. The juxtaposition of lithologies is also observed, which makes it a good structural trap. Location in Figure 11.3.

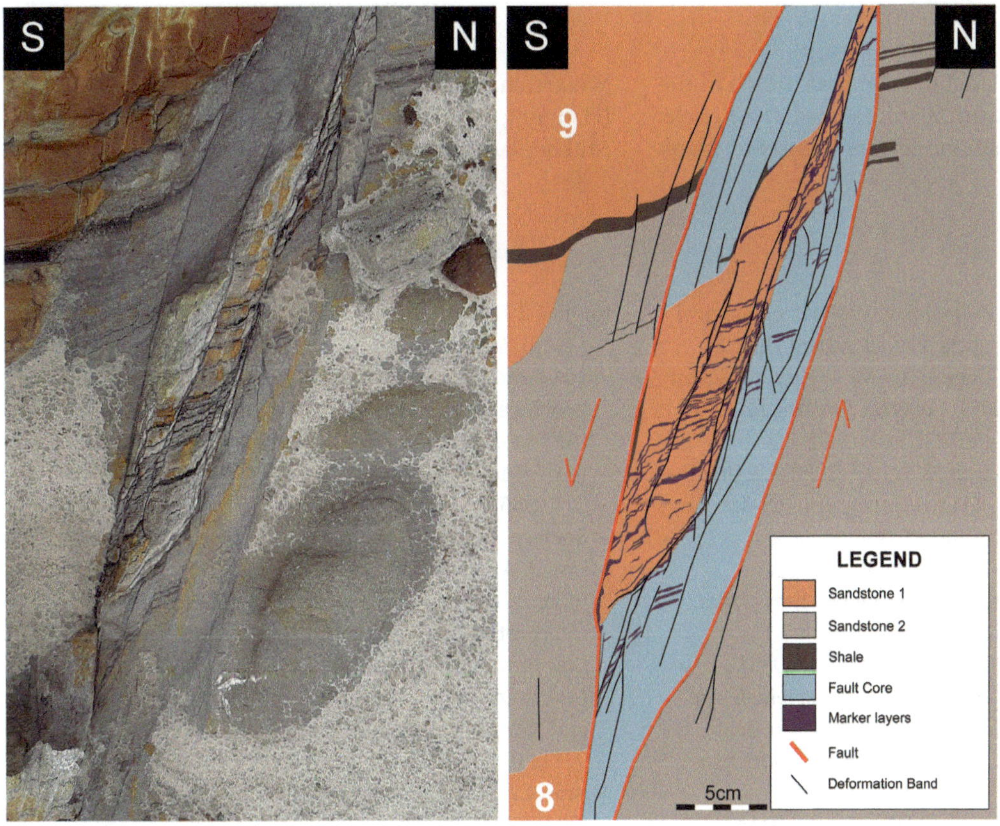

Figure 11.6 Uninterpreted (top) and interpreted (bottom) field image of the fault core showing sinistral strike-slip fault. The textbook example of fault lens is shown that contains highly strained rock and deflected layers, which are consistent with the geometry of compressional bends along the trace of the strike-slip fault. Location in Figure 11.3. Source: modified after Aaisyah et al. (2021).

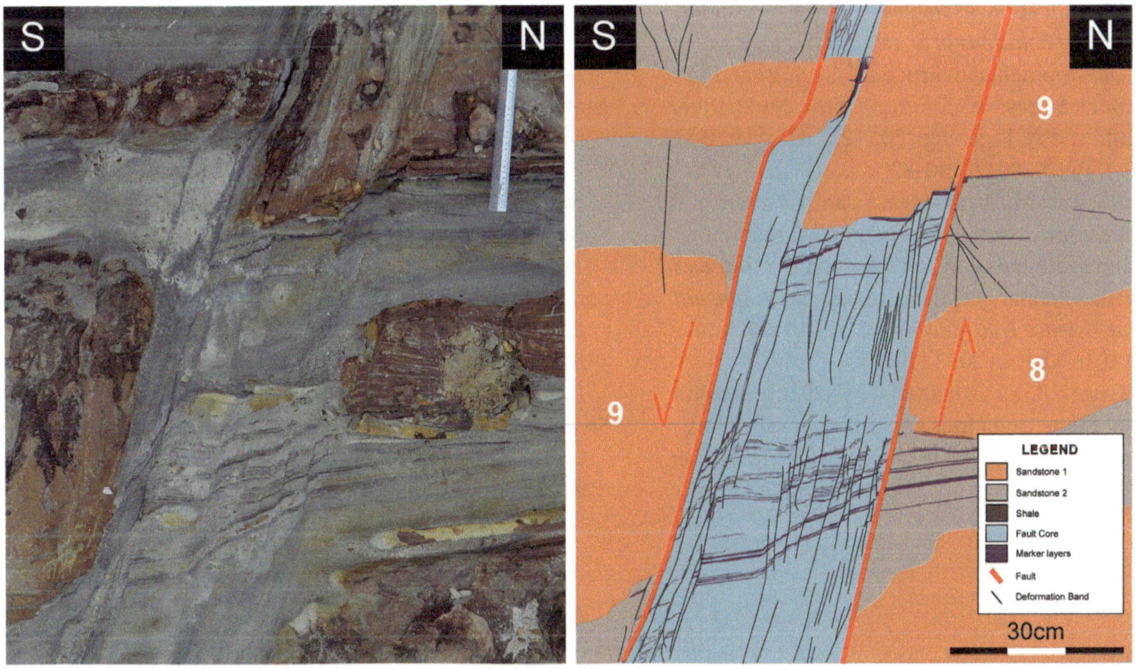

Figure 11.7 Uninterpreted (top) and interpreted (bottom) field image showing the formation of disaggregation and phyllosilicate bands within the thin layers of laminated sand and clay. The deflection of the primary sedimentary lamination is consistent with the formation during the left lateral strike slip faulting. The juxtaposition of lithologies clearly suggest a good example of a structural trap and seal. Location in Figure 11.3.

Acknowledgements

Soumyajit Mukherjee invited to submit an article in the edited book. Achyuta Ayan Misra handled this article. Wiley Blackwell Acquisition Editors (present: Frank Weinreich, past: Andrew Harrison), Stacey Woods (Handling Editor), and proofreading team. Misra and Mukherjee (2022) summarizes this article.

References

Aaisyah, D., Shah, A.A., & Zulmajdi, Z. (2020a). Geological and Structural Mapping of the Brunei-Muara District using Unmanned Aerial Vehicle (UAV). *5th International Young Earth Scientists (YES) Congress "Rocking Earth's Future"*, 2. https://doi.org/10.2312/yes19.02.

Aaisyah, D., Shah, A.A., Garcia, K., et al. (2020b). Evidence of Strike-Slip Faulting in Brunei Darussalam, NW Borneo. In: 82nd EAGE Annual Conference & Exhibition (Vol. 2020, No. 1, pp. 1–5). European Association of Geoscientists & Engineers. https://doi.org/10.3997/2214-4609.202011727.

Aaisyah, D., Sahari, S., Shah, A. et al. (2021). COVID-19 as an opportunity to make field-based earth sciences and other similar courses easily accessible and affordable. *Environmental Resilience and Transformation in Times of COVID-19* 333–342.

Dasgupta, S. and Mukherjee, S. (2017). Brittle shear tectonics in a narrow continental rift: asymmetric non-volcanic Barmer basin (Rajasthan, India). *The Journal of Geology* 125: 561–591.

Dasgupta, S. and Mukherjee, S. (2019). Remote sensing in lineament identification: examples from western India. In: *Problems and Solutions in Structural Geology and Tectonics*, Developments in Structural Geology and Tectonics Book Series, vol. 5. Series Editor: Mukherjee, S. (ed. A. Billi and A. Fagereng), 205–221. Elsevier. ISSN: 2542-9000. ISBN: 9780128140482.

Dasgupta, S., Biswas, M., Mukherjee, S., and Chatterjee, R. (2022). Depositional system, morphological signatures, tectonics and sedimentation pattern along the transform margin – Palar-Pennar basin, Indian east coast. *Journal of Petroleum Science & Engineering.* 211: 110155. https://doi.org/10.1016/j.petrol.2022.110155.

Gogoi, M.P., Gogoi, B., and Mukherjee, S. (2022). Tectonic instability of the petroliferous upper Assam valley (NE India): a geomorphic approach. *Journal of Earth System Science* 131: 18.

Kaplay, R.D., Md, B., Mukherjee, S., and Kumar, T.V. (2017). Morphotectonic expression of geological structures in eastern part of south East Deccan volcanic province (around Nanded, Maharashtra, India). In: *Tectonics of the Deccan Large Igneous Province* (ed. S. Mukherjee, A.A. Misra, G. Calvès and M. Nemčok), 317–335. London, Special Publications 445: Geological Society.

Misra, A.A. and Mukherjee, S. (2022). Introduction to "atlas of structural geological and geomorphological interpretation of remote sensing images". In: *Atlas of Structural Geological and Geomorphological Interpretation of Remote Sensing Images* (ed. A.A. Misra and S. Mukherjee). Wiley Blackwell. ISBN: 9781119813354.

Misra, A.A., Bhattacharya, G., Mukherjee, S., and Bose, N. (2014). Near N-S paleo-extension in the western Deccan region in India: does it link strike-slip tectonics with India-Seychelles rifting? *International Journal of Earth Sciences* 103: 1645–1680.

Morley, C.K., Crevello, P., and Ahmad, Z. (1998). Shale tectonics and deformation associated with active diapirism: the Jerudong Anticline, Brunei Darussalam. *Journal of the Geological Society* 115: 475–490.

Morley, C.K., Back, S., Van Rensbergen, P. et al. (2003). Characteristics of repeated, detached, Miocene–Pliocene tectonic inversion events, in a large delta province on an active margin, Brunei Darussalam, Borneo. *Journal of Structural Geology* 25 (7): 1147–1169.

Sandal, St. T. (1996). The Geology and Hydrocarbon Resources of Negara Brunei Darussalam. 1996 revision. Brunei Shell Petroleum Company / Brunei Museum. Brunei Darussalam: Syabas.

Shah, A.A., Zhafri, M.N., Delson, J., and Navakanesh, B. (2018). Major strike-slip faults identified using satellite data in Central Borneo. *SE Asia. Geosciences* 8 (5): 156.

12

Surface Deformation Along Katrol Hill Fault, Kachchh, Evidenced by Satellite and DEM Data

Atul K. Patidar[*,1], Mohamedharoon Shaikh[2], Prabhuti Tiwari[2], Deepak M. Maurya[2], and Laxman S. Chamyal[2]*

[1] *Department of Petroleum Engineering and Earth Sciences, University of Petroleum and Energy Studies, Dehradun, Uttarakhand, India*
[2] *Department of Geology, The M.S. University of Baroda, Vadodara, Gujarat, India*

Remote sensing has been a powerful tool to interpret structural geology and geomorphology of terrains (e.g. Misra et al. 2014; Kaplay et al. 2017; Dasgupta and Mukherjee 2017, 2019; Dasgupta et al. 2022; Gogoi et al. 2022). The Kachchh Rift Basin (KRB) is located at the western continental margin of India (Maurya et al. 2017; Padmalal et al. 2021; Shaikh et al. 2018, 2020). The seismically active landscape of this basin is dominantly controlled by E–W tectonic fabric and NE–SW, NW–SE oriented transverse trending fault systems, which is clearly reflected in its geomorphic configuration. The Kachchh Basin can be subdivided into five geomorphic units: (i) the Great Ranns and the Banni plain; (ii) the Island belt; (iii) the Wagad highland; (iv) the Mainland Kachchh region; and (v) the Coastal zone of Kachchh (Figure 12.1). This basin preserved 3000–4000 m thickness of Mesozoic and Cenozoic sediments, overlaid unconformably by a thin apron of Quaternary sediments. The fault-controlled geomorphic setup and series of elliptical to oval-shaped domal structures are attributed to the inversion of the basin in the Late Cretaceous (Biswas 1987). The domes of mainland Kachchh are bounded by E–W trending master faults. The Kachchh Mainland Fault (KMF) and Katrol Hill Fault (KHF) are the major structural elements controlling the rugged topography of mainland Kachchh (Figure 12.1). Both these faults are seismically active and responsible for major earthquakes that occurred in the recent past. The geomorphological mapping of seismically active regions using satellite data and its integration with field evidence of neotectonic activities may help in the delineation of landscape variability, geomorphic evaluation of terrain, and slope-stability assessments (Chowksey et al. 2010; Maurya et al. 2003, 2017; Sissakian et al. 2020; Tiwari et al. 2021). It can also contribute to understanding the chronology of paleo-seismic events and natural hazard zonation. In this chapter, we have demonstrated the application of remotely sensed satellite data to interpret the neotectonic deformation and their geomorphic responses along KHF.

12.1 Tectonic Geomorphology of KHF

In the southern part of mainland Kachchh, KHF controls the topography, where E–W trending active fault line with youthful fault scarp extends ~60 km (Figures 12.1, 12.2). Geomorphic mapping using satellite data and neotectonic field studies reveal that this intra-uplift master fault is segmented by transverse faults and is not rupturing along its entire span (Figure 12.2) (Patidar 2010). The fluvio-tectonic interpretation of the Katrol Hill Range reveals that the KHF has been dissected into four discrete segments by transverse faults (Figure 12.3). The linear profiles of fault scarp and drainage divide (Figure 12.4), local in homogeneities and variations in the attitude of fault plane along its length (Figure 12.2), systematic deflection, offsetting, and incision by younger order drainages within the vicinity of KHF, Quaternary terrace formation and surface rupture evidence along Khari river (Figure 12.5), abrupt meandering and deep incision in Quaternary sediments by Gunawri river in back-valley and retreated fault scarp of Khatrod area (Figure 12.6), and other geomorphological characters of individual segments indicate differential fault block

*Corresponding Author: apatidar@ddn.upcs.ac.in

Atlas of Structural Geological and Geomorphological Interpretation of Remote Sensing Images, First Edition.
Edited by Achyuta Ayan Misra and Soumyajit Mukherjee.

Companion website: www.wiley.com/go/misra/RemoteSensingImages

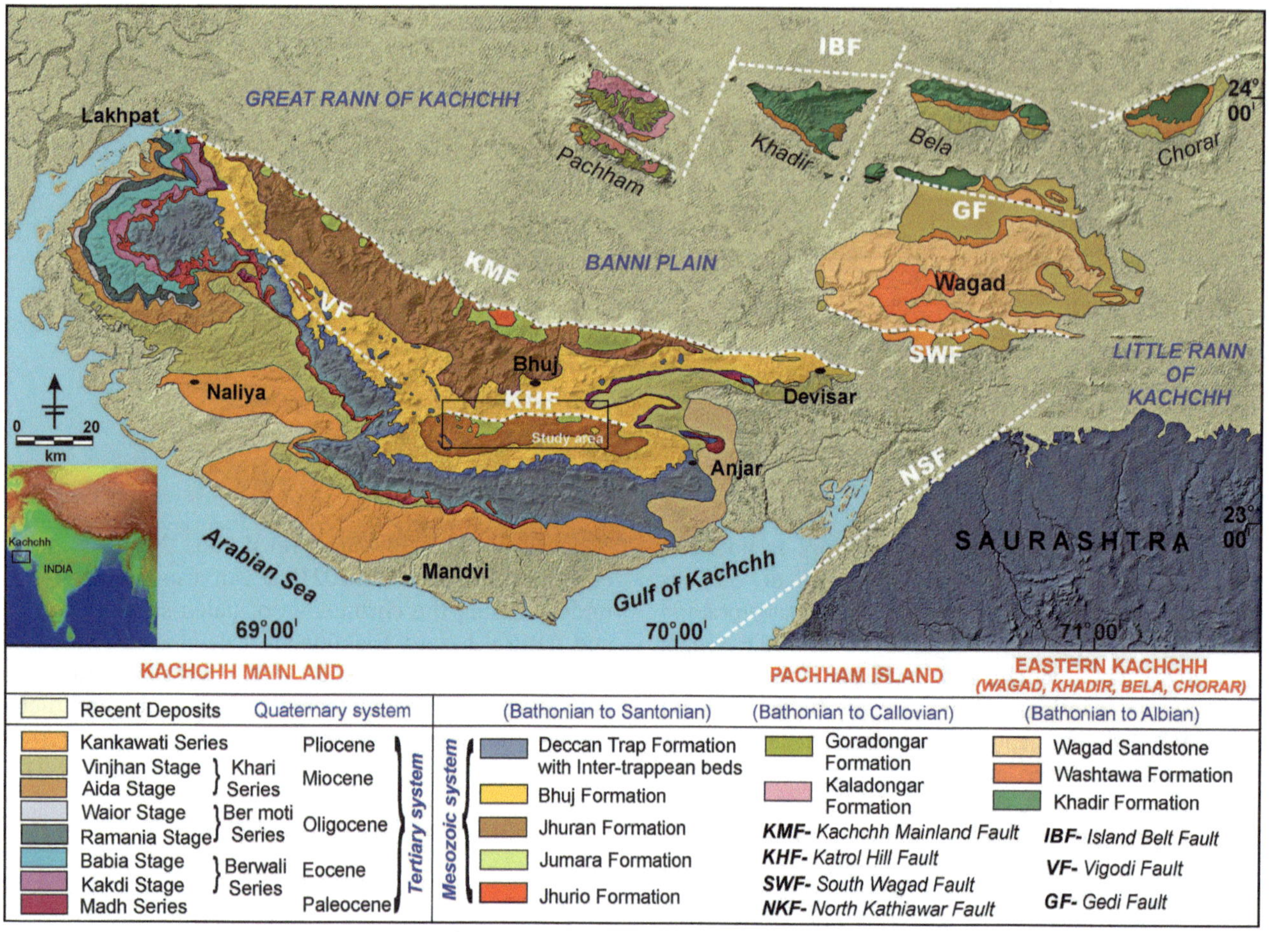

Figure 12.1 Geological map of the Kachchh Basin superimposed on digital elevation model (https://www.usgs.gov) showing the tectono-geomorphic configuration. Geological and tectonic details are after Biswas (1987) and the color map is refurbished by Patidar et al. 2008. The present study area to the south of Bhuj is shown by a rectangle along Katrol Hill Fault (KHF). An up-to-date map of the area with compiled geoscientific information can be found in Shaikh et al. (2022).

movement along the KHF (Maurya et al. 2017, 2021; Patidar et al. 2007, 2008). Systematic interpretation of remotely sensed satellite data supported by neotectonic field studies validate the persistence of KHF segments in space and time. It is also noticed that the dip of KHF varies in all four segments. The KHF is dipping gently (45–50°) due south in western segments while it becomes nearly vertical (75–80°) toward the eastern extremity (Figure 12.2).

The present work successfully demonstrated the uses of satellite and DEM data in morphostructural analysis of the Katrol Hill Range like drainage deflection and their reorganization due to neotectonic activity, back valley formation by Gunawri river due to vertical upliftment along KHF, tilted late quaternary miliolite sediments near the vicinity of transverse faults, vertically incised paired terraces and displacement of quaternary sediment along KHF in Khari river basin area (Figure 12.5), field validation and mapping of the fault plane, and retreated scarp analysis in the eastern part of the KHF. The quaternary landscape development, geomorphic observations, and neotectonic topographies indicate late quaternary tectonic reactivation of the KHF. It can be concluded that the transecting sites of KHF with transverse fault are the weakest place to transfer accumulated fault stresses and responsible for recurring seismic activities in the mainland Kachchh region, and are also susceptible to future surface rupture.

Figure 12.2 Tectonically controlled topography of the Katrol Hill Range (southern mainland Kachchh) and exposed traces of Katrol Hill Fault (KHF). (a) Uninterpreted satellite image (https://earth.google.com) showing a 3D view of the Katrol Hill Range. (b) A digital elevation model (DEM) of the southern mainland Kachchh, prepared using 90 m spatial resolution SRTM DEM (https://www.usgs.gov), showing southerly tilted rugged topography of E–W trending Katrol Hill Range and associated domal structures. The location of the KHF and transverse faults are marked based on satellite data interpretation and geological field mapping of neotectonic features (Patidar 2010). The tectono-geomorphic evaluation of the southern mainland Kachchh using field and ground-penetrating radar studies is established by Maurya et al. (2021) and Patidar et al., 2007, 2008. (c) Field photographs of the KHF showing variation in the dip of the fault plane in its entire length from east to west. The photograph shown in Figure 12.2c is taken from the SE of Samatra showing a gentle dip (45° due south). (d) Exposed KHF to the north of Bharasar dome showing a dip of 60° due south. (e) A steeply southward dipping (76°) fault plane exposed in a younger order stream to the North of Tapkeshwari. A prominent lithotectonic contact between Jhuran shale and Bhuj formation can be seen and marked as KHF. (f) Field photograph of steeply southward dipping (~80°) KHF plane exposed to the North of Khatrod peak. The location of field photographs is shown in Figure 12.2b by red color fonts. The notations in the map are as follows: 1. Sansosradungar (233 m); 2. Dholadungar (270 m); 3. Jandia hill (234 m); 4. Khatrol (349 m); 5. Marutonkdungar; 6. Tapkeshwari; 7. Bharasar; 8. Samatra; 9. Bhujiya fort (232 m); TF – Transverse fault; KHF – Katrol Hill Fault.

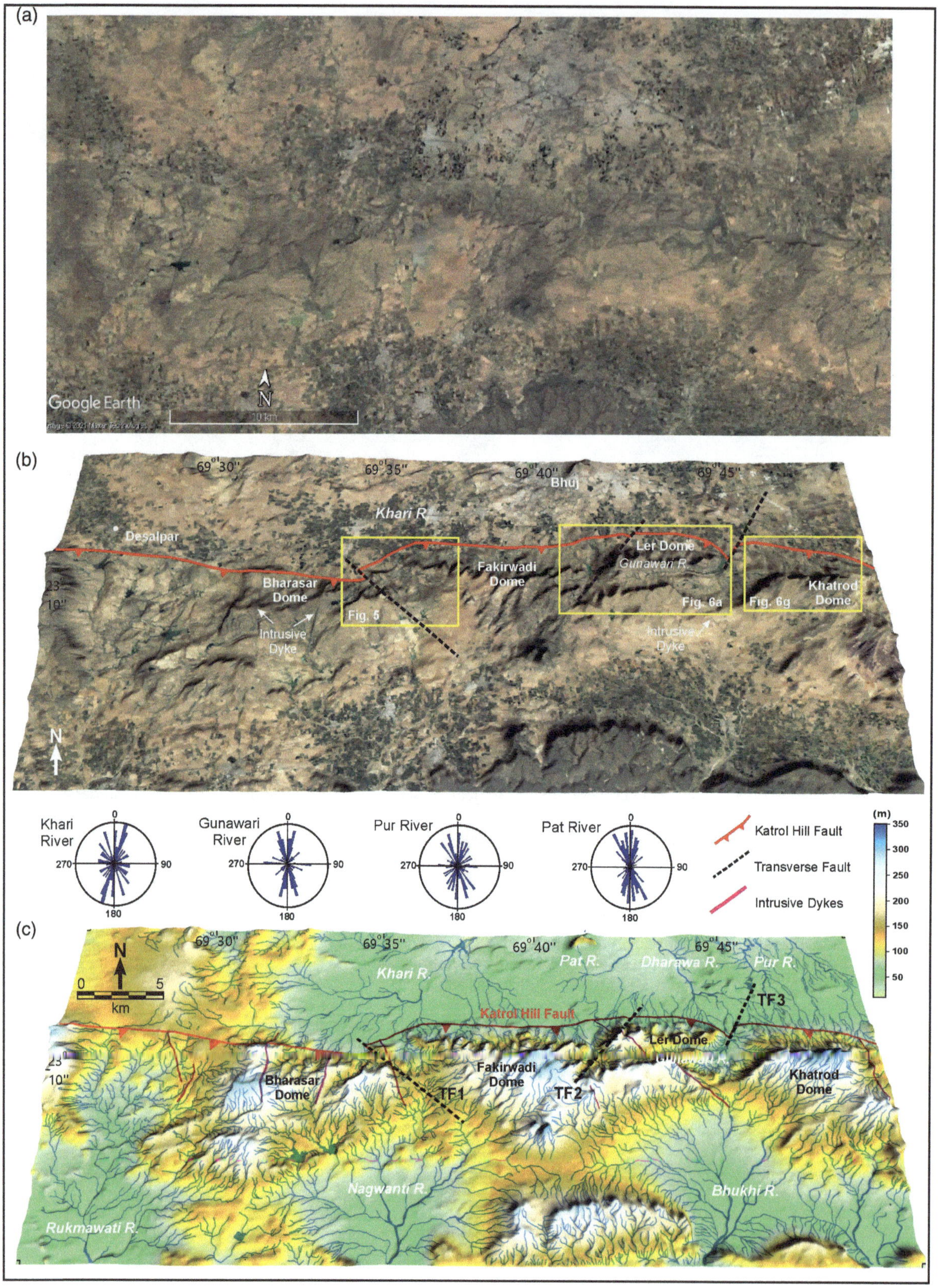

Figure 12.3 Tectono-geomorphic analysis of the Katrol Hill Range. (a) A plan view of an uninterpreted satellite image (2021, https://earth.google.com) showing rugged topography and tectonically controlled geomorphology of the Katrol Hill Range. (b) Interpreted satellite image draped over 90 m spatial resolution SRTM DEM (https://www.usgs.gov), showing a 3D view of the terrain. Note the tectonically complex and deformed structures in the southern part, while the low laying rocky plane of Bhuj is situated toward the North of KHF. The location of the KHF, transverse faults, and other structures are marked based on satellite data interpretation and geological field mapping (Patidar et al. 2007). The location of Figures 12.5 and 12.6 is shown by yellow boxes. (c) A 3D perspective view of a drainage network map of the Katrol Hill Range draped over 90 m spatial resolution SRTM DEM (https://www.usgs.gov), showing terrain elevation difference and tectono-geomorphic relation. Note the sharp geomorphic contrast between the rocky plain of Bhuj and the rugged topography of older Mesozoic formations to the south of the KHF. The rose diagrams prepared along north-flowing Khari, Pat, Dharawa, and Pur rivers show a distribution of directional data along oblique trending lineaments too. Dendritic and radial drainage patterns controlled by bedrock topography and patchy domal structures can be seen. An E–W trending Gunawari river back valley can be seen between range front scarps and drainage divide. See Figure 12.6 for more details.

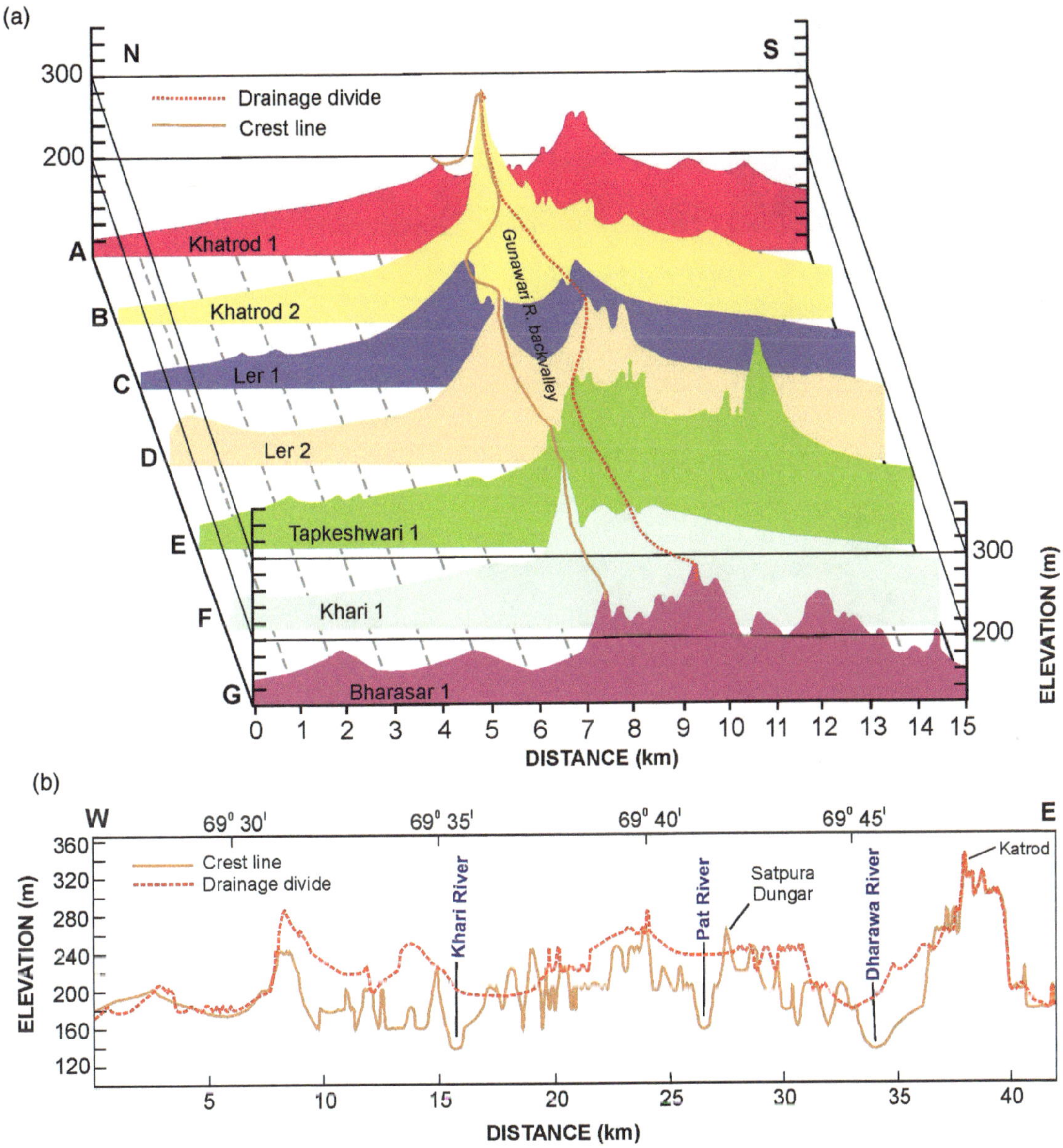

Figure 12.4 (a) N–S trending randomly selected topographic sections of the Katrol Hill Range showing topographic differences between crest line of the range front scarps and the drainage divide. Note the southward tilted rugged topography and formation of Gunawari river back valley between topographic section B and E. (b) Linear topographic sections along the E–W trending crest line and the drainage divide for the comparison of topographic changes and their relation with north-flowing rivers of the Katrol Hill Range. Note the irregular altitude along the crest line because of rapid erosion of the range front scarp due to the uplift along KHF. A remarkable observation is made that the Khari, Pat, and Dharawa rivers cross to KHF and flow northward from the areas where higher elevation differences between drainage divide and crest line are found. It is good evidence of geomorphic evaluation of surface topography, associated landforms, and younger fluvial system controlled by subsurface tectonic processes.

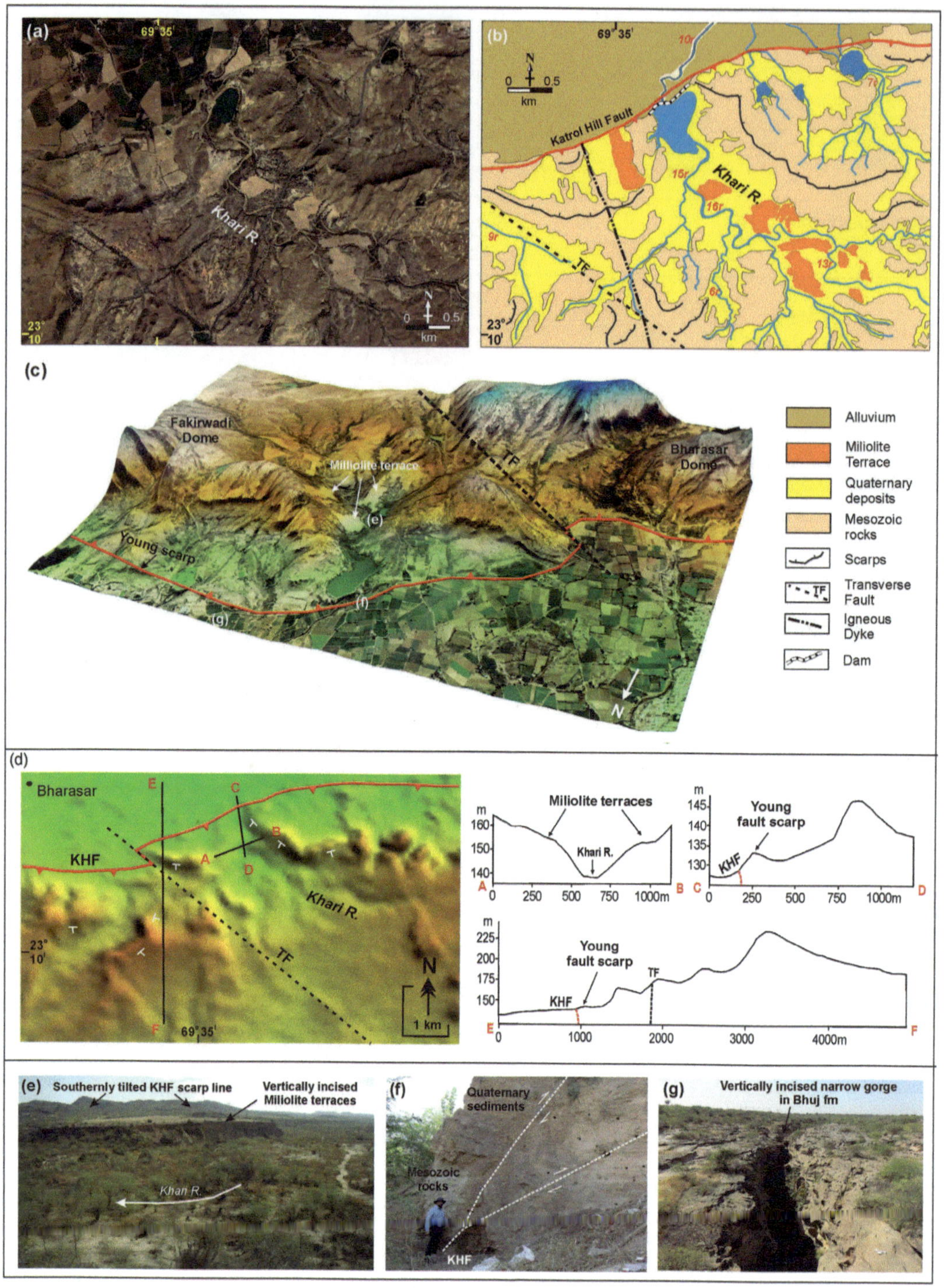

Figure 12.5 Geomorphic analysis and neotectonic field evidence from Kharriver, SW of Bhuj city. The location of this area is marked in Figure 12.3b. (a) A plan view of an uninterpreted satellite image (2021, https://earth.google.com) showing upper reaches of Quaternary filled Khari river valley. Note the rugged Mesozoic topography and formation of a deeply incised narrow valley. (b) Interpreted geological map of the area shown in (a). Mapping of various tectonic elements, KHF, transverse fault, and stratigraphy (Mesozoic/ Quaternary) is carried out based on interpretation of remote sensing data and geological field evidence. Formation of vertically incised paired terraces (field photograph shown in (e)) of eolian and valley-filled miliolite sediments (youngest quaternary sediments) by vertical incision (ranging between 6 and 16 m) by lower-order streams of Khari river indicates vertical uplift of the area due to reverse movement along KHF during its Late Quaternary reactivation. (c) Satellite image of the same area draped over 90 m spatial resolution SRTM DEM (https://www.usgs.gov), showing a 3D perspective view of Quaternary filled Khari river valley, miliolite terraces, young scarp, and other tectonic elements. (d) Geomorphic mapping of the region shows southward tilted rugged topography of Fakirwadi and Bharasar domes to the south of KHF, while the low laying rocky plane of Bhuj is located to the North. Three cross-sections are also shown. The cross-section A-B is taken across the Khari river showing vertically incised narrow valley and paired miliolite terraces. The cross-section C-D is taken across the KHF showing a rapid increase in elevation toward the south due to reverse motion and upward movement of the hanging wall side. The precise location of the KHF is marked by field evidence and near-surface investigations using GPR (Patidar et al. 2007, 2008). The cross-section E-F is taken along an N–S transect showing sharp geomorphic contrast across the KHF. Note variation in the dip of the scarp due to doming of the Mesozoic formations. (e) Field photograph of miliolite terrace incised by Khari river for ground validation of satellite image shown in (a)–(c). (f) Exposed cliff section of late Quaternary sediments along Khari river showing reverse movement along KHF. Splaying nature of the gently southward dipping KHF plane is noticed during upward propagation in Quaternary sediment (Patidar et al. 2008). (g) Field photograph showing vertically incised narrow gorge developed to the north of KHF in Bhuj formation indicating differential upliftment along KHF. The location of field photographs is shown in (c).

Figure 12.6 Geomorphic analysis and neotectonic field evidence from Gunawari back valley and Khatrod area located toward the eastern part of the Katrol Hill Range. The location of this area is marked in Figure 12.3b. (a) A plan view of an uninterpreted satellite image (2021, https://earth.google.com) showing Gunawari river valley located between the range front scarps and the drainage divide as shown in Figure 12.3a. (b) Satellite image draped over 90 m spatial resolution SRTM DEM (https://www.usgs.gov), showing an east-facing 3D perspective view of Gunawari river back valley filled with Quaternary deposits. Range front scarp of the KHF and drainage divide of the Katrol Hill Range to the south of the back valley can be seen. Mapping of E–W trending KHF and oblique trending transverse fault near Ler is marked based on interpretation of satellite data and geomorphic evidence of neotectonic deformation and GPR. (c) Field photograph of the back valley filled with Quaternary sediments. Note the E–W trending view of the drainage divide in the background. (d) Field photograph showing deeply incised (~10 m thick) Late Quaternary sediments (valley-filled miliolite, alluvium) by a meandering curve of Gunawari river in the back valley. Overall southward tilted topography of Mesozoic strata can be seen in the background. (e) Field photograph from further east showing unconformable contact between south-dipping Mesozoic rocks and overlaying Quaternary stratigraphy (valley-filled gravel and miliolite deposits). (f) Incised cliff exposing fine-grained alluvium that unconformably overlies the Mesozoic formations at the base in the eastern part of the Gunawari river valley. (g) An uninterpreted satellite image showing the Khatrod dome located at the eastern end of the KatrolHill Range. The location of this area is marked in Figure 12.3b. (h) Satellite image draped over 90 m spatial resolution SRTM DEM (https://www.usgs.gov), showing a north-facing 3D perspective view of the rang front Khatrod scarp. At this part of the Katrol Hill Range, the crest line of the range front scarp and drainage divide meets together and extends further eastward as shown in Figure 12.4a. During the geological field mapping, a steeply southward dipping (~80°) plane of KHF is found around 1 km north to the range front scarp, which indicates the southward retreat of the fault scarp due to erosion. Abrupt deflection in the north-flowing small order drainages is also noticed near the vicinity of KHF. (i) A panoramic view of north-facing range front scarp of the KHF at khatrod area. A youthful topography of a fault scarp and the fan-shaped scarp-derived colluvium deposits can be seen at the base.

Acknowledgements

Soumyajit Mukherjee invited to submit and also handled this article. Wiley Blackwell Acquisition Editors (present: Frank Weinreich, past: Andrew Harrison), Stacey Woods (Handling Editor), and proofreading team. Summaried in Misra and Mukherjee (2022). Atul Kumar Patidar gratefully acknowledges the support received from the University of Petroleum & Energy Studies, Dehradun.

References

Biswas, S.K. (1987). Regional tectonic framework, structure and evolution of the western marginal basins of India. *Tectonophysics* 135: 302–327.

Chowksey, V., Maurya, D.M., Khonde, N., and Chamyal, L.S. (2010). Tectonic geomorphology and evidence for active tilting of the Bela, Khadir and Bhanjada islands in the seismically active Kachchhpalaeoriftgraben, Western India. *ZeitschriftfürGeomorphologie* 54 (4): 467–490. doi: 10.1127/0372-8854/2010/0054-0021.

Dasgupta, S. and Mukherjee, S. (2017). Brittle shear tectonics in a narrow continental rift: asymmetric non-volcanic Barmer basin (Rajasthan, India). *The Journal of Geology* 125: 561–591.

Dasgupta, S. and Mukherjee, S. (2019). Remote sensing in lineament identification: examples from western India. In: *Problems and Solutions in Structural Geology and Tectonics*, Developments in Structural Geology and Tectonics Book Series, vol. 5. Series Editor: Mukherjee, S. (ed. A. Billi and A. Fagereng), 205–221. Elsevier. ISSN: 2542-9000. ISBN: 9780128140482.

Dasgupta, S., Biswas, M., Mukherjee, S., and Chatterjee, R. (2022). Depositional system, morphological signatures, tectonics and sedimentation pattern along the transform margin- Palar-Pennar basin, Indian east coast. *Journal of Petroleum Science & Engineering* 211: 110155. https://doi.org/10.1016/j.petrol.2022.110155.

Gogoi, M.P., Gogoi, B., and Mukherjee, S. (2022). Tectonic instability of the petroliferous upper Assam valley (NE India): a geomorphic approach. *Journal of Earth System Science* 131: 18.

Kaplay, R.D., Babar, M., Mukherjee, S., and Kumar, T.V. (2017). Morphotectonic expression of geological structures in eastern part of south east Deccan volcanic province (around Nanded, Maharashtra, India). In: *Tectonics of the Deccan large Igneous Province* (ed. S. Mukherjee, A.A. Misra, G. Calvès and M. Nemčok), 317–335. London, Special Publications 445: Geological Society.

Maurya, D.M., Thakkar, M.G., and Chamyal, L.S. (2003). Implications of transverse fault system on tectonic evolution of mainland Kachchh, western India. *Current Science* 85 (5): 661–667.

Maurya, D.M., Chowksey, V., Patidar, A.K., and Chamyal, L.S. (2017). A review and new data on neotectonic evolution of active faults in the Kachchh Basin, Western India: legacy of post-Deccan trap tectonic inversion. *Geological Society, London, Special Publications* 445 (1): 237–268. https://doi.org/10.1144/SP445.7.

Maurya, D.M., Tiwari, P., Shaikh, M.A. et al. (2021). Late quaternary drainage reorganization assisted by surface faulting: the example of the Katrol Hill fault zone, Kachchh, western India. *Earth Surface Processes and Landforms* 1–26: https://doi.org/10.1002/esp.5097.

Misra, A.A. and Mukherjee, S. (2022). Introduction to "atlas of structural geological and geomorphological interpretation of remote sensing images". In: *Atlas of Structural Geological and Geomorphological Interpretation of Remote Sensing Images* (ed. A.A. Misra and S. Mukherjee). Wiley Blackwell. ISBN: 9781119813354.

Misra, A.A., Bhattacharya, G., Mukherjee, S., and Bose, N. (2014). Near N-S paleo-extension in the western Deccan region in India: does it link strike-slip tectonics with India-Seychelles rifting? *International Journal of Earth Sciences* 103: 1645–1680.

Padmalal, A., Maurya, D.M., Vanik, N.P. et al. (2021). Impact of long term uplift on stream networks in tectonically active Northern Hill range, Kachchh palaeo-rift basin, western India. *Journal of Mountain Science* 18: 1609–1629. https://doi.org/10.1007/s11629-020-6420-9.

Patidar, A.K.(2010). Neotectonic Studies in Southern Mainland Kachchh Using GPR with Special Reference to Katrol Hill Fault. Ph.D. thesis. The Maharaja Sayajirao University of Baroda, Vadodara, Gujarat. Shodhganga. http://hdl.handle.net/10603/59101

Patidar, A.K., Maurya, D.M., Thakkar, M.G., and Chamyal, L.S. (2007). Fluvial geomorphology and neotectonic activity based on field and GPR data, Katrol hill range, Kachchh, western India. *Quaternary International* 159: 74–92. https://doi.org/10.1016/j.quaint.2006.08.013.

Patidar, A.K., Maurya, D.M., Thakkar, M.G., and Chamyal, L.S. (2008). Evidence of neotectonic reactivation of Katrol Hill Fault (KHF) during late quaternary and its

GPR characterization. *Current Science* 94: 338–346. http://www.jstor.org/stable/24100341.

Shaikh, M.A., Maurya, D.M., Vanik, N.P. et al. (2018). Uplift induced structurally controlled landscape development: example from fault bounded Jumara and Jara domes in Northern Hill Range, Kachchh, Western India. *Geosciences Journal* 1–19. https://doi.org/10.1007/s12303-018-0061-9.

Shaikh, M.A., Maurya, D.M., Mukherjee, S. et al. (2020). Tectonic evolution of the intra-uplift Vigodi-Gugriana-Khirasra-Netra Fault System in the seismically active Kachchh rift basin, India: implications for the western continental margin of the Indian plate. *Journal of Structural Geology* 140: 104124. https://doi.org/10.1016/j.jsg.2020.104124.

Shaikh, M.A., Patidar, A.K., Vanik, N.P. et al. (2022). Building tectonic framework of a blind active fault zone using field and ground-penetrating radar data. *Journal of Structural Geology* 155: 104526. https://doi.org/10.1016/j.jsg.2022.104526.

Sissakian, V.K., Al-Ansari, N., and Abdullah, L.H. (2020). Neotectonic activity using geomorphological featuresin the Iraqi Kurdistan region. *Geotechnical and Geological Engineering* 38: 4889–4904. https://doi.org/10.1007/s10706-020-01334-1.

Tiwari, P., Maurya, D.M., Shaikh, M.P. et al. (2021). Surface trace of the active Katrol Hill Fault and estimation of paleo-earthquake magnitude for seismic hazard, Western India. *Engineering Geology* 295: 106416. https://doi.org/10.1016/j.enggeo.2021.106416.

13

Tectonics, Fault Zones, and Topography in the Alaska–Canada Cordillera with a Focus on the Alaska Range and Denali Fault Zone

Jonathan Saul Caine[*,1] *and Jeff A. Benowitz*[2]

[1] *U.S. Geological Survey, Denver, CO, USA*
[2] *Fairbanks, AK, USA*

13.1 Introduction

Synergistic interactions between geologic structures and topography have long been recognized to reflect numerous Earth processes and rock properties over time (e.g. Dasgupta and Mukherjee 2017, 2019). In this review contribution we briefly highlight the scientific history and then the tectonic and geologic framework, and upper crustal structural geologic attributes that have collectively contributed to the topography of the Alaska–Canada Cordillera in existence today (Figure 13.1). Alaskan geologists and explorers of the late nineteenth century observed and recognized the intimate connections among mountainous topography, geologic structures, and rock types. Paraphrasing Walter C. Mendenhall of the U.S. Geological Survey (Mendenhall 1905): "...The general strike of the beds, [is] parallel to the axis of the adjacent parts of the Alaska Range, [and] to the structural valleys of the region. Toward the north the formation is limited by a profound fault. The faulting has crumpled the soft Permian shales, and the confusion too complex for solution by reconnaissance work." Brooks and Prindle (1911), Moffit and Pogue (1915), Capps and Moffit (1933), and many colleagues in subsequent explorations made additional improvements on their predecessor's field work. These field-based observations, over time, built upon each other resulting in important intellectual advancements in the pondering of the hypothesis that tremendous blocks of Earth could move long distances along "great faults" and ultimately control topography.

13.2 Regional Tectonics of the Northern Cordillera

It was not until the advent of plate tectonics in the mid-twentieth century that researchers began to view the nature of the northern Cordillera orogen as a quilt of foreign pieces of crust or "suspect terranes" (e.g. Coney et al. 1980; McPhee 1983). Alaska and northwestern Canada including Yukon, British Columbia, and associated continental margins reflect the record of largely additive or accretionary tectonic processes and the evolution of the western North American craton (i.e. Laurentia; Figures 13.1 and 13.2; Colpron et al. 2007; Monger and Gibson 2018). This complex (and still active) marginal region of Laurentia includes continental and oceanic terranes sculpted by multiple phases of contraction, translation, and extension from the late Proterozoic. Microcontinents and various fragments have been adjoined to the stable cratonic margin, and possibly pushed out of the way (i.e. extruded); ocean basins have opened and closed; ancient and modern volcanic island arcs with associated subduction zone plutonic systems have stitched these pieces together; and rocks were buried-metamorphosed-exhumed (Plafker and Berg 1994; Dusel-Bacon et al. 2006; Bradley et al. 2014). The entirety of this crustal patchwork was reincorporated into the combined contractional and translational conveyor system via continental scale shear zones with hundreds of kilometers of displacement and, in the present landscape, host some of the world's highest and longest terrestrial mountain ranges thus directly controlling adjacent topography (Figures 13.1–13.3; e.g. St. Amand 1954, 1957; Till et al. 2007; Trop et al. 2019).

*Corresponding Author: jscaine@usgs.gov

Atlas of Structural Geological and Geomorphological Interpretation of Remote Sensing Images, First Edition.
Edited by Achyuta Ayan Misra and Soumyajit Mukherjee.

Companion website: www.wiley.com/go/misra/RemoteSensingImages

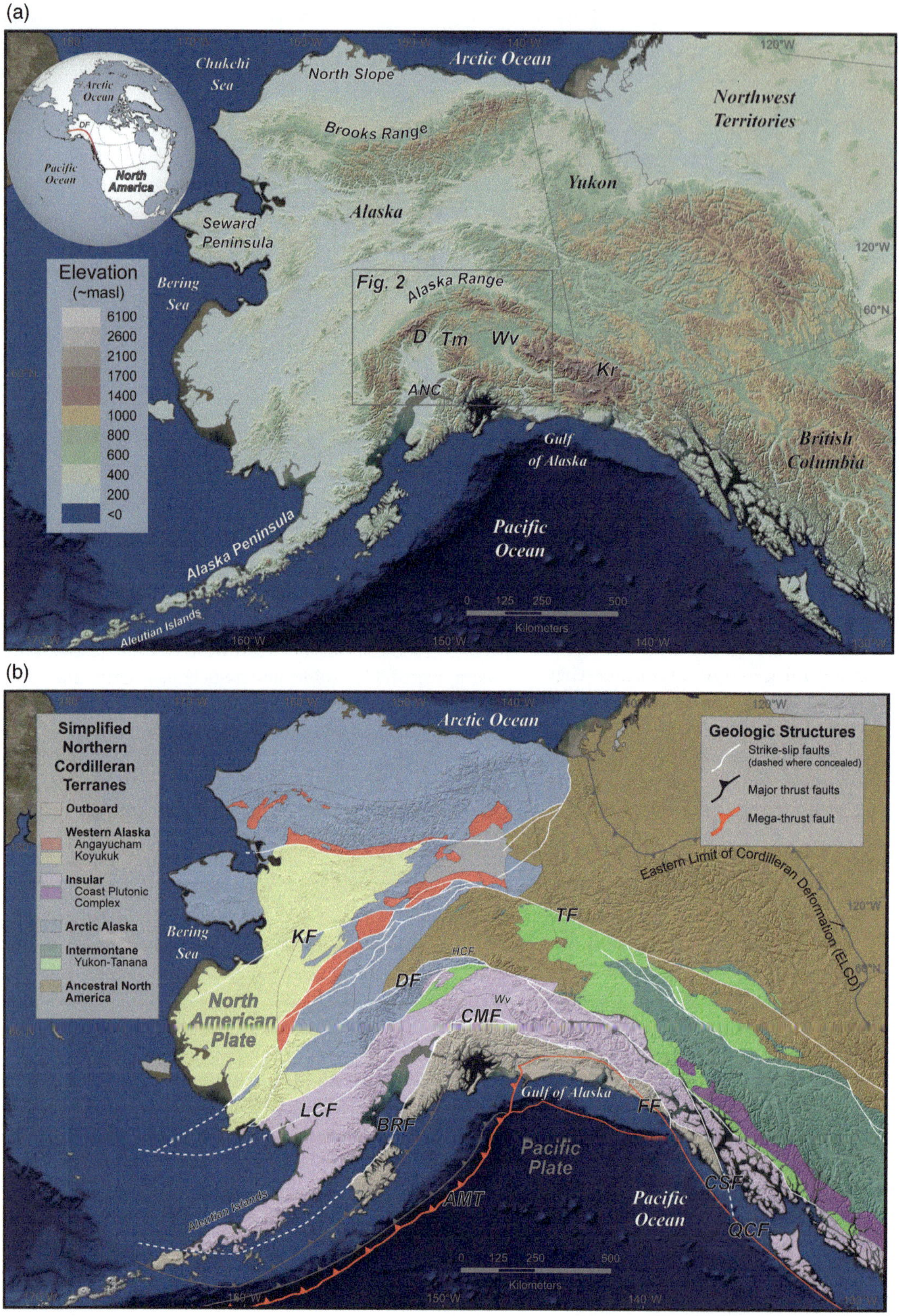

Figure 13.1 (a) 30-m digital elevation model (DEM) hillshade image showing major continental topography and bathymetry. The locations of Denali (D), formerly named Mount McKinley; Talkeetna Mountains (Tm); and the Wrangell volcanic arc (Wv) are shown in the inset map area for Figure 13.2. Kr denotes the location of the Kluane Ranges and the eastern Denali fault (Source: Image and DEM is from Airbus, USGS, NASA, CGIAR, NCEAS, NLS, OS, NMA, Geodatastyrelsen, GS, and the GIS User Community). (b) Major terranes grouped by tectonic affinity, tectonic features, and fault map of the northern Cordillera overlaid on DEM (Source: modified from Colpron and Nelson 2011). Major faults are shown as white, dark gray, and red traces and include AMT, Aleutian megathrust; BRF, Border Ranges; LCF, Lake Clark; CMF, Castle Mountain; CSF, Chatham Strait; DF, Denali; HCF: Hines Creek; FF, Fair weather; KF, Kaltag; QCF, Queen Charlotte; TF, Tintina.

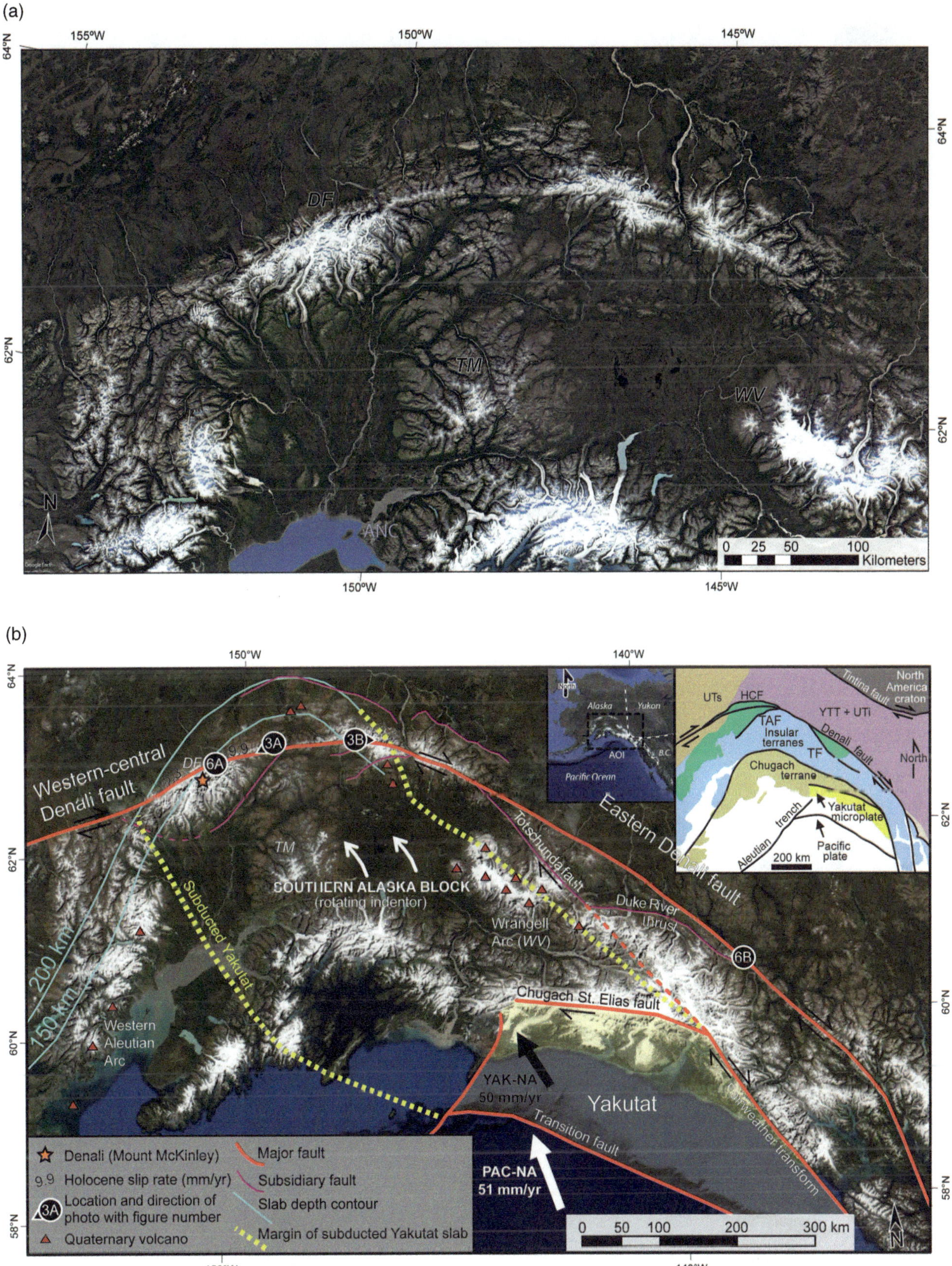

Figure 13.2 (a) Google Earth™ image showing topographic details of the Alaska Range, abbreviations as in Figure 13.1. (b) Image in A with major structural geologic, tectonic, and geodynamic features superimposed. Insets show overall location in the northern Cordillera of Alaska (USA) and Canada and major terranes.

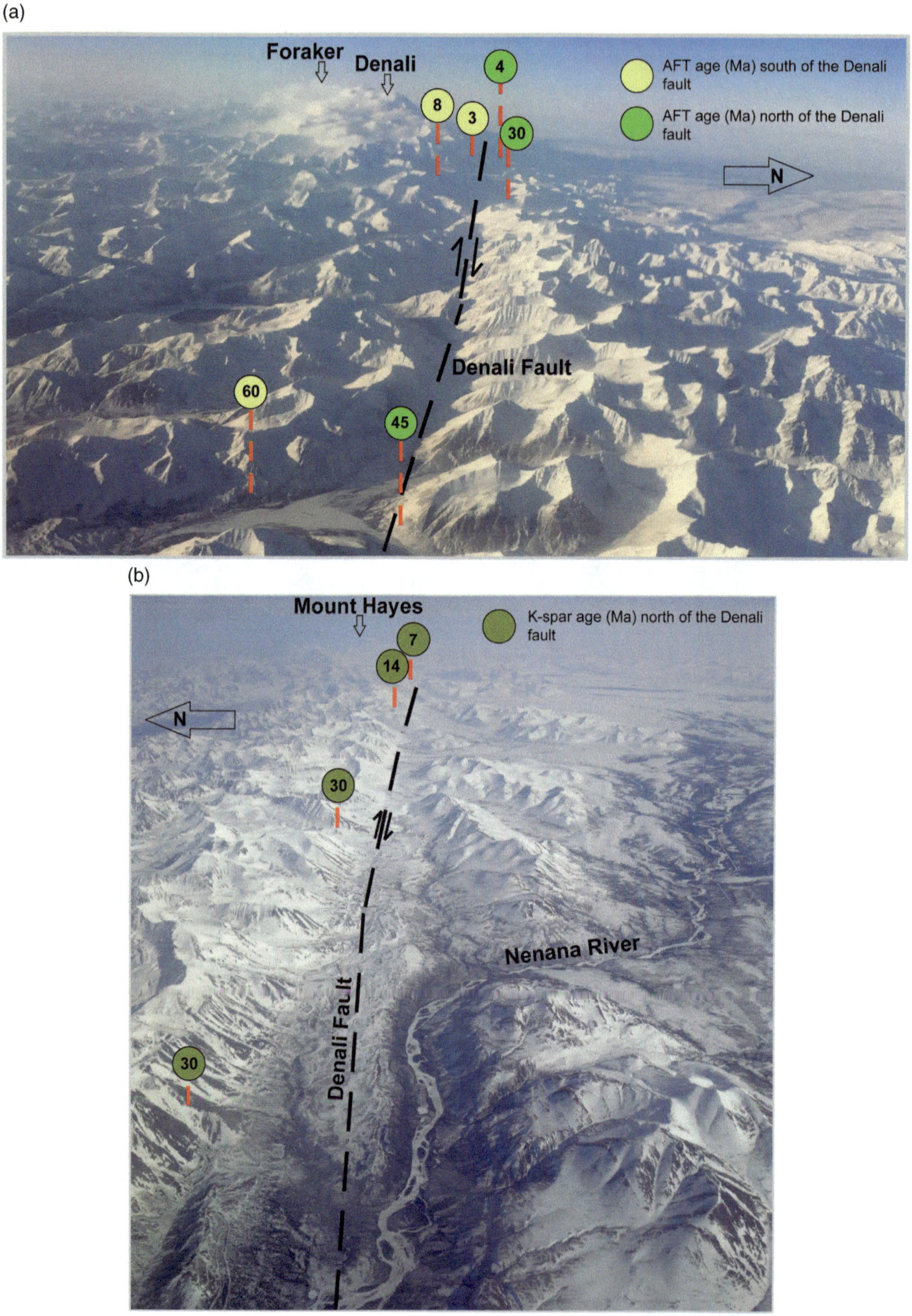

Figure 13.3 Oblique aerial photographs looking along the Denali fault in the Alaska Range (Source: photos by Jeff Benowitz). Thermochronometric ages and associated mineral chronometers with respective ages are also shown on either side of the Denali fault with red dashed lines pointing to the sample locations. Note the lack of age continuity with position along the fault and with topography (see Benowitz et al. 2011, 2014; Burkett et al. 2016). Photo locations shown in Figure 13.2. View directions are (a) West to Denali (Mount McKinley) and (b) East to Mounts Deborah and Hayes.

13.3 The Denali Fault and the Alaska Range: Topography, Geophysics, and Crustal Processes

Like a giant scar sliced through the Earth's upper crust, the ~2000 km long Denali fault has puckered up the Alaska Range as one of the paramount examples of a mountain range-fault system (Figures 13.1–13.4). The mountain range-fault system is arcuate in shape, covered in alpine glaciers, and is segmented with displacements and topography that are not uniformly distributed along its trace. The Eastern and Central Denali fault segments have accommodated at least 400 km of right-lateral strike-slip separation and in places kilometer-scale reverse slip since ca. 57 Ma (Riccio et al. 2014). The far western segment has ~130 km of dextral separation since ~85–70 Ma (e.g. Eisbacher 1976; Nokleberg et al. 1985; Plafker and Berg 1994; Lowey 1998; Cole et al. 1999; Miller et al. 2002; Ridgway et al. 2002). Ancient to modern components of reverse, oblique, and strike slip are reflective of the dominantly transpressional nature of the fault system (e.g. Plafker and Berg 1994; Eberhart-Phillips et al. 2003; Bemis et al. 2015).

The Alaska Range shows complexity in topographic, geometric, and exhumational age asymmetry along and across the strike of the Denali fault zone attributable to several factors (Figures 13.4 and 13.5; Benowitz et al. 2011, 2012). These factors include orogen-scale transpressional tectonics; fault geometric elements such as the major restraining bend along the Denali fault where the extreme topography of Denali occurs (formerly known as Mount McKinley, the highest peak in North America at 6190 masl); variations of juxtaposed rock types of different intrinsic strengths; as well as climate and weathering, particularly due to glacial processes (Benowitz et al. 2011; Fitzgerald et al. 2014; Lease et al. 2016; Burkett et al. 2016; Regan et al. 2020). The modern topographic signature of the Denali fault zone reflects a combination of linked vertical and horizontal slip since at least 45 million years, northward convergence of southern Alaska against the Denali fault since at least 30 million years, and late Cenozoic global cooling and related glaciation rearranging mass distribution in these sub-arctic mountains (Benowitz et al. 2011; Riccio et al. 2014; Lease et al. 2016; Haeussler et al. 2017).

To illustrate fault-related topography and asymmetry associated with the Denali fault, simple topographic profiles showing average elevation versus distance along fault strike are presented in Figure 13.4. Figure 13.4 shows topographic profiles along the fault trace and subparallel profiles connecting many of the highest summits on either side that fall within 5–10 km of the fault trace (see Data Supplement for profile locations and other metadata). From the west end of the fault to Denali (Mount McKinley, see point *DF* on Figure 13.4) there is symmetry in the locations of topographic highs among the profiles, moderate variations in the magnitudes of the highs relative to the fault trace, and a switch in the topographic highs where the north side has a higher high compared with the south side relative to the

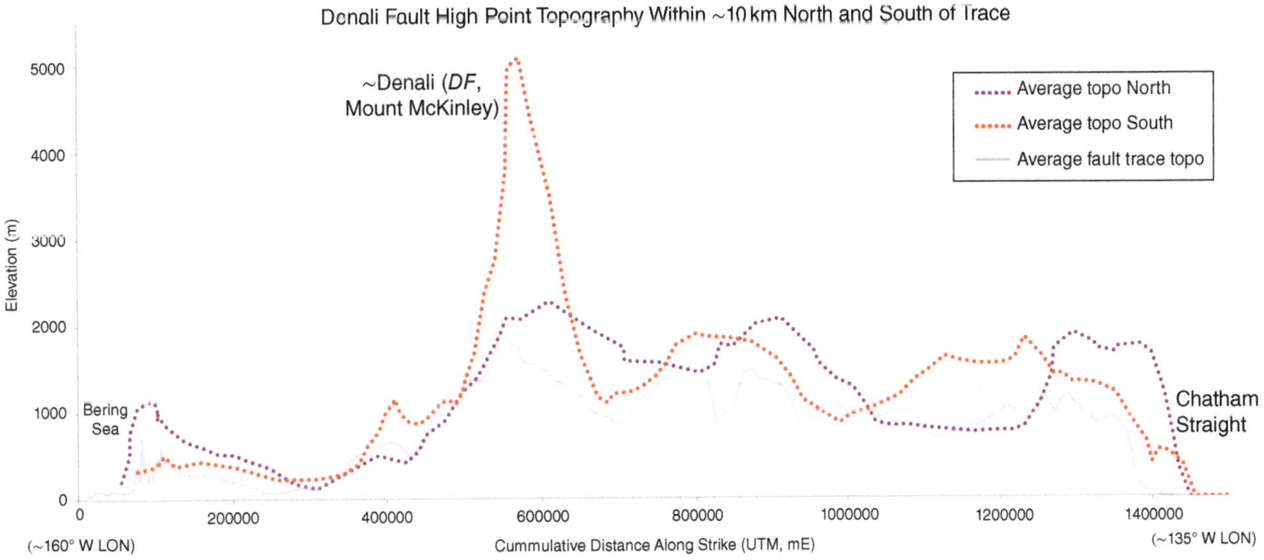

Figure 13.4 Plot of averaged elevation along the surface trace of the Denali fault and along two arbitrary subparallel lines on either side of the trace that connect many of the high summits. (Note that the profiles are smoothed with the use of a 4-point moving average.) The fault trace is the solid and faint lavender line, the south line is red and the north magenta, each are dotted and within about 5–10 km of the trace. The distance along strike is in units of cumulative meters from longitude ~160° west to ~135° west using a Universal Transverse Mercator projection (UTM, World Geodetic Survey 1984 datum).

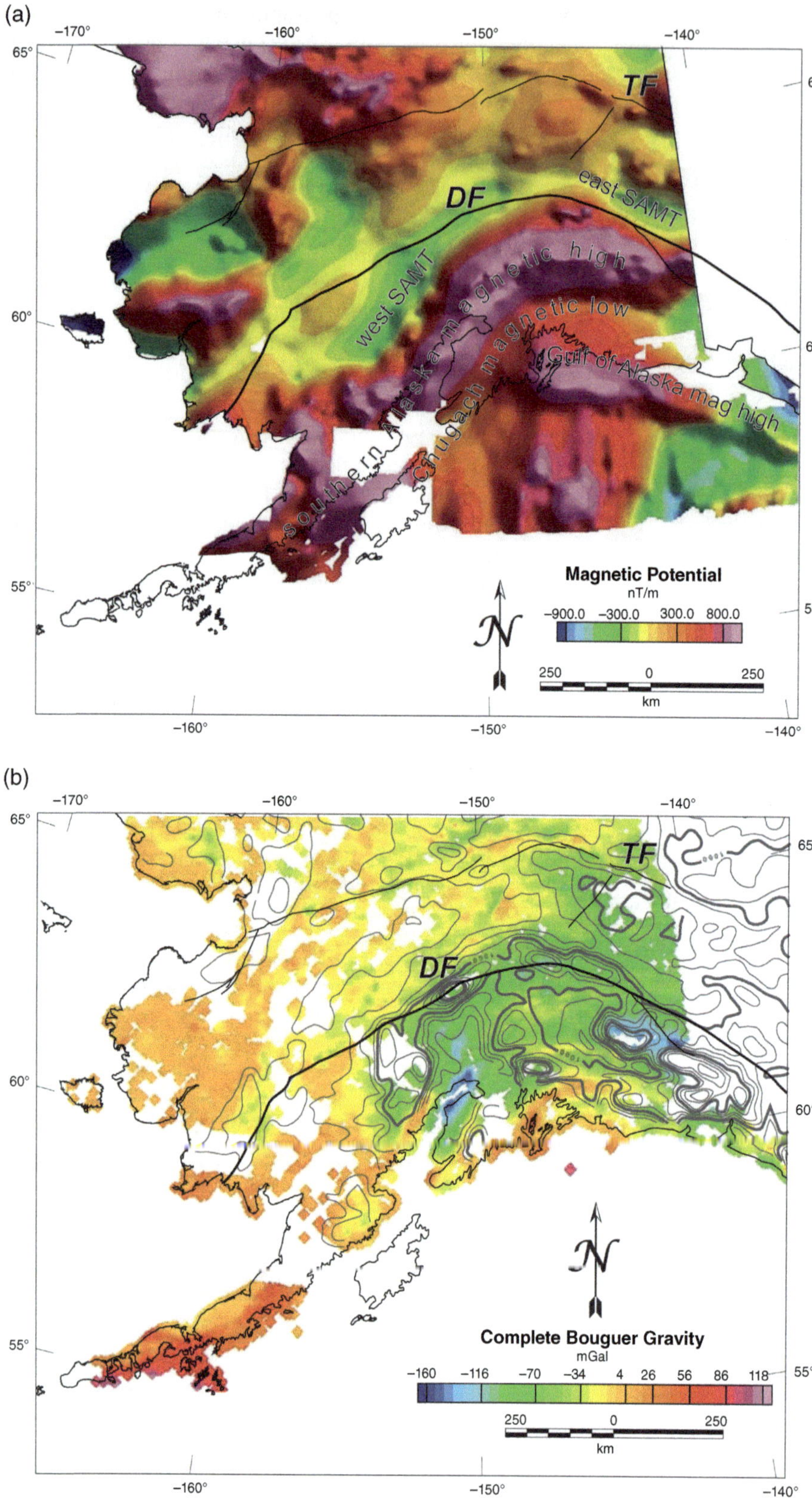

Figure 13.5 Geophysical data associated with the northern Cordillera of Alaska showing major faults (DF, Denali fault; TF, Tintina fault; modified from Saltus et al. 2007). (a) Magnetic potential derived from merged aeromagnetic data are shown as colored regions with regional aeromagnetic domains noted with text. (b) Complete Bouguer gravity (colored regions) and topography (gray contour lines have 1000-m intervals and the thicker contours are 1000 and 5000 m).

trace (Figure 13.4). Just west of Denali (Mount McKinley) the fault trace, the south and north sides of the fault, all show roughly co-located increases in topography. However, the magnitude of the high on the south side is about three times greater than that along the trace and along the north side reflecting the immense topography of Denali (Mount McKinley) on the concave-south side of the bend in the fault trace (Figure 13.4). Additionally, the overall possible effects of the fault on local topography are found within a narrow region on either side of the fault, particularly when compared with the fault trace length (e.g. Figure 13.3). Directly east of Denali (Mount McKinley) the profiles show strike-parallel broadening in the topography with fault trace and south-side profile highs being reasonably co-located (Figure 13.4). These topographic features are consistent with overall right-lateral strike-slip but also vertical components of slip; topography in the restraining bend at Denali (Mount McKinley); vertical extrusion between the Hines Creek and Denali faults (Figure 13.2); the complexities in exhumational age; and possibly the relative variations in resistance to weathering of various rock types within and adjacent to the Denali fault itself (e.g. Benowitz et al. 2011, 2014).

The tectonic-scale processes associated with slip, deformation, and topographic change along the Denali fault are tied to oblique convergence of the ancient Kula and Pacific plates with the North American plate, and more recently with flat slab subduction of the Yakutat microplate resulting in modern seismic activity such as the 2002 M7.9 Denali fault earthquake (Figure 13.2; e.g. Eberhart-Phillips et al. 2003; Eberhart-Phillips et al. 2006; Haeussler et al. 2003). The arcuate, oroclinal shape of the Denali fault and the Alaska Range are controversial but have been in place since ca. 45 Ma and likely earlier (Murphy 2018; Benowitz et al. 2021). The Oligocene to present subduction of the over-thickened Yakutat flat slab involved a change in plate motion at ~6 Ma (e.g. Fitzgerald et al. 1995). This change caused several geodynamic effects along much of the Alaska Range and the Denali fault including the onset of rapid exhumation of Denali (Mount McKinley) as well as significant volcanism in the Wrangell volcanic arc (Figures 13.1 and 13.2; e.g. Brueseke et al. 2018; McDermott et al. 2021).

Magnetic and gravity data show distinctive, continental-scale domanial anomalies (Figure 13.5; Saltus et al. 2007; cf. Schulte-Pelkum et al. 2020). These anomalies are geometrically aligned with the southern Alaska plate margin, the orogenic tectonic grain, and the Alaska Range, and thus indicate that subsurface crustal features may play a role in controlling tectonic inheritance, reactivation of geologic structures, and topography. Alaska's gravity signature is challenging to model, particularly due to the dynamic nature of subduction along the southern margin and Aleutian megathrust (Saltus et al. 2007). Although the commonly observed inverse relation between topography and the complete Bouguer gravity anomaly is absent in southern Alaska and is also not strong along the Alaska range in central Alaska, particularly at Denali (Mount McKinley), there is a residual gravity gradient along the northern edge of the Alaska Range (Figure 13.5; Saltus et al. 2007). This gravity gradient from the western to central Denali fault disappears along and north of the eastern Denali fault. The domains also include regional scale alternating magnetic highs and troughs. The northeastern margin of the southern Alaska magnetic high is co-located with the Alaska range but diverges to the southwest at the central and western segments of the Denali fault zone. The northern boundary of this magnetic high tracks with the northern boundaries of the Wrangellian and Peninsular terranes (Insular terranes in Figure 13.1). The magnetic high has been attributed to Late Triassic emplacement of deep crustal mafic magmas, associated tectonism, and over-thickened crust (Saltus et al. 2007) prior to the inception of the Denali fault. However, the divergence of the magnetic boundary from the central to western Denali fault may indicate that the factors controlling tectonic inheritance may not be the same along any one terrane boundary at any given time. Additionally, the possible northwestern gravity gradient may also be consistent with post-Late Triassic processes such as the location of inception, geometry, and exhumation associated with the younger evolution of the Alaska Range and Denali fault zone.

13.4 Exceptional Bedrock Exposures Reveal Strain Localization Along the Denali Fault

Although direct exposures of the Denali fault zone in bedrock are exceptionally rare, regional to outcrop scale observations show the common internal structure consisting of some degree of strain localization in one or more, and presumably relatively weak, fault cores and an associated, commonly hydrothermally altered, damage zone (e.g. Caine and Forster 1999; Faulkner et al. 2003). In the field, the influence of the Denali fault and its geometry are easily recognized controls on local topography. Figures 13.3, 13.4, and 13.6 show that, in general, the fault core and possibly damage zone regions, comprising the entire fault, are weathered out and in many areas host glaciers, rivers, and in the southeastern segment of the Denali fault, major fjords. Within about 5–10 km on either side of the fault zone the major vertical component of high topography, directly related to the fault itself, begins to fall back to background

Figure 13.6 Photographs of outcrop-scale, bedrock exposures of the Denali fault (photos by the authors). (a) View looking west at the Gunsight Pass-Muldrow Glacier exposure in Denali National Park (Benowitz et al. 2011). The fault juxtaposes ca. 40 Ma McGonagall Granite (light gray rocks, EMg) against Mesozoic meta sediments (dark gray rocks, MzMs). The fault trace in magenta strikes west–east and the dip is subvertical. Fault block motion toward the reader is shown by the magenta circled dot and motion away by the circled plus sign. The inset was taken from Gunsight Pass and looks east to where the enclosing photo was taken from on the Muldrow glacier. (b) Oblique aerial view of the Eastern Denali fault in the Kluane Ranges of southwest Yukon, Canada (Caine et al. 2015). The view is looking northwest showing the juxtaposition of Permian Hasen Creek formation metapelites (PH) against Upper Triassic Bear Creek assemblage metaplutonic and metavolcanic rocks (uTrBv). The fault strikes northwest–southeast and is subvertical shown by the magenta trace with arrows showing relative motion of each fault block). The reddish, fault parallel bands are iron-oxide rich hydrothermally altered zones associated with faulting.

elevations (Figures 13.4 and 13.5; cf. Benowitz et al. 2011; Schulte-Pelkum et al. 2020). Along the eastern Denali fault in southwest Yukon, the entire fault zone is exposed near Kluane Lake (Figure 13.6). Here the fault zone is characterized by the juxtaposition of highly weatherable carbonaceous metapelites against more resistant metavolcanic rocks (Caine et al. 2015). The entire fault zone here is on the order of only 100 m wide indicating extreme and long-lived strain localization, and it also forms an important modern drainage feature. Thus, from regional scale topographic and geophysical data to outcrop scale observations, the Denali fault zone appears to be a highly strain-localized feature. Ongoing research that integrates multi-scale geologic mapping, topographic analyses, advances in thermochronology, geophysics, and geodynamics will continue to refine the original hypotheses made over a century ago toward an understanding of the fully integrated scope of tectonics, fault zones, and landscape evolution in the northern Cordilleran orogen.

Acknowledgements

Soumyajit Mukherjee invited to submit an article in the edited book. Achyuta Ayan Misra handled this article. Wiley Blackwell Acquisition Editors (present: Frank Weinreich, past: Andrew Harrison), Stacey Woods (Handling Editor), and proofreading team. Summarized in Misra and Mukherjee (2022).

This work was funded by the U.S. Geological Survey, Mineral Resources Program. We thank Rick Saltus and Adrian Bender for discussion and review of an early version. Reviews by Evan Thoms and Cal Ruleman improved the contribution. Any use of trade, firm, or product names is for descriptive purposes only and does not imply endorsement by the U.S. Government.

References

Bemis, S.P., Weldon, R.J., and Carver, G.A. (2015). Slip partitioning along a continuously curved fault: quaternary geologic controls on Denali fault system slip partitioning, growth of the Alaska range, and the tectonics of south-Central Alaska. *Lithosphere* 7 (3): 235–246. http://lithosphere.gsapubs.org/cgi/doi/10.1130/L352.1.

Benowitz, J.A., Layer, P.W., Armstrong, P. et al. (2011). Spatial variations in focused exhumation along a continental-scale strike-slip fault: the Denali fault of the eastern Alaska range. *Geosphere* 7 (2): 455–467.

Benowitz, J.A., Haeussler, P.J., Layer, P.W. et al. (2012). Cenozoic tectono-thermal history of the Tordrillo Mountains, Alaska: Paleocene-Eocene ridge subduction, decreasing relief, and late Neogene faulting. *Geochemistry, Geophysics, Geosystems* 13 (4): https://doi.org/10.1029/2011GC003951.

Benowitz, J.A., Layer, P.W., and Vanlaningham, S. (2014). Persistent long-term (c. 24 Ma) exhumation in the eastern Alaska range constrained by stacked thermochronology. *Geological Society, London, Special Publications* 378 (1): 225–243.

Benowitz, J., Roeske, S., Regan, S., et al. (2021) Large Scale Crustal-Block Vertical Extrusion Between the Hines Creek and Denali Faults Coeval with Slip Localization on the Denali Fault since ca. 45 Ma: Hayes Range, Alaska, Geological Society of America Abstracts with Programs, vol. 53. http://dx.doi.org/10.1130/abs/2021AM-365121.

Bradley, D.C., McClelland, W.C., Friedman, R.M. et al. (2014). Proterozoic geochronological links between the farewell, Kilbuck, and arctic Alaska terranes. *Journal of Geology* 122 (3): 237–258. https://doi.org/10.1086/675663.

Brooks, A.H. and Prindle, L.M. (1911). The Mount McKinley region, Alaska, with descriptions of the igneous rocks and of the Bonnifield and Kantishna districts: U.S. Geological Survey, Professional Paper 70, 234 p.

Brueseke, M.E., Benowitz, J.A., Trop, J.M. et al. (2018). The Alaska Wrangell arc: ~30 Ma of subduction-related magmatism along a still active arc-transform junction. *Terra Nova* 31 (1): 59–66. https://doi.org/10.1111/ter.12369.

Burkett, C.A., Bemis, S.P., and Benowitz, J.A. (2016). Along-fault migration of the Mount McKinley restraining bend of the Denali fault defined by late quaternary fault patterns and seismicity, Denali National Park & Preserve, Alaska. *Tectonophysics* 693: 489–506.

Caine, J.S. and Forster, C.B. (1999). Fault zone architecture and fluid flow: insights from field data and numerical modeling. In: *Faults and Subsurface Fluid Flow in the Shallow Crust*, Geophysical Monograph Series, vol. 113 (ed. W.C. Haneberg, P.S. Mozley, J.C. Moore and L.B. Goodwin). American Geophysical Union https://doi.org/10.1029/GM113p0101.

Caine, J.S., Israel, S.A., Benowitz, J.A., and Murphy, D.A. (2015). A fault rock record of late Eocene strain partitioning along the Denali fault zone in southwestern Yukon. *Geological Society of America Abstracts with Programs* 47 (4): 51.

Capps, S.R. and Moffit, F.H. (1933). Mineral resources of Alaska, report on progress of investigations in 1930. The eastern portion of Mount McKinley National Park, Alaska. The Kantishna District, Alaska. Mining development in the Tatlanika and Totatlanika basins, Alaska: U.S. *Geological Survey Bulletin* 836-D: 219–300.

Cole, R.B., Ridgway, K.D., Layer, P.W., and Drake, J. (1999). Kinematics of basin development during the transition from terrane accretion to strike-slip tectonics. Late cretaceous-early tertiary Cantwell formation, south Central Alaska. *Tectonics* 18 (6): 1224–1244. https://doi.org/10.1029/1999TC900033.

Colpron, M. and Nelson, J.L. (2011). A digital atlas of terranes for the northern Cordillera, Yukon Geological Survey. http://data.geology.gov.yk.ca/Compilation/2.

Colpron, M., Nelson, J.L., and Murphy, D.C. (2007). Northern cordilleran terranes and their interactions through time. *GSA Today* 17: 4–7.

Coney, P.J., Jones, D.L., and Monger, J.W.H. (1980). Cordilleran suspect terranes. *Nature* 288 (7): 329–333.

Dasgupta, S. and Mukherjee, S. (2017). Brittle shear tectonics in a narrow continental rift: asymmetric non-volcanic Barmer basin (Rajasthan, India). *The Journal of Geology* 125: 561–591.

Dasgupta, S. and Mukherjee, S. (2019). Remote sensing in lineament identification: examples from western India. In: *Problems and Solutions in Structural Geology and Tectonics*, Developments in Structural Geology and Tectonics Book Series, vol. 5. Series Editor: Mukherjee, S. (ed. A. Billi and A. Fagereng), 205–221. Elsevier. ISSN: 2542-9000. ISBN: 9780128140482.

Dusel-Bacon, C., Hopkins, M.J., Mortensen, J.K. et al. (2006). Paleozoic tectonic and metallogenic evolution of the pericratonic rocks of east-central Alaska and adjacent Yukon, in Paleozoic Evolution and Metallogeny of Pericratonic Terranes at the Ancient Pacific Margin of North America. In: *Canadian and Alaskan Cordillera* (ed. M. Colpron and J.L. Nelson), 25–74. St. Johns, Newfoundland, Canada, Special Paper 45: Geological Association of Canada.

Eberhart-Phillips, D., Haeussler, P.J., Freymueller, J.T. et al. (2003). The 2002 Denali fault earthquake, Alaska: a large magnitude, slip-partitioned event. *Science* 300 (5622): 1113–1118. https://doi.org/10.1126/science.1082703.

Eberhart-Phillips, D., Christensen, D.H., Brocher, T.M. et al. (2006). Imaging the transition from Aleutian subduction to Yakutat collision in Central Alaska, with local earthquakes and active source data. *Journal of Geophysical Research: Solid Earth* 111 (B11).

Eisbacher, G. (1976). Sedimentology of the Dezadeash flysch and its implications for strike-slip faulting along the Denali fault, Yukon territory and Alaska. *Canadian Journal of Earth Sciences* 13 (11): 1495–1513. https://doi.org/10.1139/e76-157.

Faulkner, D.R., Lewis, A.C., and Rutter, E.H. (2003). On the internal structure and mechanics of large strike-slip fault zones: field observations of the Carboneras fault in southeastern Spain. *Tectonophysics* 367 (3–4): 235–251.

Fitzgerald, P.G., Sorkhabi, R.B., Redfield, T.F., and Stump, E. (1995). Uplift and denudation of the Central Alaska range: a case study in the use of apatite fission track thermochronology to determine absolute uplift parameters. *Journal of Geophysical Research* 100 (B10): 20118–20175. https://doi.org/10.1029/95JB02150.

Fitzgerald, P.G., Roeske, S.M., Benowitz, J.A. et al. (2014). Alternating asymmetric topography of the Alaska range along the strike-slip Denali fault: strain partitioning and lithospheric control across a terrane suture zone. *Tectonics* 33 (8): 1519–1533.

Haeussler, P.J., Bradley, D.C., Wells, R.E., and Miller, M.L. (2003). Life and death of the resurrection and subduction plate: evidence for its existence in in the northeastern time Pacific. *Geological Society of America Bulletin* 115 (7): 867–880. https://doi.org/10.1130/0016-7606(2003)115<0867.

Haeussler, P.J., Matmon, A., Schwartz, D.P., and Seitz, G.G. (2017). Neotectonics of interior Alaska and the late quaternary slip rate along the Denali fault system. *Geosphere* 13 (5): 1445–1463.

Lease, R.O., Haeussler, P.J., and O'Sullivan, P. (2016). Changing exhumation patterns during Cenozoic growth and glaciation of the Alaska range: insights from detrital thermochronology and geochronology. *Tectonics* 35 (4): 934–955. https://doi.org/10.1002/2015TC004067.

Lowey, G.W. (1998). A new estimate of the amount of displacement on the Denali fault system based on the occurrence of carbonate megaboulders in the Dezadeash formation (Jura-Cretaceous), Yukon, and the Nutzotin Mountains sequence (Jura-Cretaceous), Alaska. *Bulletin of Canadian Petroleum Geology* 46 (3): 379–386.

McDermott, R.G., Ault, A.K., and Caine, J.S. (2021). Dating fault damage along the eastern Denali fault zone with hematite (U-Th)/He thermochronometry. *Earth and Planetary Science Letters* 563: 116872. https://doi.org/10.1016/j.epsl.2021.116872.

McPhee, J. (1983). *Suspect Terrain*, 209. Farrar, Straus and Giroux.

Mendenhall, W.C. (1905). The geology of the central Copper River region, Alaska: U.S. Geological Survey Professional Paper 41, 133 p.

Miller, M.L., Bradley, D.C., Bundtzen, T.K., and McClelland, W. (2002). Late cretaceous through Cenozoic strike-slip tectonics of southwestern Alaska. *The Journal of Geology* 110 (3): 247–270. https://doi.org/10.1086/339531.

Misra, A.A. and Mukherjee, S. (2022). Introduction to "atlas of structural geological and geomorphological interpretation of remote sensing images". In: *Atlas of Structural Geological and Geomorphological Interpretation of Remote Sensing Images* (ed. A.A. Misra and S. Mukherjee). Wiley Blackwell. ISBN: 9781119813354.

Moffit, F.H. and Pogue, J.E. (1915). The broad pass region, Alaska, with sections on quaternary deposits, igneous rocks, and glaciation: U.S. Geological Survey, Bulletin 608, 80 p

Monger, J.W.H. and Gibson, H.D. (2018). Mesozoic-Cenozoic deformation in the Canadian cordillera: the record of a "continental bulldozer"? *Tectonophysics* 757: 153–169. https://doi.org/10.1016/j.tecto.2018.12.023.

Murphy, D.C. (2018). Latest cretaceous-early Eocene Pacific-Arctic?-Atlantic connection: co-evolution of strike-slip fault systems, oroclines, and transverse fold-and-thrust belts in the northwestern north American cordillera. In: *Circum-Arctic Structural Events: Tectonic Evolution of the Arctic Margins and Trans-Arctic Links with Adjacent Orogens*, vol. 541, Issue

28 (ed. K. Piepjohn, J.V. Strauss, L. Reinhardt and W.C. McClelland), 1–22. Geological Society of America Special Paper 541.

Nokleberg, W.J., Jones, D.L., and Silberling, N.J. (1985). Origin and tectonic evolution of the Maclaren and Wrangellia terranes, eastern Alaska range, Alaska. *Geological Society of America Bulletin* 96 (10): 1251–1270. https://doi.org/10.1130/0016-7606(1985)96<1251:OA-TEOT>2.0.CO;2.

Plafker, G. and Berg, H.C. (1994). Overview of the geology and tectonic evolution of Alaska. In: *The Geology of North America* (ed. G. Plafker and H.C. Berg), 989–1021. Boulder, Colorado: v. G-1.

Regan, S.P., Benowitz, J.A., and Holland, M.E. (2020). A plutonic brother from another magma mother: disproving the Eocene Foraker-McGonagall pluton piercing point and implications for long-term slip on the Denali fault. *Terra Nova* 32 (1): 66–74. https://doi.org/10.1111/ter.12437.

Riccio, S.J., Fitzgerald, P.G., Benowitz, J.A., and Roeske, S.M. (2014). The role of thrust faulting in the formation of the eastern Alaska range: Thermochronological constraints from the Susitna glacier thrust fault region of the intracontinental strike-slip Denali fault system. *Tectonics* 33 (11): 2195–2217. https://doi.org/10.1002/2014TC003646.

Ridgway, K.D., Trop, J.M., Nokleberg, W.J. et al. (2002). Mesozoic and Cenozoic tectonics of the eastern and Central Alaska range: progressive basin development and deformation in a suture zone. *Geological Society of America Bulletin* 114 (12): 1480–1504. https://doi.org/10.1130/0016-7606(2002)114<1480:MACTOT>2.0.CO;2.

Saltus, R.W., Hudson, T.L., and Wilson, F.H. (2007). The geophysical character of southern Alaska-implications for crustal evolution. *Special Paper of the Geological Society of America* 431 (01): 1–20. https://doi.org/10.1130/2007.2431(01).

Schulte-Pelkum, V., Caine, J.S., Jones, J.V. III et al. (2020). Imaging the tectonic grain of the northern cordillera Orogen using transportable Array receiver functions. *Seismological Research Letters* 91 (6): 3086–3105. https://doi.org/10.1785/0220200182.Supplemental.

St. Amand, P. (1954). Tectonics of Alaska as deduced from seismic data. *Bulletin of the Geological Society of America* 65: 1350.

St. Amand, P. (1957). Geological and geophysical synthesis of the tectonics of portions of British Columbia, the Yukon territory, and Alaska. *Geological Society of America Bulletin* 68 (10): 1343–1370. http://gsabulletin.gsapubs.org/content/68/10/1343.short.

Till, A.B., Roeske, S.M., Bradley, D.C. et al. (2007). Early tertiary transtension-related deformation and magmatism along the Tintina fault system, Alaska. *Special Papers-Geological Society of America* 434: 233.

Trop, J.M., Benowitz, J., Cole, R.B., and Sullivan, P.O. (2019). Cretaceous to Miocene magmatism, sedimentation, and exhumation within the Alaska Range suture zone: a polyphase reactivated terrane boundary. *Geosphere* 15 (4): 1066–1101.

14

Use of Remote Sensing in Lineament Analysis

Exploring its Potentials in a Humid Subtropical and Semi-Arid Environment

Swakangkha Ghosh[1], *Thota Sivasankar**[,2], *and Gokul Anand*[3]

[1] *Cactus Communications Pvt. Ltd., Mumbai, Maharashtra, India*
[2] *NIIT University, Neemrana, Rajasthan, India*
[3] *North Eastern Space Applications Centre, Shillong, Meghalaya, India*

The demarcation and analysis of linear geomorphological features from satellite images occupies a significant place in several geologic investigations (e.g. Dasgupta et al. 2022; Dasgupta and Mukherjee 2017, 2019; Gogoi et al. 2022; Kaplay et al. 2017; Misra et al. 2014). Lineaments are used to infer mineral prospects of a region, analyze structural deformation pattern/trends, identify geological boundaries, and infer crustal structure and various subsurface phenomena in areas of unexposed lithology (Anand and Rajaram 2004; Tiren 2010). In addition, lineaments help in understanding the hydrogeological conditions of a region along with its extensive use in seismic and landslide hazard assessment studies (Ahmadi and Pekkan 2021; Azar et al. 2019; Marghany and Hashim 2010; Masoud and Koike 2006; Mountrakis and Luo 2011; Pour and Hashim 2012; Qari et al. 2008; Ramli et al. 2010). Remotely sensed imagery plays a vital role in lineament mapping over large areas and particularly in inaccessible regions of the Earth's surface. Imagery acquired using sensors in different spectral regions like optical and Synthetic Aperture Radar (SAR) has complementary information in this regard and has shown significant improvement in both the quantitative and qualitative aspects. Taking into account the varied aspects, we present here a comparative analysis of geological lineaments derived from SAR and optical data taken from two different climatic regimes pertaining to humid sub-tropics and semi-arid regions. The humid subtropical region is mostly covered with moderate to dense vegetation, whereas the semi-arid region contains sparse vegetation due to dry weather conditions. The extracted lineaments from satellite data along with rose diagrams representing its trend and length are demonstrated in figures for an improved understanding regarding the different advanced semi-automatic algorithms implemented.

We utilized freely available satellite imagery, i.e. 10 m pixel size C-band Sentinel-1 SAR ground range detected (GRD) products acquired in IW mode (https://developers.google.com/earth-engine/datasets/catalog/COPERNICUS_S1_GRD#description), 25 m spatial resolution L-band ALOS PALSAR global mosaic of 2017 (https://developers.google.com/earth-engine/datasets/catalog/JAXA_ALOS_PALSAR_YEARLY_SAR#description), and Level-2A Sentinel-2 multi-spectral surface reflectance products (https://developers.google.com/earth-engine/datasets/catalog/COPERNICUS_S2_SR) datasets from Google Earth Engine (GEE), a cloud computing platform. In this study, Lineament Extraction (LINE) tool inbuilt in the software PCI Geomatica is used for initial lineament extraction. The LINE tool parameters, i.e. RADI, GTHR, LTHR, FTHR, ATHR, and DTHR are optimized based on geological lineaments provided by the Geological Survey of India (GSI) through Bhukosh portal (https://bhukosh.gsi.gov.in/Bhukosh/Public). Furthermore, automatic extracted lineaments are further refined, based on the ArcGIS true color composite image and Digital Elevation Model (DEM) database over the study area.

14.1 A Case Study from Humid Subtropical Region

The study area is a part of the Indo-Burma Range with an extent between 93.50° E and 93.89° E longitudes and 24.38° N to 24.94° N latitudes as shown in Figure 14.1. Overall, the study area is encompassed by several neotectonic features, viz. thrust faults, anticlines, and right and left lateral active faults in its vicinity. Geologically, the rocks in the

*Corresponding Author: siva.iirs@gmail.com

Atlas of Structural Geological and Geomorphological Interpretation of Remote Sensing Images, First Edition.
Edited by Achyuta Ayan Misra and Soumyajit Mukherjee.

Companion website: www.wiley.com/go/misra/RemoteSensingImages

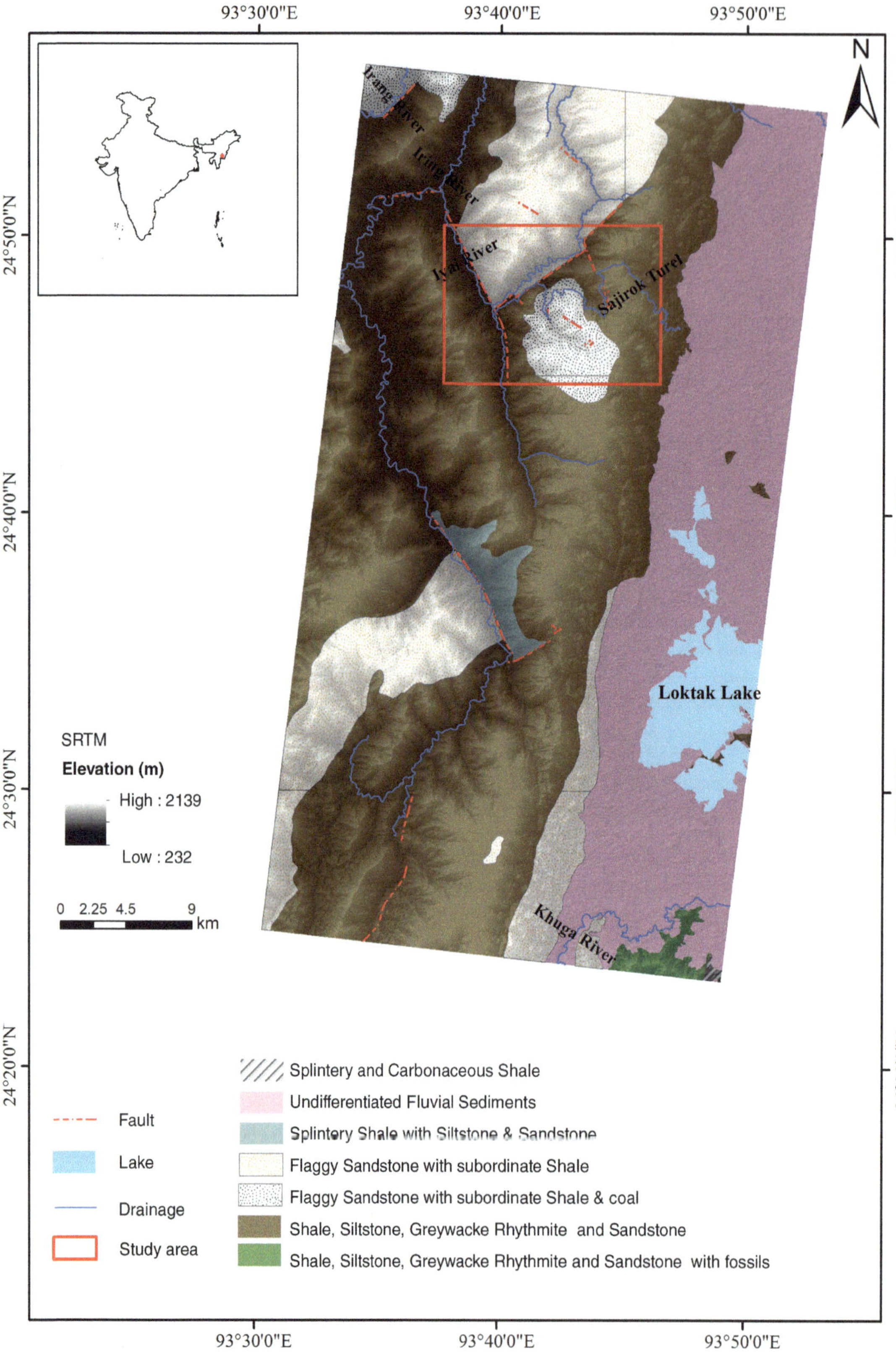

Figure 14.1 Geological map of the study area showing the major lithotectonic units adapted from GSI (2020); accessed on 3 February 2020. The area is encompassed by several neotectonic features like thrust faults, anticlines, and right and left lateral active faults. The Loktak lake located in Moirang is one of the largest and most important freshwater lakes (287 km^2) in northeast India. The red square denotes the area of interest for lineament extraction demonstration.

Indo-Burman ranges are made up of Neogene molasses, Tertiary, and Cretaceous Flysch sediments, which occupy 70% of Manipur with older igneous and metamorphic rocks (Singh et al. 2013). The terrain in this region is highly undulating with slope ranging from 0 to 73°.

In this section we analyze the performance of multi-parametric SAR data for lineament extraction. The median image of Sentinel-1 SAR data collection acquired during January to December 2018 and ALOS mosaic data of 2017 are considered for this case. In addition to these, cloud-free composite generated from Sentinel-2 multi-spectral data acquired during February to April 2018 is also considered. The multi-frequency (C and L-bands) multi-polarization (like and cross) provided in Figure 14.2 clearly shows that C-band is more sensitive toward topographic variations than the

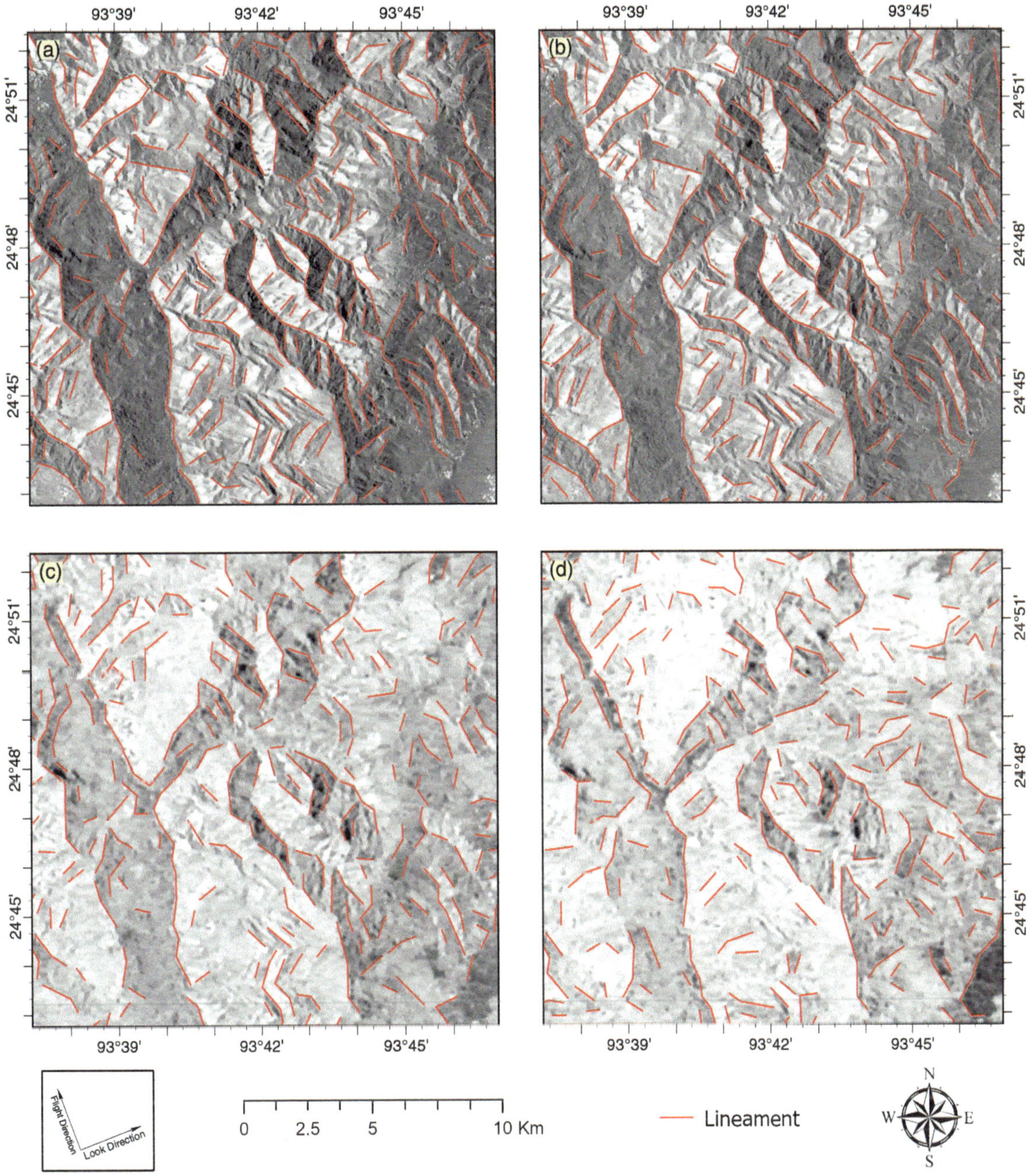

Figure 14.2 Analysis of multi-frequency multi-polarized SAR data for lineaments in a humid sub-tropical region. Lineaments are extracted from part of the Indo-Burma ranges: (a) C-band VV (vertical transmitting, vertical receiving); (b) C-band VH (vertical transmitting, horizontal receiving); (c) L-band HH (horizontal transmitting, horizontal receiving); and (d) L-band HV (horizontal transmitting, vertical receiving). Results are enlarged to the red square extent denoted in Figure 14.1 for demonstration purposes.

L-band. It is also observed that the like-polarization (VV) provided slightly improved results in comparison with cross-polarization (VH) data. Figure 14.3 exemplifies that the lineaments oriented perpendicular to the SAR sensor look direction are more prominently extracted than those in parallel direction. Thus, integration of lineaments extracted from multi-look SAR data has been carried out to improve the performance (see Figure 14.4). The detailed procedure for integration of lineaments extracted from multi-look SAR data

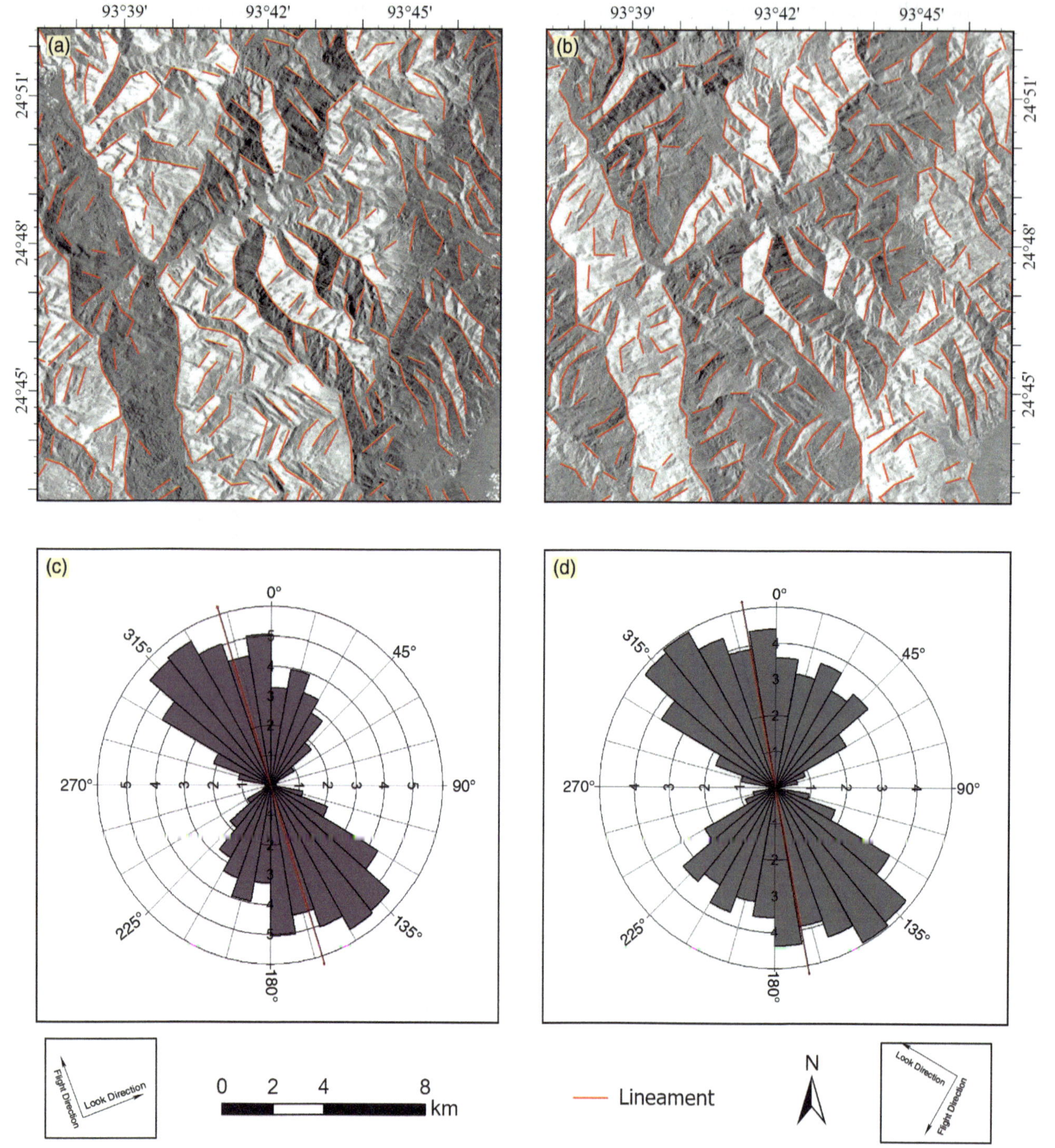

Figure 14.3 Analysis and superposition of lineaments extracted from C-band VV backscatter data for (a) ascending pass and (b) descending pass for a region in the humid tropics. Results are enlarged to the red square extent denoted in Figure 14.1 for demonstration purposes; (c) and (d) represent the bidirectional rose diagrams showing the length of the lineaments along with its major trend for entire study area. The red line in the rose diagrams denotes the mean lineation vector.

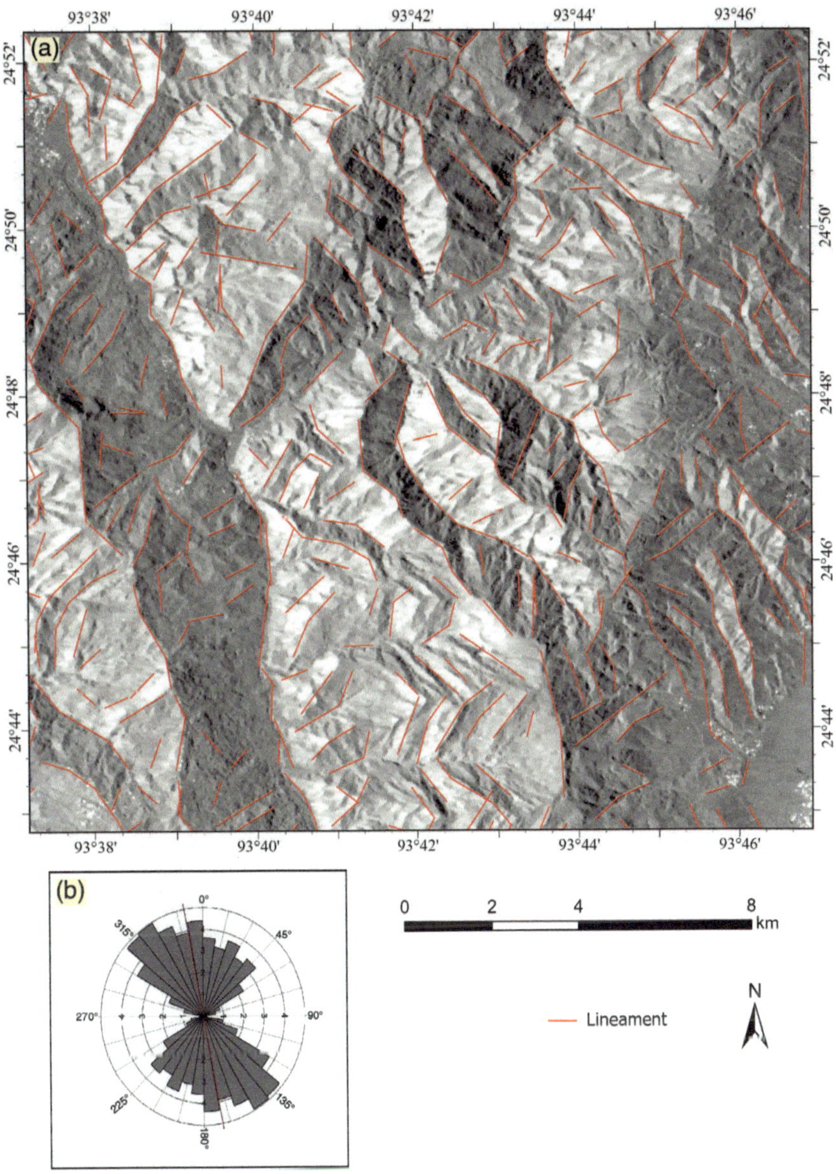

Figure 14.4 (a) Analysis and superposition of lineaments extracted from multi-look C-band SAR VV backscatter data in the humid subtropical region. Results are enlarged to the red square extent denoted in Figure 14.1 for demonstration purposes; (b) Bidirectional rose diagram representing length for lineaments along with its major trend for entire study area. The red line in the rose diagram denotes the mean lineation vector.

is provided in Ghosh et al. (2021). In addition to this, Sentinel-2 optical data has also been explored for this purpose. In this regard, 10 bands acquired in blue, green, red, red edge, near infrared, and shortwave infrared regions are considered. These bands are initially processed using principal component analysis (PCA) and identified as the best band for lineament extraction. The lineaments derived from Sentinel-2 optical data is given in Figure 14.5. The study results, as observed from Figures 14.4 and 14.5, demonstrated that the lineaments derived from SAR and optical datasets show significant difference in pattern/trends in vegetation-covered regions. It is observed that the multi-look SAR data derived lineaments are numerous as well as matching with the pattern/trends with the lineaments provided in Bhukosh portal of GSI.

14.2 A Study from an Arid Region

An area south of the Khetri copper belt in the northwestern part of India has been chosen for extraction and analysis of lineaments using satellite images. Copper deposits in the Khetri copper belt are hosted in the Proterozoic sequence of rocks, which belong to the Ajabgarh group of Delhi Supergroup situated in the foothill zone of Aravalli Mountain range North-West, India. The copper belt is a 100 km-long metallogenic province trending in the NE–SW direction, extending from Singhana in the north to Sangarava in the south. The area is located in a sub-tropical, semi-arid region with mean annual rainfall around 500 m. The mid-Proterozoic Delhi Supergroup comprises shallow-water, sedimentary rocks, with lesser mafic and felsic volcanic rocks hosting

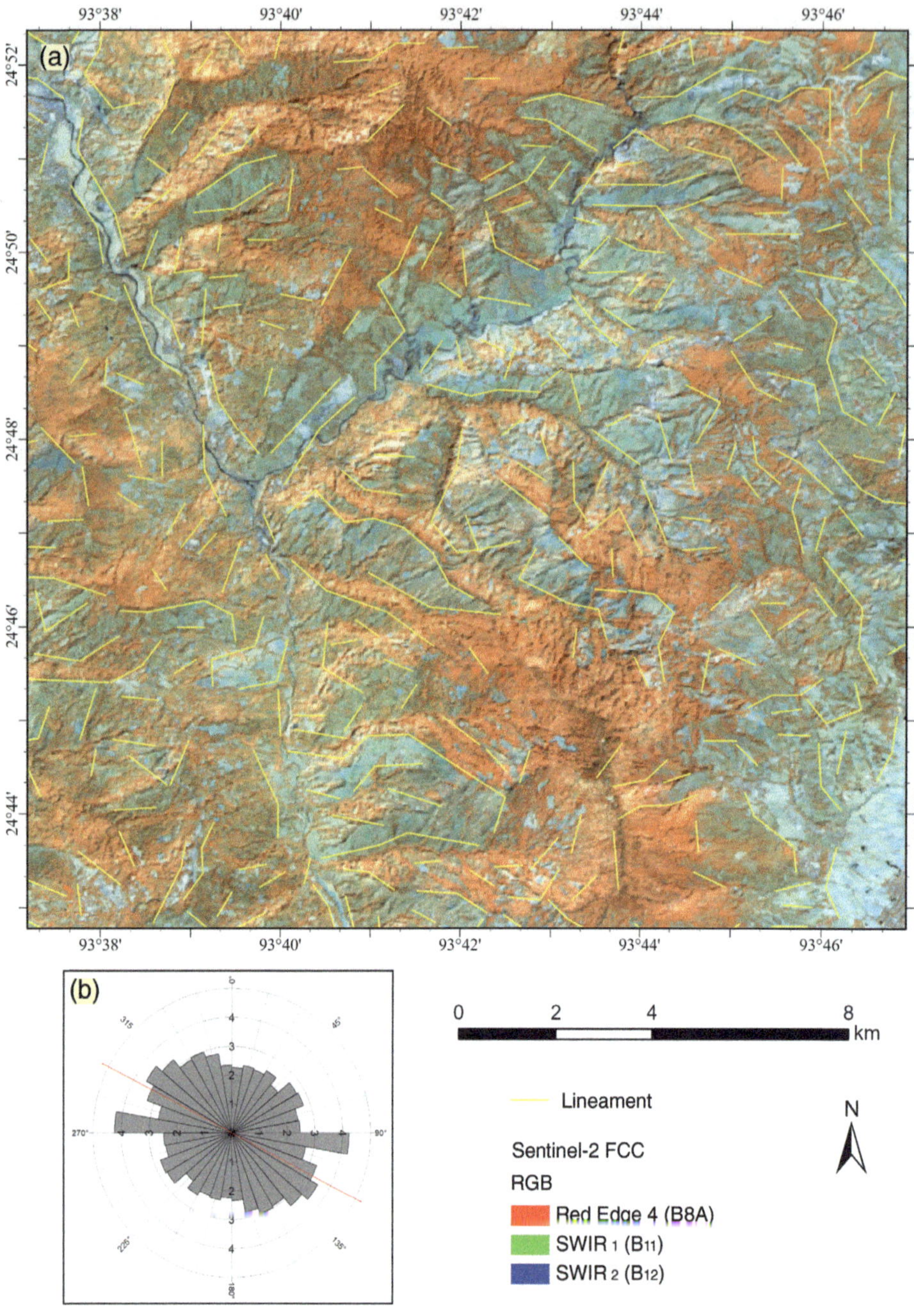

Figure 14.5 (a) Analysis and superposition of lineaments extracted from optical data (Sentinel-2) in the humid subtropical region (false color composite is considered to enhance geology). Results are enlarged to the red square extent denoted in Figure 14.1 for demonstration purposes; (b) Bidirectional rose diagram representing length for lineaments along with its major trend for entire study area. The red line in the rose diagrams denotes mean lineation vector.

the copper mineralization (Knight 2002). These rocks have been deformed into numerous NE–SW striking, doubly plunging folds as seen in Figure 14.6.

In this section, the median image of dual-polarized Sentinel-1 SAR data collection acquired during January to December 2020 and cloud-free composite of Sentinel-2 optical image collection acquired during April to May 2020 are considered for analysis. Figures 14.7 and 14.8 demonstrate the lineaments extracted from multi-look Sentinel-1 SAR VV backscatter and Sentinel-2 optical

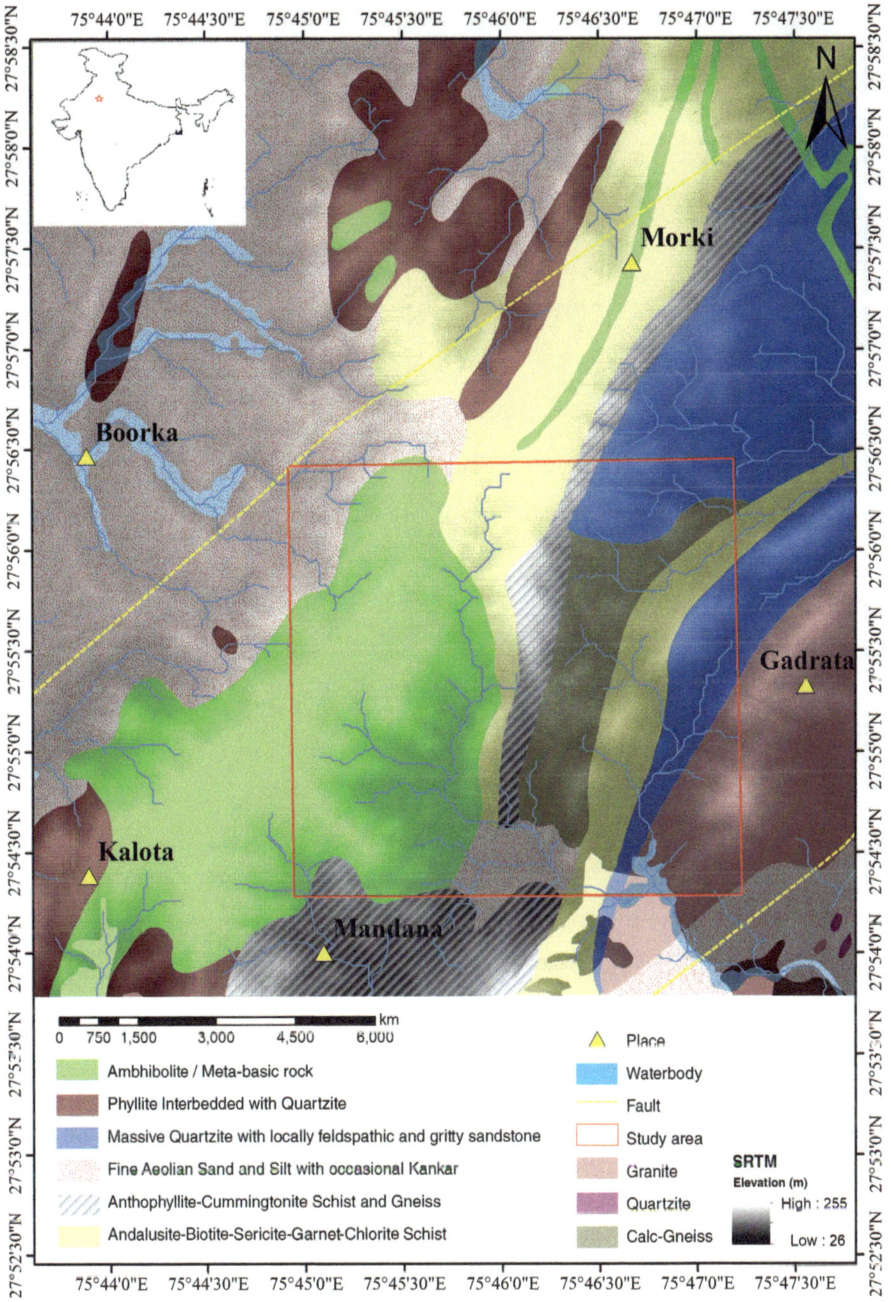

Figure 14.6 Figure 14.1 Geological map of the study area showing the major lithotectonic units adapted from GSI (2020 [accessed 3 February 2020]). The red square denotes the area of interest for lineament extraction demonstration from remotely sensed imagery. The entire region is located south of the Khetri Copper belt, NW India.

imagery, respectively. The rose diagrams denote the major trend of the lineaments extracted from satellite images, in the NE–SW direction, which is complementary with the trend of the lineaments formed in this region. The study results observed that the multi-look SAR and optical derived lineaments are showing similar pattern/trends in the arid region. However, SAR data-derived lineaments are numerous in comparison to those derived from optical data. Therefore, it is indicated that the use of multi-look SAR data is preferable to the optical data for lineament extraction purposes in arid regions.

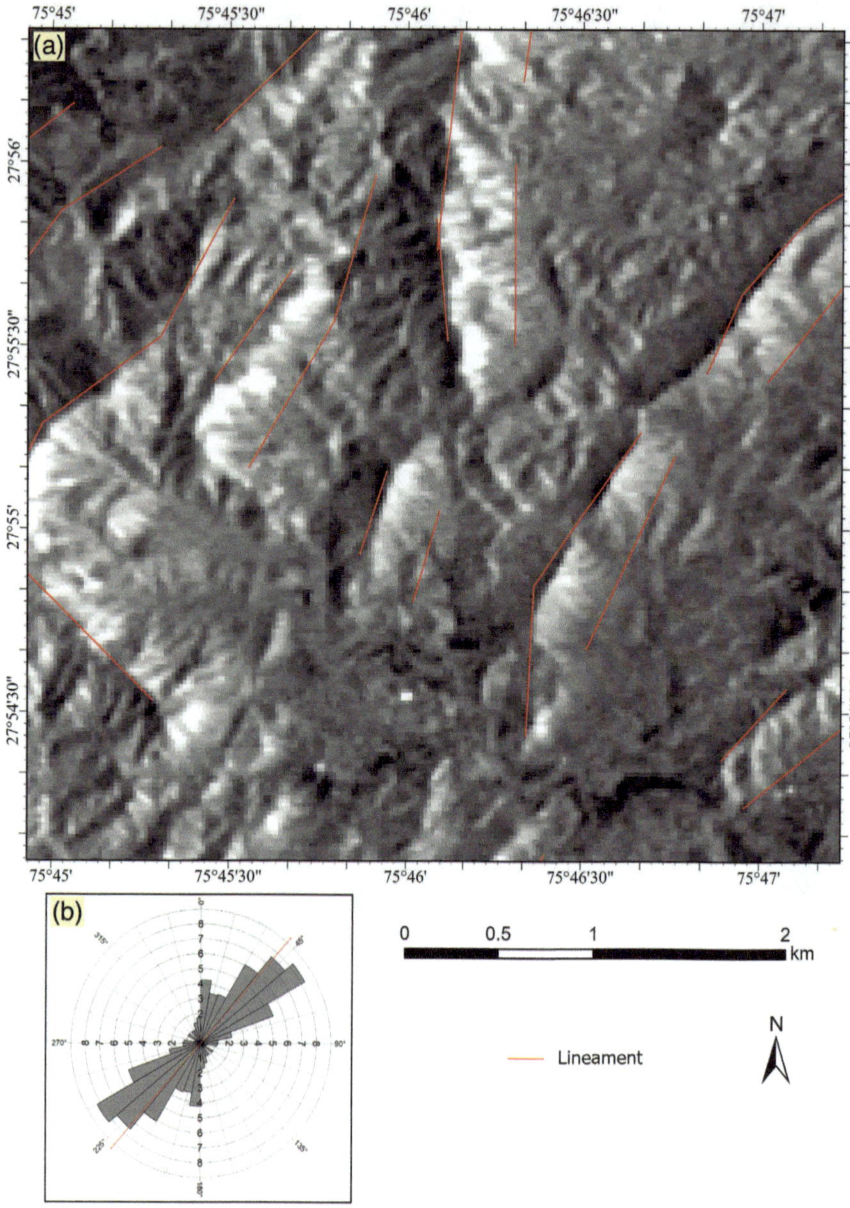

Figure 14.7 (a) Analysis and superposition of lineaments extracted from multi-look Sentinel-1 SAR data in an arid region. Results are enlarged to the red square extent denoted in Figure 14.6 for demonstration purposes; (b) Bidirectional rose diagram representing length for lineaments along with its major trend. Red line in rose diagram: mean lineation vector.

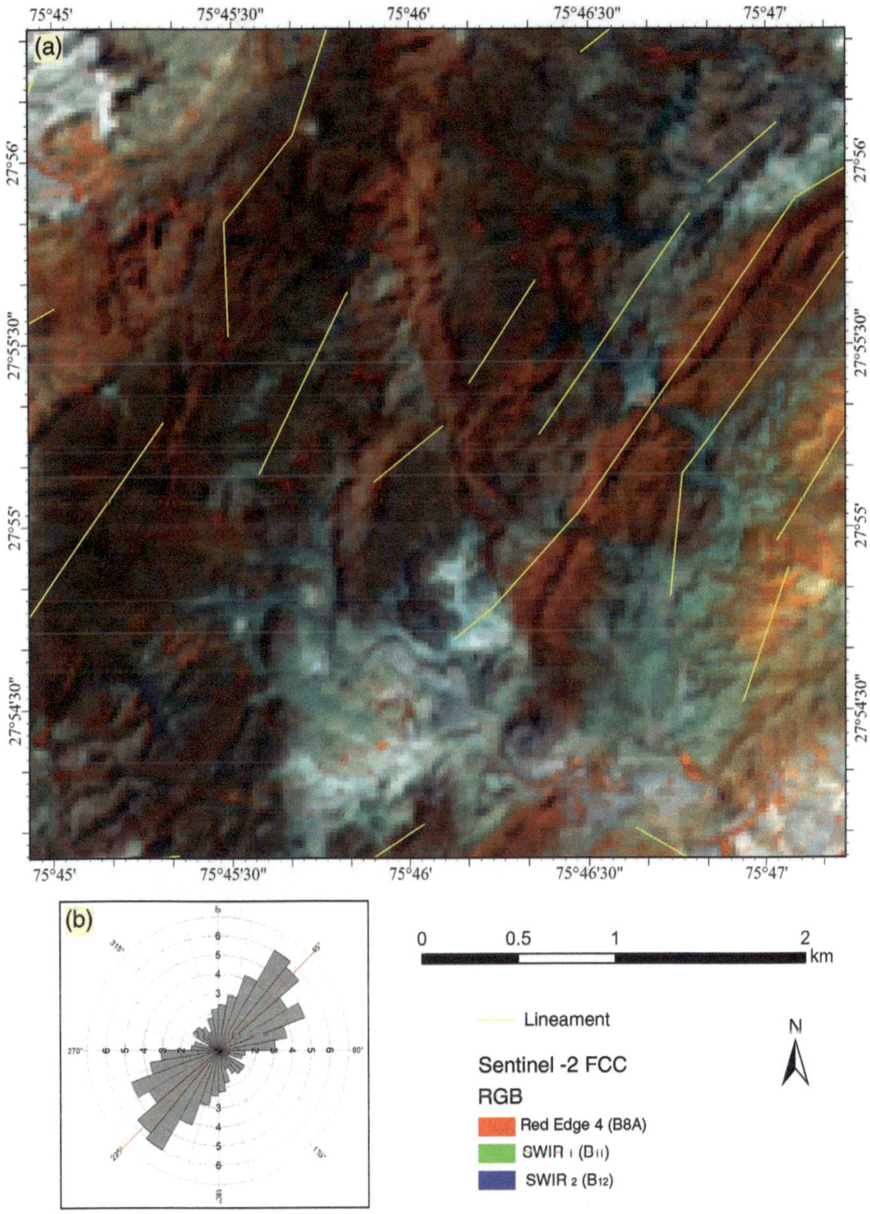

Figure 14.8 (a) The analysis and superposition of lineaments extracted from Sentinel 2 optical Image. Results are enlarged to the red square extent denoted in Figure 14.1 for demonstration purposes; (b) Bidirectional rose diagram representing length for lineaments along with its major trend. The red line in the rose diagram denotes mean lineation vector.

Acknowledgements

Soumyajit Mukherjee invited to submit an article in the edited book. Achyuta Ayan Misra handled this article. Wiley Blackwell Acquisition Editors (present: Frank Weinreich, past: Andrew Harrison), Stacey Woods (Handling Editor), and proofreading team. Summarized in Misra and Mukherjee (2022).

References

Ahmadi, H. and Pekkan, E. (2021). Fault-based geological lineaments extraction using remote sensing and GIS – a review. *Geosciences* 11 (5): 183.

Anand, S.P. and Rajaram, M. (2004). Crustal structure of Narmada-Son lineament: an aeromagnetic perspective. *Earth, Planets and Space* 56 (5): e9–e12.

Azar, A.P.K., Askari, G., Crispini, L. et al. (2019). Field and spaceborne imagery data for evaluation of the paleo-stress regime during formation of the Jurassic dike swarms in the KalatehAlaeddin Mountain area, Shahrood, North Iran. *Arabian Journal of Geosciences* 12 (17): 1–12.

Dasgupta, S. and Mukherjee, S. (2017). Brittle shear tectonics in a narrow continental rift: asymmetric non-volcanic Barmer basin (Rajasthan, India). *The Journal of Geology* 125: 561–591.

Dasgupta, S. and Mukherjee, S. (2019). Remote sensing in lineament identification: examples from western India. In: *Problems and Solutions in Structural Geology and Tectonics*, Developments in Structural Geology and Tectonics Book Series, vol. 5. Series Editor: Mukherjee, S. (ed. A. Billi and A. Fagereng), 205–221. Elsevier. ISSN: 2542-9000. ISBN: 9780128140482.

Dasgupta, S., Biswas, M., Mukherjee, S., and Chatterjee, R. (2022). Depositional system, morphological signatures, tectonics and sedimentation pattern along the transform margin- Palar-Pennar basin, Indian east coast. *Journal of Petroleum Science & Engineering* 211: 110155. https://doi.org/10.1016/j.petrol.2022.110155.

Ghosh, S., Sivasankar, T., and Anand, G. (2021). Performance evaluation of multi-parametric synthetic aperture radar data for geological lineament extraction. *International Journal of Remote Sensing* 42 (7): 2574–2593.

Gogoi, M.P., Gogoi, B., and Mukherjee, S. (2022). Tectonic instability of the petroliferous upper Assam valley (NE India): a geomorphic approach. *Journal of Earth System Science* 131: 18. https://doi.org/10.1007/s12040-021-01752-6.

GSI. (2020). Bhukosh Geoportal. Geological Society of India. www.bhukosh.gsi.gov.in (accessed 03 February 2020).

Kaplay, R.D., Md, B., Mukherjee, S., and Kumar, T.V. (2017). Morphotectonic expression of geological structures in eastern part of south east Deccan volcanic province (around Nanded, Maharashtra, India). In: *Tectonics of the Deccan Large Igneous Province* (ed. S. Mukherjee, A.A. Misra, G. Calvès and M. Nemčok), 317–335. London, Special Publications 445: Geological Society.

Knight, J. (2002). The Khetri copper belt, Rajasthan: iron-oxide copper-gold terrane in the Proterozoic of NW India. *Hydrothermal Iron Oxide Copper-Gold and Related Deposits: A Global Perspective* 2, AMF: 321–341.

Marghany, M. and Hashim, M. (2010). Lineament mapping using multispectral remote sensing satellite data. *International Journal of Physical Sciences* 5 (10): 1501–1507.

Masoud, A.A. and Koike, K. (2006). Arid land salinization detected by remotely-sensed landcover changes: a case study in the Siwa region, NW Egypt. *Journal of Arid Environments* 66 (1): 151–167.

Misra, A.A. and Mukherjee, S. (2022). Introduction to "atlas of structural geological and geomorphological interpretation of remote sensing images". In: *Atlas of Structural Geological and Geomorphological Interpretation of Remote Sensing Images* (ed. A.A. Misra and S. Mukherjee). Wiley Blackwell. ISBN: 9781119813354.

Misra, A.A., Bhattacharya, G., Mukherjee, S., and Bose, N. (2014). Near N-S paleo-extension in the western Deccan region in India: does it link strike-slip tectonics with India-Seychelles rifting? *International Journal of Earth Sciences* 103: 1645–1680.

Mountrakis, G. and Luo, L. (2011). Enhancing and replacing spectral information with intermediate structural inputs: a case study on impervious surface detection. *Remote Sensing of Environment* 115 (5): 1162–1170.

Pour, A.B. and Hashim, M. (2012). The application of ASTER remote sensing data to porphyry copper and epithermal gold deposits. *Ore Geology Reviews* 44: 1–9.

Qari, M.H.T., Madani, A.A., Matsah, M.I.M., and Hamimi, Z. (2008). Utilization of Aster and Landsat data in geologic mapping of basement rocks of Arafat area, Saudi Arabia. *The Arabian Journal for Science and Engineering* 33 (1C): 99–116.

Ramli, M.F., Yusof, N., Yusoff, M.K. et al. (2010). Lineament mapping and its application in landslide hazard assessment: a review. *Bulletin of Engineering Geology and the Environment* 69 (2): 215–233.

Singh, P., Thakur, J.K., and Kumar, S. (2013). Delineating groundwater potential zones in a hard-rock terrain using geospatial tool. *Hydrological Sciences Journal* 58 (1): 213–223.

Tiren, S. (2010). Lineament interpretation, short review and methodology. In: *Swedish Radiation Safety Methodology*. Stockholm, Sweden: Swedish Radiation Safety Authority.

15

Tectonic Structures Interpretation Using Airborne-Based LiDAR DEM on the Examples from the Polish Outer Carpathians

Maciej Kania and Mateusz Szczęch*

Faculty of Geography and Geology, Institute of Geological Sciences, Jagiellonian University, Kraków, Poland

15.1 Introduction

The application of remote sensing has greatly enhanced structural and geomorphic interpretations of terrains (e.g. Misra et al. 2014; Kaplay et al. 2017; Dasgupta and Mukherjee 2017, 2019; Dasgupta et al. 2022; Gogoi et al. 2022). Airborne LiDAR (Light Detection and Ranging) is currently one of the most advanced remote sensing methods of collecting data to produce digital elevation models (DEM). The main advantage of LiDAR is detecting series of reflections, which after filtering allows us to get high-resolution data of the elevation even in areas with a lush vegetation cover, which is the case in the Carpathians. LiDAR data has become a great source of structural information, especially in geological mapping (e.g.: Cieszkowski et al. 2017; Jagodnik et al. 2020; Kania and Szczęch 2020).

The study area is located in the Polish part of the Outer Carpathians (Figure 15.1). The Carpathians are the mountain-arc located in central Europe between Austria and Romania. From the geological point of view, there are two main zones: The Outer and the Inner Carpathians separated by Pieniny Klippen Belt (Książkiewicz 1977; Plašienka 2018; Golonka et al. 2019). The selected examples are in the Polish sector of the Outer Carpathians, known also as Flysh Carpathians (Mahel 1974; Książkiewicz 1977). They are built mainly of the flysh deposits of the Late Jurassic to Middle Miocene age. The total thickness of these deposits is 6000 m (Mahel 1974; Książkiewicz 1977; Golonka et al. 2005, 2019; Ślączka et al. 2006). The sediments were deposited in the deep basins of the northern Tethys, which were separated by ridges. During the Miocene, these sediments were detached from the basement, folded, and overthrusted as a series of nappes (from the south): Magura Nappe; Dukla and Foremagura nappes; and Silesian, Subsilesian, and Skole nappes (Książkiewicz 1977). The folding, overthrusting, and later faulting were results of the collision of Inner Carpathians, which are part of the ALCAPA microplate with the North European Platform (Plašienka 2018; Golonka et al. 2021). The nappes of the Outer Carpathians were overthrusted on the autochton of the Carpathian foredeep Miocene deposits with the underlying southern margin of the North European Platform (Mahel 1974; Golonka et al. 2005, 2019; Plašienka 2018).

We present some examples of how geological structures such as faults and joints, overthrusts, bedding, and folds can be interpreted from the high-resolution airborne LiDAR DEMs. The data were obtained during the Polish state program of the digital mapping of geohazards (Wężyk 2015). All models are in 1 m/pixel resolution.

The basic image for interpretation presented in this chapter is a multidirectional hill-shade model. However, we were using the following auxiliary images derived from the DEMs: slope angle maps, topographic position index, and vertical and horizontal Sobel filter (Nixon and Aguado 2019). We observe that the most effective method of interpreting the model is often changing the visualization method, however with a constant working map scale. Additionally, on the interpretation map, we added topographic contours to show the bedding dip.

15.2 Faults and Joints

These structures are the most common ones. The interpretation bases on the presence of the lineaments (O'Leary et al. 1976). The lineaments can be: (i) clear straight features present in morphology, like stream valleys;

*Corresponding Author: maciej.kania@uj.edu.pl

Atlas of Structural Geological and Geomorphological Interpretation of Remote Sensing Images, First Edition.
Edited by Achyuta Ayan Misra and Soumyajit Mukherjee.

Companion website: www.wiley.com/go/misra/RemoteSensingImages

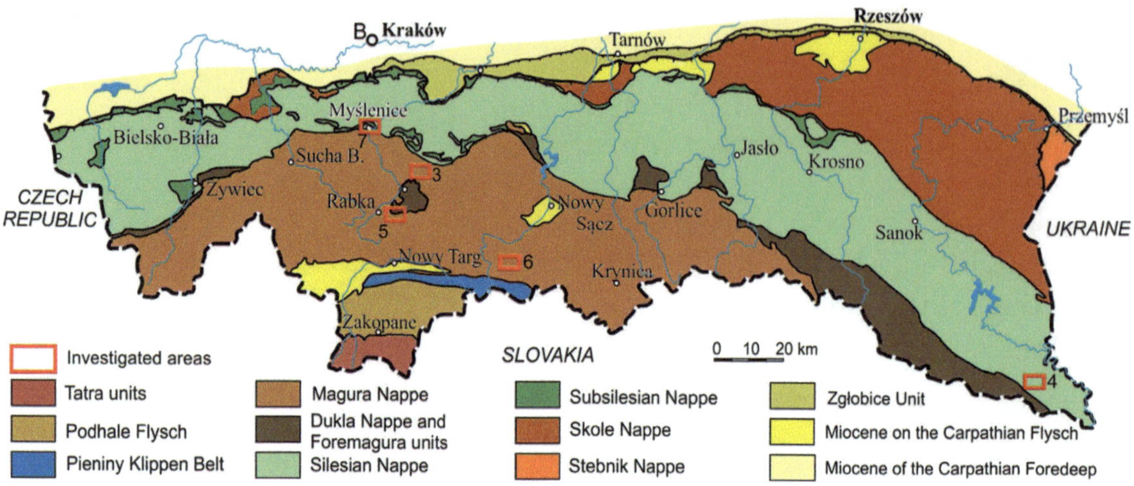

Figure 15.1 Localization of the presented models in the Polish Carpathians (Cieszkowski et al. 2017).

(ii) zones of visible relative layers displacement; and (iii) weakly visible lines in the image, in most cases verified by using other than the hill-shade image. Only type 2 can be directly interpreted as a fault; however, in most cases of type 1 and 3, field investigation allows the interpretation of lineaments as tectonolineaments, especially fault zones (Figure 15.2). In some cases, the presence of the faults is also enhanced by tearing and relative displacement of mountain ridges.

We confirmed tectonic interpretation of the lineaments in many places, by finding displacements or fault breccia zones in a field (Figure 15.2). Moreover, our interpretation fits to published geological maps as in the case of northern part of the Gorce Mts (Mszana Tectonic Window, Figure 15.5) which was mapped by Szczęch and Cieszkowski (2021).

15.3 Overthrusts

The presence of overthrust is inferred from the geological mapping of the area. However, the detailed interpretation of DEM allows us to determine the position of the overthrust even in the areas without outcrops. This is possible especially if there is a strong contrast in rocks lithology on both sides of the overthrust trace. On the other hand, the presence of the faults cutting the overthrust can often be visible on the model.

15.4 Bedding

As the Outer Carpathians are built of flysh rocks, the common situation is alternating of the layers of shales and sandstones. The contrast in the weathering and erosion susceptibility of these two lithologies creates a pattern of slight changes in the slope angle, creating the natural intersection lines on the DEMs. Some of the most prominent beds were marked on the interpretation images by wide semi-transparent lines. The layers dip can be interpreted by observing the angle between layer traces and topographic contours. In some places, the bodies of the thick-bedded sandstones (like Magura sandstone) forms clear thresholds in the morphology.

15.5 Folds

The folds axes can be interpreted mainly by detailed analyses of the pattern of the layers. This is in most cases backed by traditional geological mapping data.

We selected five cases from the different geological units of the Outer Carpathians to show examples of the structural interpretation.

15.6 The Lubogoszcz Mountain (Figure 15.3)

This is one of the Beskid Wyspowy range culminations. The characteristic feature of the morphology in this area is the presence of short, isolated mountain ridges. The mountains massifs are, from a structural point of view, cores of synclines formed by thick-bedded Oligocene-Miocene sandstones of the Magura Formation, gently dipping to the syncline axis. This dipping results in low-angle cross-cutting of the intersection lines and morphological contours. The syncline axis is marked, trending SW–NE, turning to W–E in the eastern part. Note two prominent faults, cutting the ridge with NNW–SSE trends. These faults are accompanied by some less visible, however also likely of tectonic genesis, lineaments. The NE corner of the mountain as well as its SE part are big landslides. The body of thick-bedded sandstones is separated from an underlying packet of thin-and medium-bedded sandstone-shales and shales by distinguishing morphological threshold.

15.7 Ustrzyki Górne Area (Figure 15.4)

The trending of mountain ridges in this area is a direct consequence of the geological structure. Steeply dipping beds are trending NW–SE and are composed of alternating layers of the Krosno sandstone and shales, which is visible in the DEM. In particular, relatively thin layers of the resistant sandstones are visible as sharp steps. The fault system is dominated by NE–SW faults. The second set is developed mainly in the northern part of the area. The model allows us to see relationships between two systems, showing the second one (N–S) to be older, cut and displaced by the first one. Note that the gorge of a stream in the centre of the image is also probably based on the fault of the first system.

15.8 Mszana Tectonic Window Area (Figure 15.5)

This example shows the trace of overthrust of the Magura Nappe, which is reflected as contrast in morphology. The overthrust is cut by a younger series of NW–SE trending faults, and a subsidiary set of NE–SW faults. The secondary thrust line can be spotted in the easter part of the area. A morphological threshold is a boundary between thick-bedded Magura sandstones in the south and a less resistant packet of thin- and medium-bedded sandstone-shales and shales of different lithostratigraphic formations in the north. The presence of shale packets between Magura sandstones allows tracing bedding on the model, showing in general intermediate dipping (30–40°).

15.9 Dzwonkówka (Beskid Sądecki) (Figure 15.6)

This area is located near the southern boundary of the Outer Carpathians. The dominating set of lineaments (NNW–SSE) is characterized by long, continuous forms as visible in the central part of the model. These two forms are probably conjugate with NNE–SSW short faults visible in-between forming major dislocation zone. The dominating set is continued far to the south, to the Pieniny Klippen Belt, which is a suture zone between Outer and Inner Carpathians. A relationship between two sets of lineaments that can be spotted in the NE part of the map suggests the NE–SW set to be probably an older one. The massif and steeply dipping Magura sandstones make interpretation of bedding difficult, especially in the eastern part of the area.

(a) (b) (c) (d)

Figure 15.2 Field examples of structures in the DEM lineaments localities. (a) – tectonic melange in a fault zone; (b) – fault breccia in the sandstone; (c) – damage zone in the flysch; (d) – fault surfaces in the thick-bedded sandstone. Source: All photos taken in the GorceMts (near or in the area presented on the Figure 15.5).

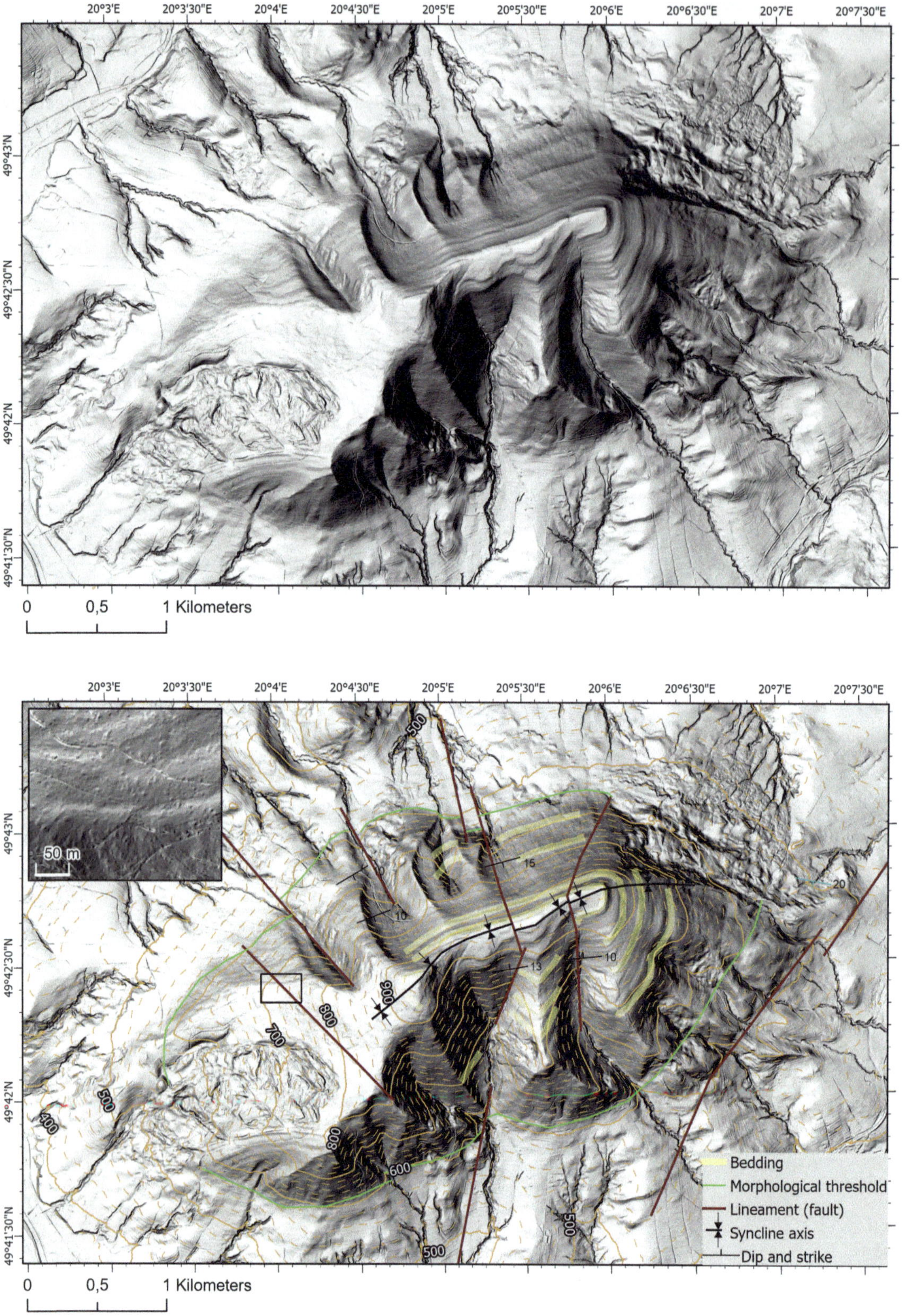

Figure 15.3 The Lubogoszcz Mountain. The massif is in fact core of the syncline built of thick bedded sandstones. The inset present slope angle image showing how the beds are broken on the fault line. More details in the text.

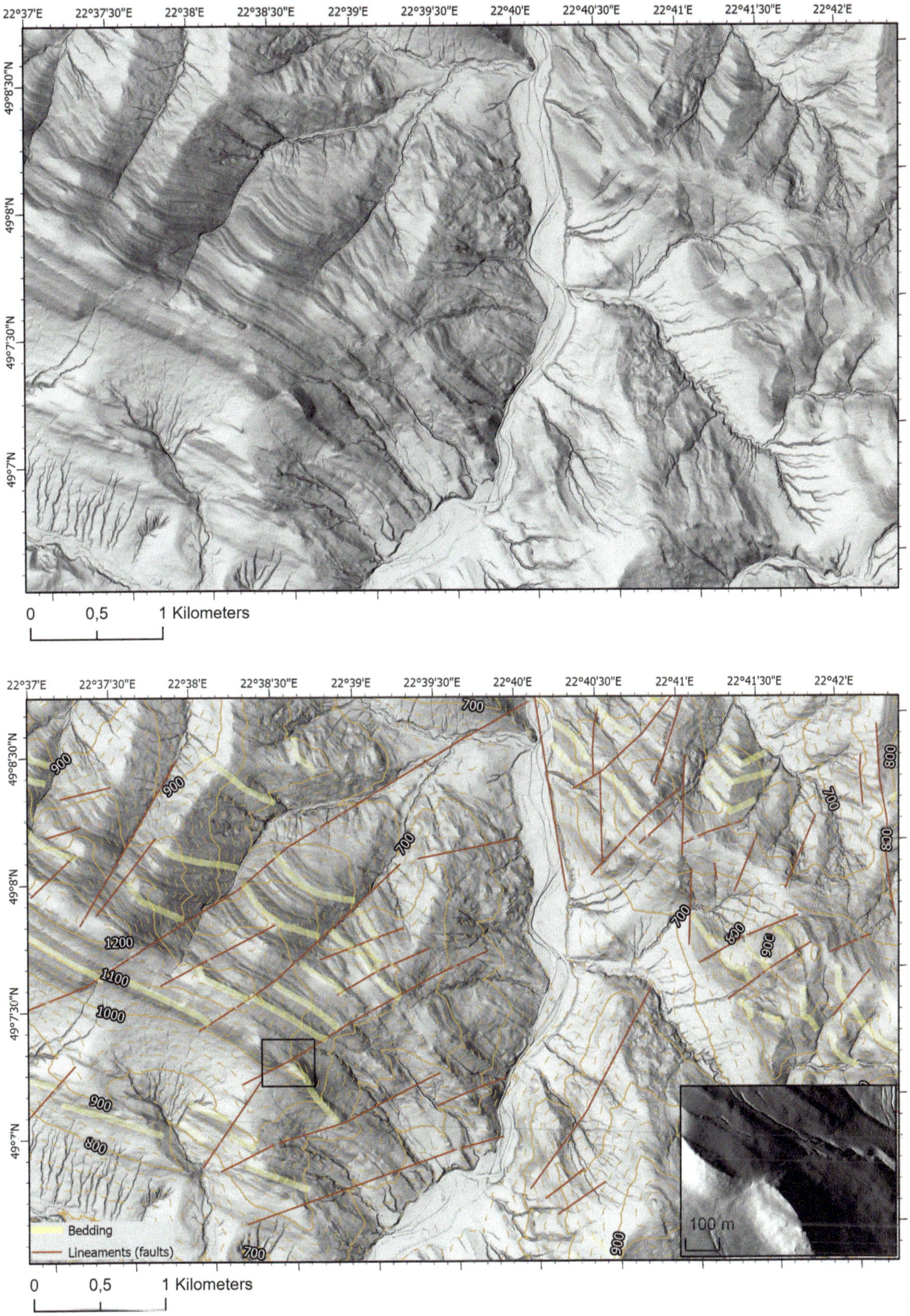

Figure 15.4 Ustrzyki Górne area in the Bieszczady Mts. Note the regular bedding cut by lineaments system. Cross-cutting faults in the NE quarter allows to interpretate fault sets relationships. An inset shows details of the one of the fault zone related lineament visualized with Sobel filter enhancing linear features. More details in the text.

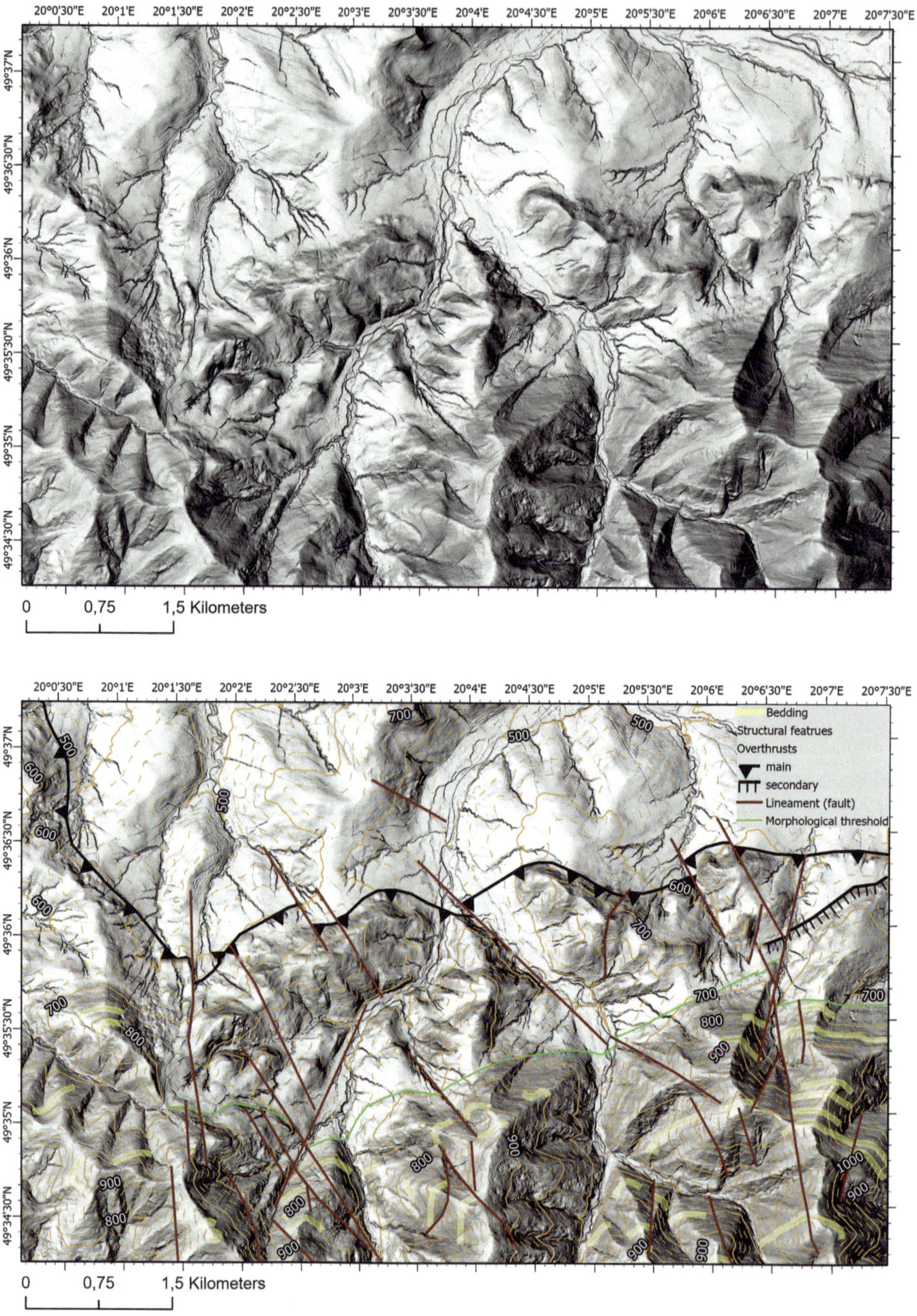

Figure 15.5 The Mszana Tectonic Window area. Note the overthrust reflecting in the morphology as well as boundary of the thick-bedded sandstones visible as morphological threshold. More details in the text.

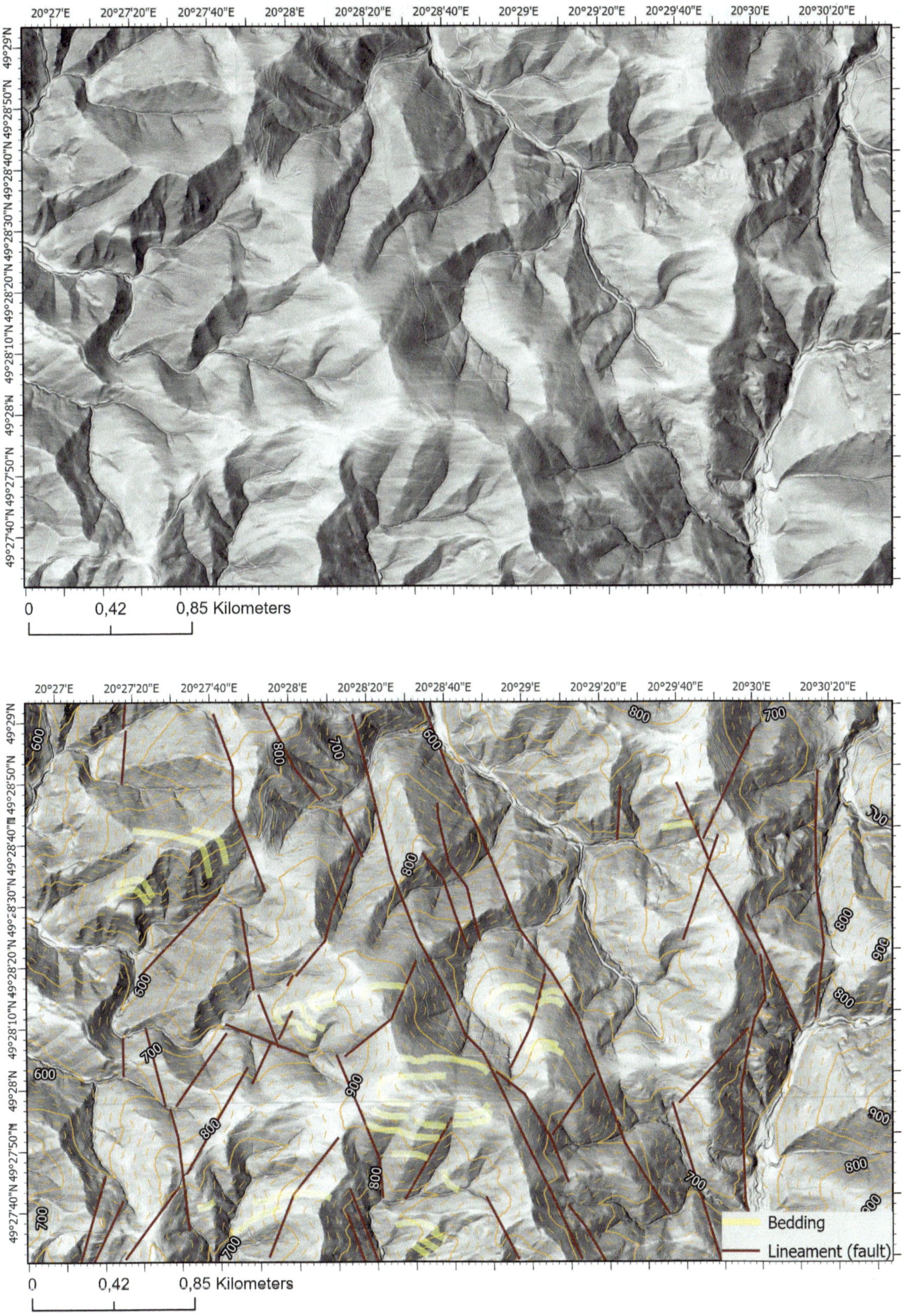

Figure 15.6 Dzwonkówka Mountain. Note the system of NNW–SSE faults forming the fault zone. More details in the text.

15.10 The Barnasiówka Ridge (Figure 15.7)

This mountain ridge, built of thick-bedded sandstones of the Godula beds surrounded by more eroded softer layers of shales and tin-thin and medium-bedded sandstone-shales flysch is an excellent opportunity to see how the bedding is reflected in the DEM. Changing in the lithology is visible in the mode as morphological thresholds. The ridge is trending parallel to the beds and is cut on the west by a prominent fault. There are two cross-cutting sets of faults that are visible in the Barnasiówka Ridge; in many cases short stream valleys are developed on these faults. The Barnasiówka Ridge's zigzag shape is also probably an effect of the tectonic pattern.

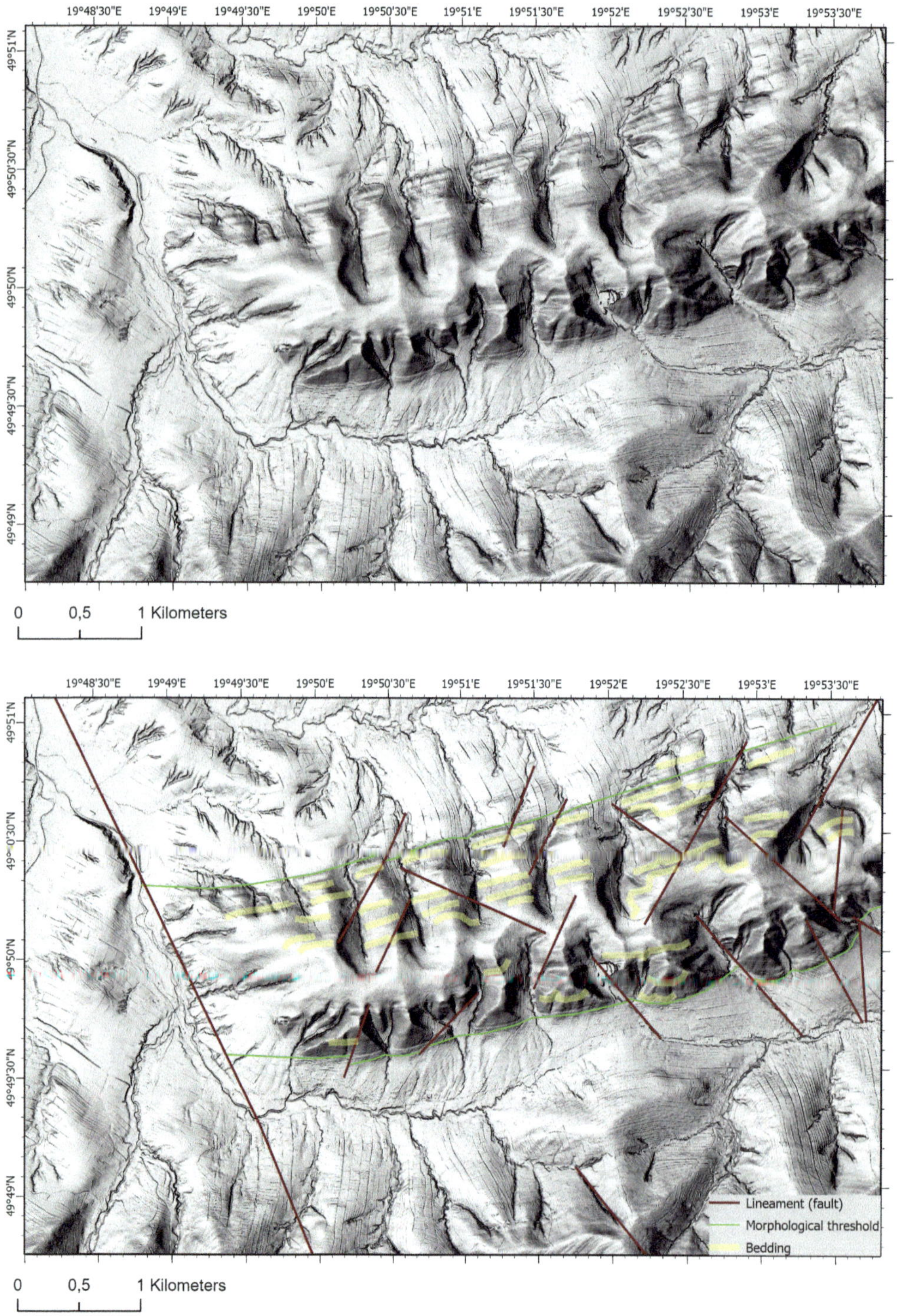

Figure 15.7 The Barnasiówka Ridge. Note the morphological thresholds on the contact of different lithologies. More details in the text.

Acknowledgements

Soumyajit Mukherjee invited to submit an article in the edited book. Achyuta Ayan Misra handled this article. Wiley Blackwell Acquisition Editors (present: Frank Weinreich, past: Andrew Harrison), Stacey Woods (Handling Editor), and proofreading team. Summarized in Misra and Mukherjee (2022).

References

Cieszkowski, M., Kysiak, T., Szczęch, M., and Wolska, A. (2017). Geology of the Magura Nappe in the Osielec area with emphasis on an Eocene olistostrome with metabasite olistoliths (outer Carpathians, Poland). *Annales Societatis Geologorum Poloniae* 87: 169–182.

Dasgupta, S. and Mukherjee, S. (2017). Brittle shear tectonics in a narrow continental rift: asymmetric non-volcanic Barmer basin (Rajasthan, India). *The Journal of Geology* 125: 561–591.

Dasgupta, S. and Mukherjee, S. (2019). Remote sensing in lineament identification: examples from western India. In: *Problems and Solutions in Structural Geology and Tectonics*, Developments in Structural Geology and Tectonics Book Series, vol. 5. Series Editor: Mukherjee, S. (ed. A. Billi and A. Fagereng), 205–221. Elsevier. ISSN: 2542-9000. ISBN: 9780128140482.

Dasgupta, S., Biswas, M., Mukherjee, S., and Chatterjee, R. (2022). Depositional system, morphological signatures, tectonics and sedimentation pattern along the transform margin- Palar-Pennar basin, Indian east coast. *Journal of Petroleum Science & Engineering* 211: 110155. https://doi.org/10.1016/j.petrol.2022.110155.

Gogoi, M.P., Gogoi, B., and Mukherjee, S. (2022). Tectonic instability of the petroliferous upper Assam valley (NE India): a geomorphic approach. *Journal of Earth System Science* 131: 18.

Golonka, J., Aleksandrowski, P., Aubrecht, R. et al. (2005). The Orava deep drilling project and post-palaeogene tectonics of the northern Carpathians. *Annales Societatis Geologorum Poloniae* 75: 211–248.

Golonka, J., Waśkowska, A., and Ślączka, A. (2019). The Western outer Carpathians: origin and evolution. *Zeitschrift Der Deutschen Gesellschaft FürGeowissenschaften* 229–254.

Golonka, J., Gawęda, A., and Waśkowska, A. (2021). Carpathians. In: *Encyclopedia of Geology*, 2e (ed. D. Alderton and S.A. Elias), 372–381. Academic Press https://doi.org/10.1016/B978-0-12-409548-9.12384-X.

Jagodnik, P., Bernat Gazibara, S., Arbanas, Ž., and MihalićArbanas, S. (2020). Engineering geological mapping using airborne LiDAR datasets–an example from the Vinodol Valley, Croatia. *Journal of Maps* 16: 856–867. https://doi.org/10.1080/17445647.2020.1831980.

Kania, M. and Szczęch, M. (2020). Geometry and topology of tectonolineaments in the Gorce Mts. (outer Carpathians) in Poland. *Journal of Structural Geology* 141: 104186. https://doi.org/10.1016/j.jsg.2020.104186.

Kaplay, R.D., Md, B., Mukherjee, S., and Kumar, T.V. (2017). Morphotectonic expression of geological structures in eastern part of south East Deccan volcanic province (around Nanded, Maharashtra, India). In: *Tectonics of the Deccan Large Igneous Province*, vol. 445 (ed. S. Mukherjee, A.A. Misra, G. Calvès and M. Nemčok), 317–335. London, Special Publications: Geological Society.

Książkiewicz, M. (1977). *The AlpineTectonicEpoch. The Tectonics of the Carpathians. Geology of Poland*, vol. 4, 476–608. Warszawa: Tectonics. Geological Institute.

Mahel, M. (1974). *Tectonics of the Carpathian–Balkan Regions, Explanations to the Tectonic Map of the Carpathian-Balkan Regions and their Foreland.* Bratislava: Štátny Geologický Ústav Dionýza Štúra.

Misra, A.A. and Mukherjee, S. (2022). Introduction to "atlas of structural geological and geomorphological interpretation of remote sensing images". In: *Atlas of Structural Geological and Geomorphological Interpretation of Remote Sensing Images* (ed. A.A. Misra and S. Mukherjee). Wiley Blackwell. ISBN: 9781119813354.

Misra, A.A., Bhattacharya, G., Mukherjee, S., and Bose, N. (2014). Near N-S paleo-extension in the western Deccan region in India: does it link strike-slip tectonics with India-Seychelles rifting? *International Journal of Earth Sciences* 103: 1645–1680.

Nixon, M.S. and Aguado, A.S. (2019). *Feature Extraction and Image Processing for Computer Vision, Feature Extraction and Image Processing for Computer Vision*. Elsevier https://doi.org/10.1016/C2017-0-02153-5.

O'Leary, D.W., Friedman, J.D., and Pohn, H.A. (1976). Lineament, linear, lineation: some proposed new standards for old terms. *Geological Society of America Bulletin 1* 87: 1463–1469.

Plašienka, D. (2018). Continuity and episodicity in the early Alpine tectonic evolution of the Western Carpathians: how large-scale processes are expressed by the orogenic architecture and rock record data. *Tectonics* 37: 2029–2079.

Ślączka, A., Kruglov, S., Golonka, J. et al. (2006). Geology and hydrocarbon resources of the outer Carpathians, Poland, Slovakia, and Ukraine: general geology. *The Carpathians and Their Foreland: Geology and Hydrocarbon Resources: AAPG Memoir* 84: 221–258.

Szczęch, M. and Cieszkowski, M. (2021). Geology of the Magura Nappe, south-western Gorce Mountains (outer Carpathians, Poland). *Journal of Maps* 17 (2): https://doi.org/10.1080/17445647.2021.1950579.

Wężyk, P. ed. (2015). *Manual for Participants of LiDAR Products Applications Training (in Polish)*, 2e. Warszawa: Główny Urząd Geodezji i Kartografii.

16

Spatial Variability of Tectonic Influences on Drainage Networks

Examples from the Narmada-Tapi Interfluve in Gujarat State, Western India

Swarali Vasaikar, Deepak M. Maurya, Prabhuti Tiwari, and Laxman S. Chamyal*

Department of Geology, The M.S. University of Baroda, Vadodara, Gujarat, India

16.1 Introduction

Structural geology and geomorphology of terrains can be explained using remote sensing images (e.g. Misra et al. 2014; Kaplay et al. 2017; Dasgupta and Mukherjee 2017, 2019; Dasgupta et al. 2022; Gogoi et al. 2022).We provide examples from the area between the west flowing Narmada and Tapi Rivers in the central part of the Indian plate that is variously known as the Central Indian Tectonic Zone (CITZ) and Son-Narmada-Tapi (SONATA) zone. This ENE–WSW trending zone comprises the Narmada Son fault (NSF) and the Tapi fault (TF) systems, in addition to several related faults (Acharyya et al. 1998; Ravi Shankar 1992). The zone has undergone intense tectonic activity several times during the geological history of the Indian plate, including the Quaternary time period (Chamyal et al. 2002). In its western part in the Gujarat state, the zone shows evidence of intense deformation under compressive stress regime during the Quaternary (Maurya et al. 2000; Chamyal et al. 2002). This is indicated by the ENE–WSW trending anticlines and reverse faults in the exposed Tertiary rocks, reverse nature of the NSF and TF, and the southward tilted trappean lava flows (Agarwal 1986; Roy 1990; Chamyal et al. 2002; Chamyal et al. 2022; Raj et al. 2003). The alluvial basin of the Narmada River to the north of NSF shows geomorphic evidences of uplift during the early and late Holocene, suggesting differential uplift along the NSF under compression (Chamyal et al. 2002; Joshi et al. 2013a) (Figure 16.1).

16.2 Narmada–Tapi Interfluve

Drainages in neotectonically and seismically active regions show remarkable influence on drainage basin shapes and channel morphology (Maurya et al. 2021). The drainage configuration in the region between the Narmada and Tapi Rivers shows a variety of strong tectonic controls. The ENE–WSW trending NSF zone is characterized by north flowing transverse drainages that arise in the trappean uplands to the south of the NSF. These drainages to the north of NSF incise the alluvial deposits by 40–50 m and show parallel drainage pattern developed in response to the prominent northward slope of the alluvial plain (Joshi et al. 2013a). In contrast, these drainages in the trappean uplands show strong trellis drainage pattern influenced by the ENE–WSW and NNW–SSE trending conjugate fracture system evidenced by the dyke swarm and joints along with linear ridges and valleys. Within the uplands, the drainages show several reaches that are straight as an arrow, suggesting the presence of faults and fracture zones that are not mapped so far. Deeply incised bedrock channels and multiple waterfalls indicate neotectonic uplift of the area.

The drainages arising from the western fringe of the trappean uplands flow westward and follow valleys controlled by the ENE–WSW trending anticlines and reverse faults in Tertiary rocks. Within the deformed Tertiary domain, a linear alignment of Kosamba, Dungari, Dinod, and Talodara anticlines bounded by subvertical reverse faults in their respective southern limbs form a major

*Corresponding Author:dmmaurya@yahoo.com

Atlas of Structural Geological and Geomorphological Interpretation of Remote Sensing Images, First Edition.
Edited by Achyuta Ayan Misra and Soumyajit Mukherjee.

Companion website: www.wiley.com/go/misra/RemoteSensingImages

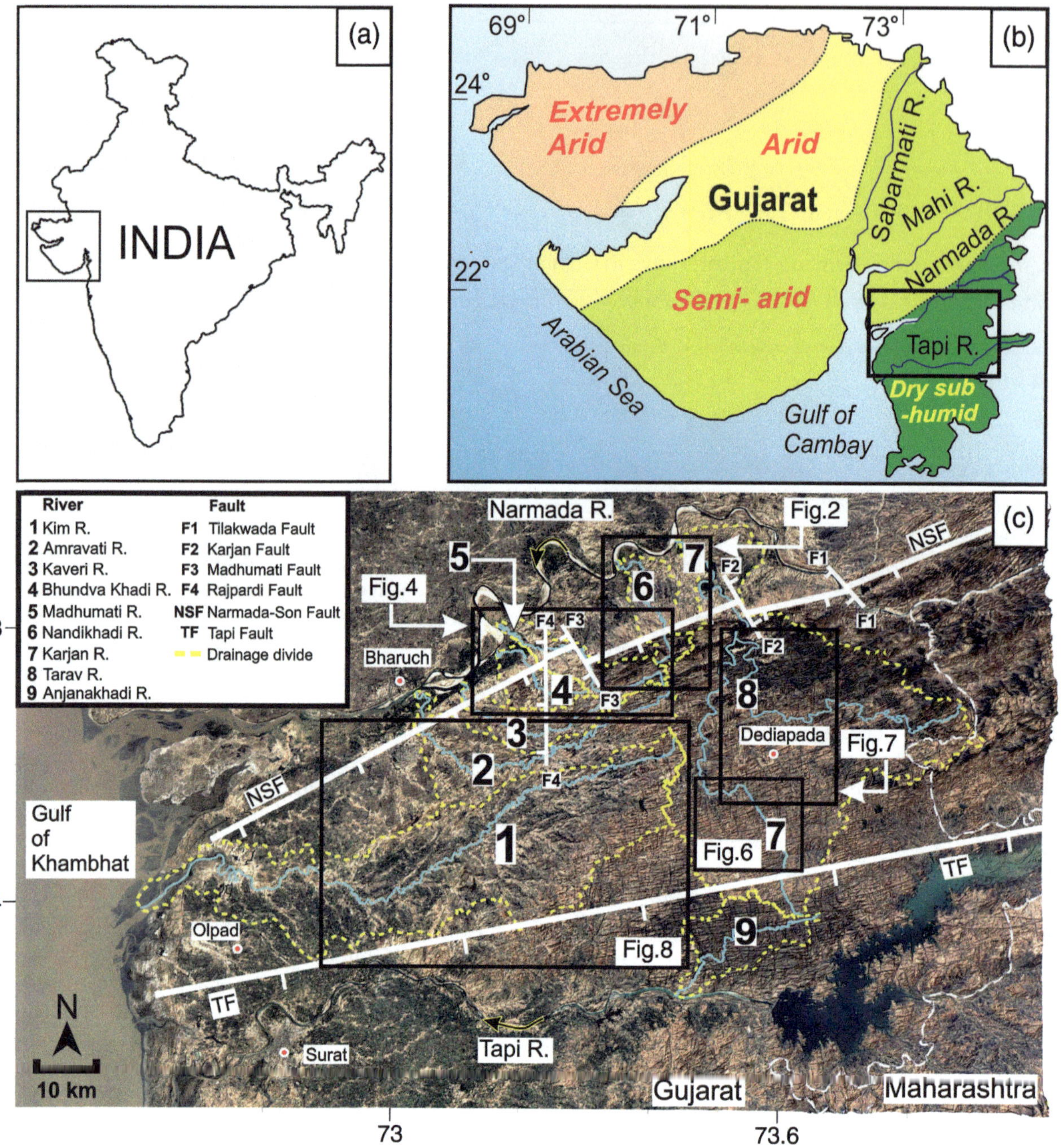

Figure 16.1 (a) Map of India showing the location of Gujarat. (b) Map of Gujarat showing major climatic zones (from https://mausam.imd.gov.in). Black box shows the area covered in (c). (c) Satellite image of the area draped over 30 m resolution DEM. Location of the Narmada-Son Fault (NSF) and the Tapi Fault (TF) is based on Joshi et al. (2013a) and Kaila et al. (1981). Black boxes show the areas enclosed in Figures 16.2–16.8. Note the influence of ENE–WSW structural trend on the landscape induced by the Narmada-Son Fault (NSF) and the Tapi Fault (TF). The majority of the drainages meet the Narmada River while there is only one tributary that meets the Tapi River on its northern bank. The Kim River basin is the only independent drainage basin in the Narmada-Tapi interfluve. (Source: Google Earth Pro V 7.3 (31 December 1990). Image Landsat/Copernicus, Resolution of the image: 4800 × 2275. http://www.earth.google.com and www.opentopography.org).

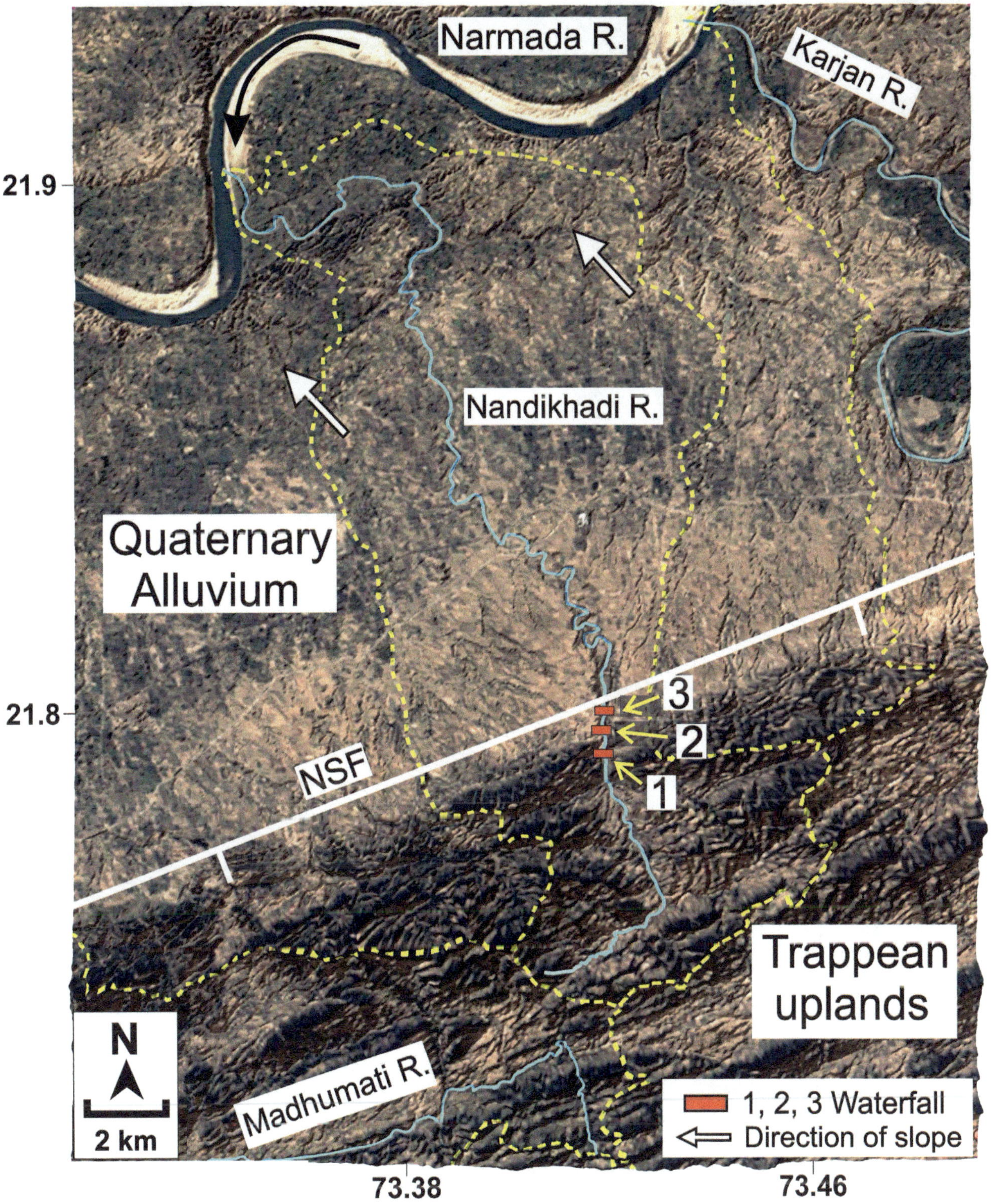

Figure 16.2 Satellite image of the Nandikhadi River, a northward flowing tributary of the Narmada River draped over 30 m resolution DEM. Note the north facing scarps along the NSF and trappean uplands to the south showing strong ENE–WSW structural fabric. The north sloping alluvial basin to the north comprises Late Pleistocene bajada fan surface in the piedmont zone and alluvial plain deposits further north (Joshi et al. 2013b). Location of the image is shown in Figure 16.1. Note multiple waterfalls in Narmada-Son Fault (NSF) zone. Field photographs of the waterfalls are shown in Figure 16.3. (Source: Google Earth Pro V 7.3 (31 December 1990). Image Landsat/Copernicus, Resolution of the image: 4800 × 2275. http://www.earth.google.com and www.opentopography.org).

Figure 16.3 Waterfalls in Nandikhadi River suggesting neotectonic activity along the Narmada-Son Fault (NSF) zone. Location of all waterfalls are shown in Figure 16.2. Yellow arrows show upstream dips of Deccan Trap lava flows while white arrows denote the flow direction of river. (a) Distant view of the waterfall 1 and rapids. (b) Close view of the waterfall 1 formed over upstream dipping trappean lava flows. (c) Panoramic view of the waterfall 2. (d) Close view of waterfall 2. (e) View of waterfall 3.

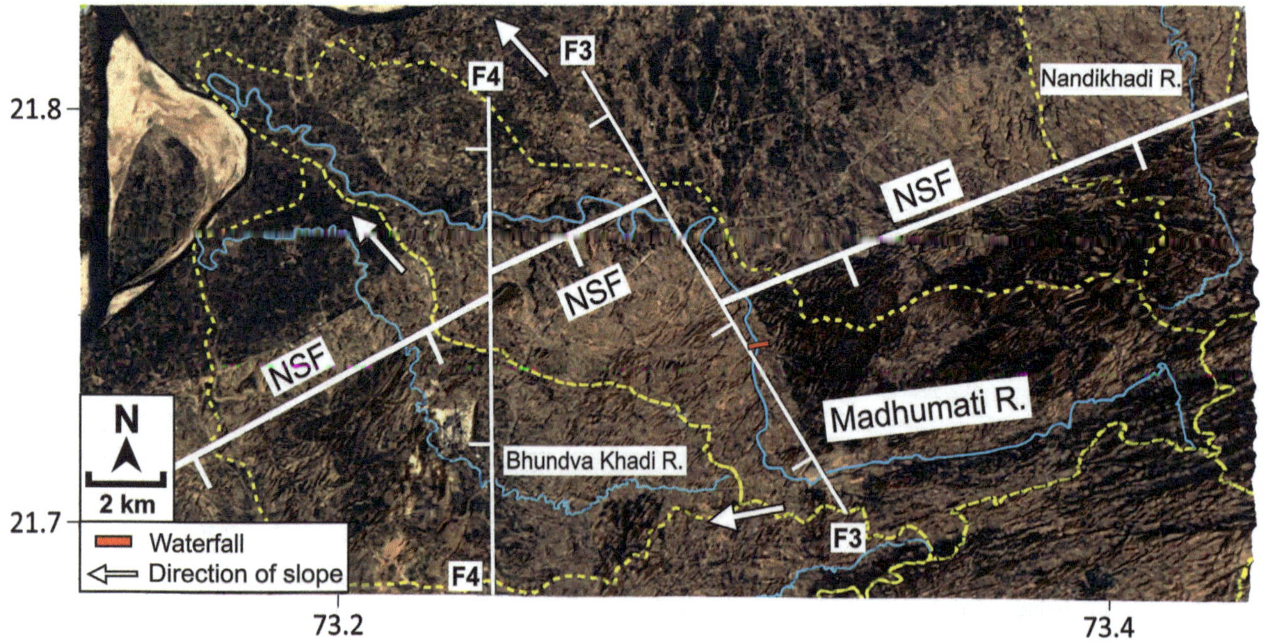

Figure 16.4 Satellite image of the Madhumati River, a north flowing tributary of the Narmada River, superimposed over 30 m resolution DEM. The NSF marks the trappean uplands to the south of it and alluvial plain to the north. Note the anomalous northward swing of the Madhumati River along the cross fault F3. The river course shows three segments: ENE–WSW trending structurally controlled channel in upper part; NNW–SSE along F3 in the middle part; and the meandering channel in the alluvial plain to the north of the NSF. Location of the image is shown in Figure 16.1. Field photographs of the incision and waterfall are shown in Figure 16.5. (Source: Google Earth Pro V 7.3 (31 December 1990). Image Landsat/Copernicus, Resolution of the image: 4800 × 2275. http://www.earth.google.com and www.opentopography.org).

Figure 16.5 (a) Upstream view of the straight canyon-like bedrock channel of Madhumati River controlled by cross-fault of NSF named as Madhumati Fault shown as F3. Note the ~40 m deep incision in the Deccan Trap lava flows forming vertical rocky cliffs along the banks. (b) ~8 m deep waterfall in Madhumati River suggesting neotectonically active nature of the Madhumati Fault. Location of the waterfall is shown in Figure 16.4.

ENE–WSW trending drainage divide. The drainages to the north of these highs swing northward around other anticlinal highs to join the Narmada River. To south of these highs, the Kim River continues to flow westward to form the only independent drainage basin between the Narmada and Tapi Rivers. The continued westward course of the Kim drainage basin is also attributed to the presence of anticlinal highs to the south of it including the Tarkeshwar high. Except for the Anjanakhadi stream, there is no major tributary joining the Tapi River on its northern bank.

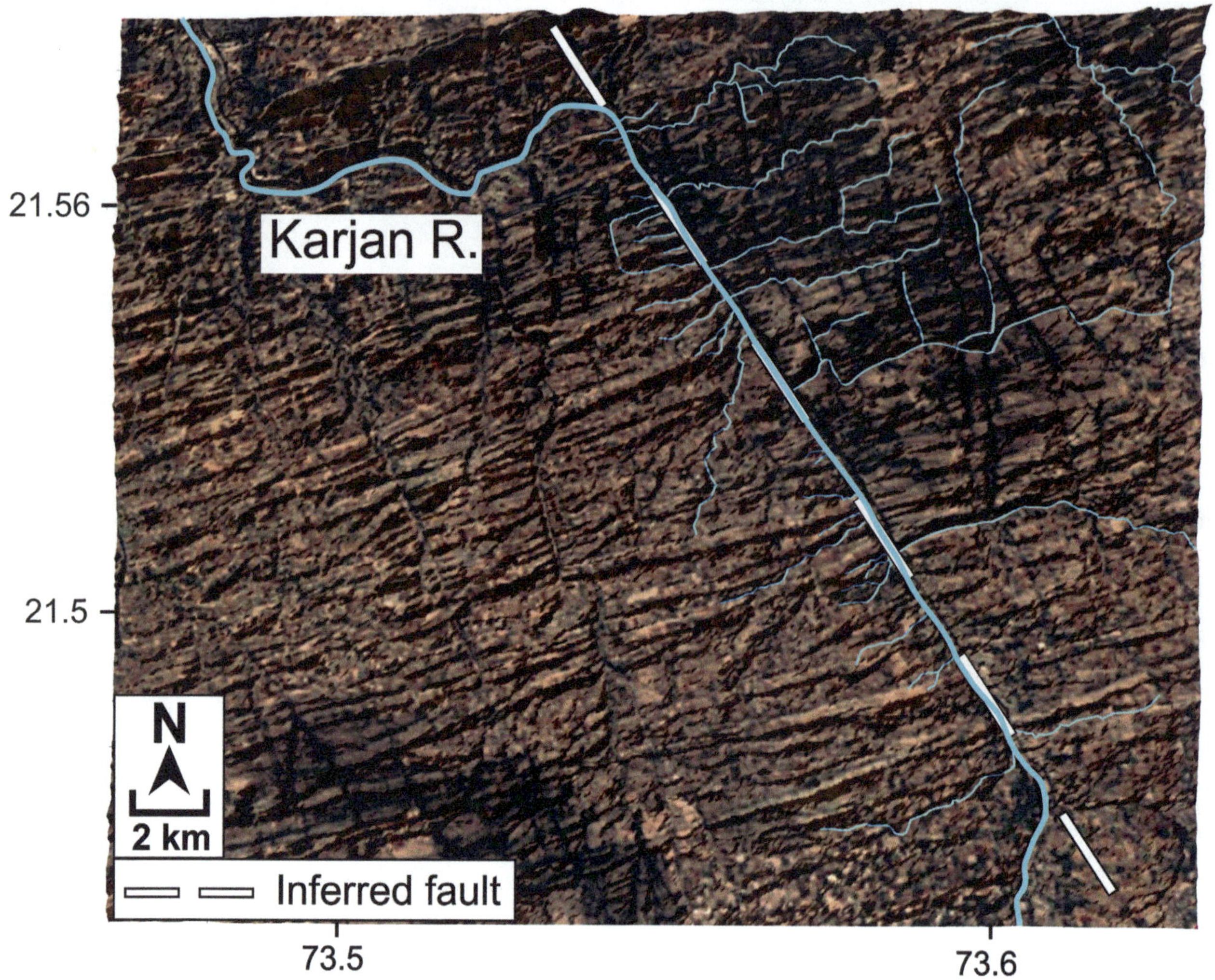

Figure 16.6 Satellite image draped over 30 m resolution DEM showing an anomalous straight segment of the Karjan River in the trappean uplands. Note the fracture controlled trellis drainage pattern and abrupt swing of the river at either ends of the straight segment of the river. The straight as arrow channel suggests the possible influence of fault. (Source: Google Earth Pro V 7.3 (31 December 1990). Image Landsat/Copernicus, Resolution of the image: 4800 × 2275. http://www.earth.google.com and www.opentopography.org).

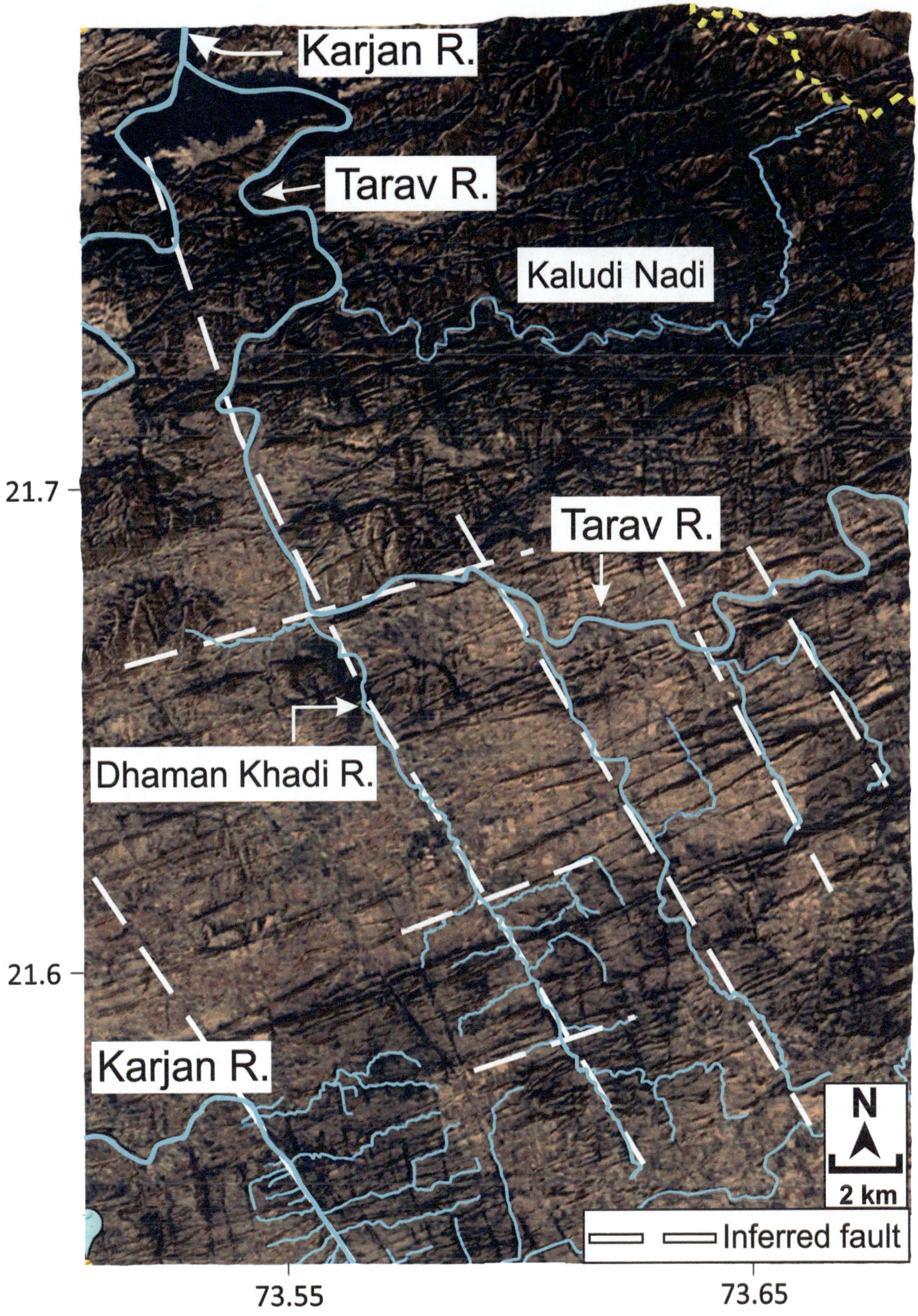

Figure 16.7 Satellite image draped over 30 m resolution showing strongly trellis drainage network in the trappean uplands indicating existence of conjugate fracture system and dykes. Persistent and straight course of the rivers suggest possible influence of faults. Notice the four-stream junction at the confluence of the Dhamankhadi River with the Tarav River. Location of the image is shown in Figure 16.1. (Source: Google Earth Pro V 7.3 (31 December 1990). Image Landsat/Copernicus, Resolution of the image: 4800 × 2275. http://www.earth.google.com and www.opentopography.org).

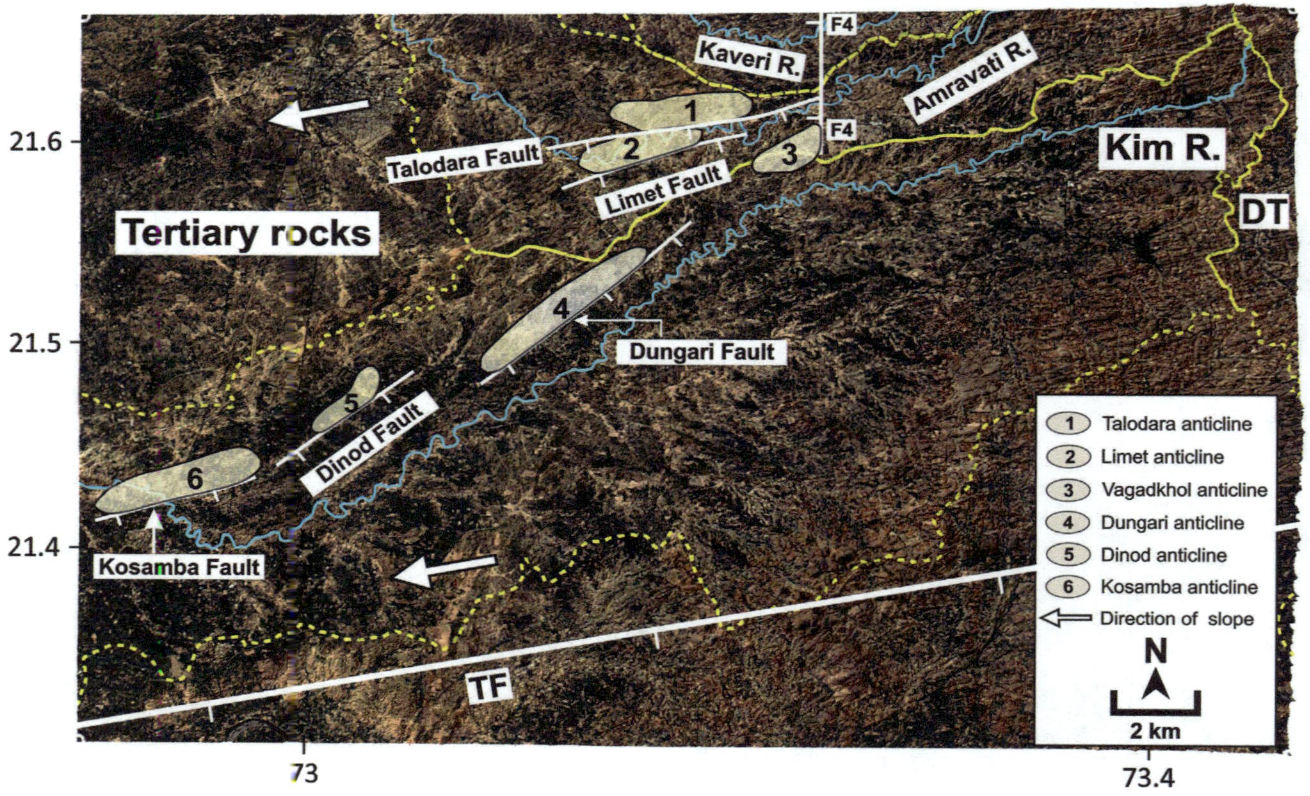

Figure 16.8 Satellite image showing central and upper part of the Kim drainage basin superimposed over 30 m resolution DEM. The right half of the image comprises Deccan traps with strong ENE–WSW tectonic fabric. The structural highs shown in the western half are formed in Palaeogene and Neogene rocks under compression. Note that the Amravati River and Kaveri River swing north due to structural highs to meet the Narmada River further north. In contrast, the Kim River continues to flow westward due to the series of ENE–WSW trending anticlinal highs bounded by reverse faults located on its northern bank. (Source: Google Earth Pro V 7.3 (31 December 1990). Image Landsat/Copernicus, Resolution of the image: 4800 × 2275. http://www.earth.google.com and www.opentopography.org).

Acknowledgements

Section of the work presented here forms part of the ongoing Ph.D. research of Ms. Swarali Vasaikar. We are grateful to Prof. Soumyajit Mukherjee for inviting us to submit an article in the edited book. Achyuta Ayan Misra efficiently handled this article. We are thankful to Wiley Blackwell Acquisition Editors (present: Frank Weinreich, past: Andrew Harrison), Stacey Woods (Handling Editor), and proofreading team. This chapter is summarized in Misra and Mukherjee (2022).

References

Acharyya, S.K., Kayal, J.R., Roy, A., and Chaturvedi, R.K. (1998). Jabalpur earthquake of May 22, 1997: constraint from aftershock study. *Journal of the Geological Society of India* 51 (3): 295–304.

Agarwal, G.C. (1986). Structure and tectonics of exposed tertiary rocks between Narmada and Kim rivers in South Gujarat. *Journal of the Geological Society of India* 27 (6): 531–542.

Chamyal, L.S., Maurya, D.M., Bhandari, S., and Raj, R. (2002). Late quaternary geomorphic evolution of the lower Narmada valley, Western India: implications for neotectonic activity along the Narmada–Son Fault. *Geomorphology* 46 (3–4): 177–202.

Chamyal, L.S., Joshi, P., Vasaikar Swarali, and Maurya, D.M. (2022). Neotectonic characterization of the Narmada-Son Fault (NSF) using field and GPR data, Gujarat, western India. *Journal of the Palaeontological Society of India* 61(1): 2–84.

Dasgupta, S. and Mukherjee, S. (2017). Brittle shear tectonics in a narrow continental rift: asymmetric nonvolcanic Barmer Basin (Rajasthan, India). *The Journal of Geology* 125 (5): 561–591.

Dasgupta, S. and Mukherjee, S. (2019). Remote sensing in lineament identification: examples from western India. In: *Problems and Solutions in Structural Geology and Tectonics*, Developments in Structural Geology and Tectonics Book Series, vol. 5. Series Editor: Mukherjee, S. (ed. A. Billi and A. Fagereng), 205–221. Elsevier.

Dasgupta, S., Biswas, M., Mukherjee, S., and Chatterjee, R. (2022). Depositional system, morphological signatures, tectonics and sedimentation pattern along the transform margin – Palar-Pennar basin, Indian east coast. *Journal of Petroleum Science & Engineering* 211: 110155. https://doi.org/10.1016/j.petrol.2022.110155.

Gogoi, M.P., Gogoi, B., and Mukherjee, S. (2022). Tectonic instability of the petroliferous upper Assam valley (NE India): a geomorphic approach. *Journal of Earth System Science* 131: 18. https://doi.org/10.1007/s12040-021-01752-6.

Joshi, P.N., Maurya, D.M., and Chamyal, L.S. (2013a). Morphotectonic segmentation and spatial variability of neotectonic activity along the Narmada–Son Fault, Western India: remote sensing and GIS analysis. *Geomorphology* 180: 292–306.

Joshi, P.N., Maurya, D.M., and Chamyal, L.S. (2013b). Tectonic and climatic controls on late Quaternary bajada sedimentation along Narmada-Son Fault (NSF), Gujarat, Western India. *International Journal of Sediment Research* 28 (1): 66–76.

Kaila, K.L., Krishna, V.G., and Mall, D.M. (1981). Crustal structure along Mehmadabad-Billimora profile in the Cambay basin, India, from deep seismic soundings. *Tectonophysics* 76 (1–2): 99–130.

Kaplay, R.D., Babar, M.D., Mukherjee, S., and Kumar, T.V. (2017). Morphotectonic expression of geological structures in eastern part of south East Deccan volcanic province (around Nanded, Maharashtra, India). In: *Tectonics of the Deccan Large Igneous Province* (ed. S. Mukherjee, A.A. Misra, G. Calvès and M. Nemčok). *Geological Society, London, Special Publication* 445 (1): 317–335.

Maurya, D.M., Raj, R., and Chamyal, L.S. (2000). History of tectonic evolution of Gujarat alluvial plains, Western India during quaternary: a review. *Journal of the Geological Society of India* 55: 343–366.

Maurya, D.M., Tiwari, P., Shaikh, M.A. et al. (2021). Late quaternary drainage reorganization assisted by surface faulting: the example of the Katrol Hill Fault zone, Kachchh, Western India. *Earth Surface Processes and Landforms* 46 (7): 1268–1293.

Misra, A.A. and Mukherjee, S. (2022). Introduction to "atlas of structural geological and geomorphological interpretation of remote sensing images". In: *Atlas of Structural Geological and Geomorphological Interpretation of Remote Sensing Images* (ed. A.A. Misra and S. Mukherjee). Wiley Blackwell. ISBN: 9781119813354.

Misra, A.A., Bhattacharya, G., Mukherjee, S., and Bose, N. (2014). Near N-S paleo-extension in the western Deccan region in India: does it link strike-slip tectonics with India-Seychelles rifting? *International Journal of Earth Sciences* 103: 1645–1680.

Raj, R., Bhandari, S., Maurya, D.M., and Chamyal, L.S. (2003). Geomorphic indicators of active tectonics in the Karjan river basin, lower Narmada Valley, western India. *Journal of the Geological Society of India* 62 (6): 739–752.

Ravi Shankar (1992). Structural and geomorphological evolution of SONATA rift in Central India in response to the Himalayan uplift. *The Palaeontological Society of India* 38: 17–30.

Roy, T.K. (1990). Structural styles in southern Cambay Basin, India and role of Narmada Geofracture in formation of giant hydrocarbon accumulations. *Bulletin Oil and Natural Gas Commission* 27: 15–38.

17

Archival Airborne Visible/Infrared Imaging Spectrometer (AVIRIS) Image of Faults in a Mixed Carbonate-Clastic Succession, Northwestern Spring Mountains, Nevada, USA

*Mark Abolins**

Department of Geosciences, Middle Tennessee State University, Murfreesboro, TN, USA

17.1 Introduction

Remote sensing has been a powerful tool to study structural geology, tectonics, and geomorphology in different terrains and rocks types (e.g. Misra et al. 2014; Dasgupta and Mukherjee 2017, 2019; Dasgupta et al. 2022; Gogoi et al. 2022). Remote sensing can reveal faults within lithologically variable sedimentary rock units that are not subdivided on geologic maps. For example, the Ediacaran Johnnie Formation is widely exposed within the southwestern United States, but the formation is not subdivided on the highly generalized Geologic Map of Nevada (Stewart and Carlson 1978) and more recent compilations (e.g. Slate et al. 1999). However, the Johnnie Fm. contains numerous carbonate, siltstone, sandstone, and quartzite intervals. These intervals are up to tens of meters thick and are mappable on the ground over distances of up to a few kilometers (e.g. Abolins 1999; Witkosky and Wernicke 2018), indicating that it should be possible to use Airborne Visible/Infrared Imaging Spectrometer (AVIRIS) images to identify locations where strata are offset across macroscale faults. AVIRIS Classic images are used because the images have a smaller cell size (~20 m) than Landsat multispectral images (~30 m cell size). Also, AVIRIS has a higher spectral resolution (224 contiguous bands from 400 to 2500 nm) than Landsat and commercial multispectral systems. The AVIRIS Classic system has acquired data since 1986. Although the newer AVIRIS Next Generation (NG) system can acquire images having a higher spatial resolution (typically 0.3–4 m), AVIRIS Classic data has already been acquired at select locations around the world, and is available at no cost upon request (URL: https://aviris.jpl.nasa.gov/data/get_aviris_data.html).

17.2 Geologic Setting of the Northwestern Spring Mountains, Nevada

The Spring Mountains are a relatively unextended block between the Death Valley and Lake Mead extended domains in the Basin and Range geomorphic province of the southwestern United States (Figure 17.1). Although the Spring Mountains are less extended than adjoining areas, both extensional and contractile structures are exposed within the range (e.g. Burchfiel et al. 1974; Abolins 1999; Giallorenzo et al. 2017). As shown in Figure 17.2, the Johnnie Fm. outcrops at elevations of ~830–2050 m along the boundary between the northwestern Spring Mountains and the adjoining extended domains. For example, the contact between the Johnnie Fm. and the overlying Ediacaran Stirling Quartzite is at an elevation of ~1820 m at Point A and ~1240 m at Point B on Figure 17.2. Because of this difference in elevation (~580 m) and the east, northeast, and southeast dip of the Johnnie-Stirling contact throughout most of the northwestern Spring Mountains, there is probably a major dip-slip fault between Points A and B. This chapter describes an AVIRIS image of lithologic units cut by two faults between points A and B. These faults are Patrick's Fault (~1.9 km long) and the Hill 5482 Fault (~17.3 km long). Another fault, the Gold Springs Fault (~9.6 km long), is also imaged. All three faults appear on the Stewart and Carlson (1978) compilation, which is based primarily on field mapping.

*Corresponding Author: Mark.Abolins@mtsu.edu

Atlas of Structural Geological and Geomorphological Interpretation of Remote Sensing Images, First Edition.
Edited by Achyuta Ayan Misra and Soumyajit Mukherjee.

Companion website: www.wiley.com/go/misra/RemoteSensingImages

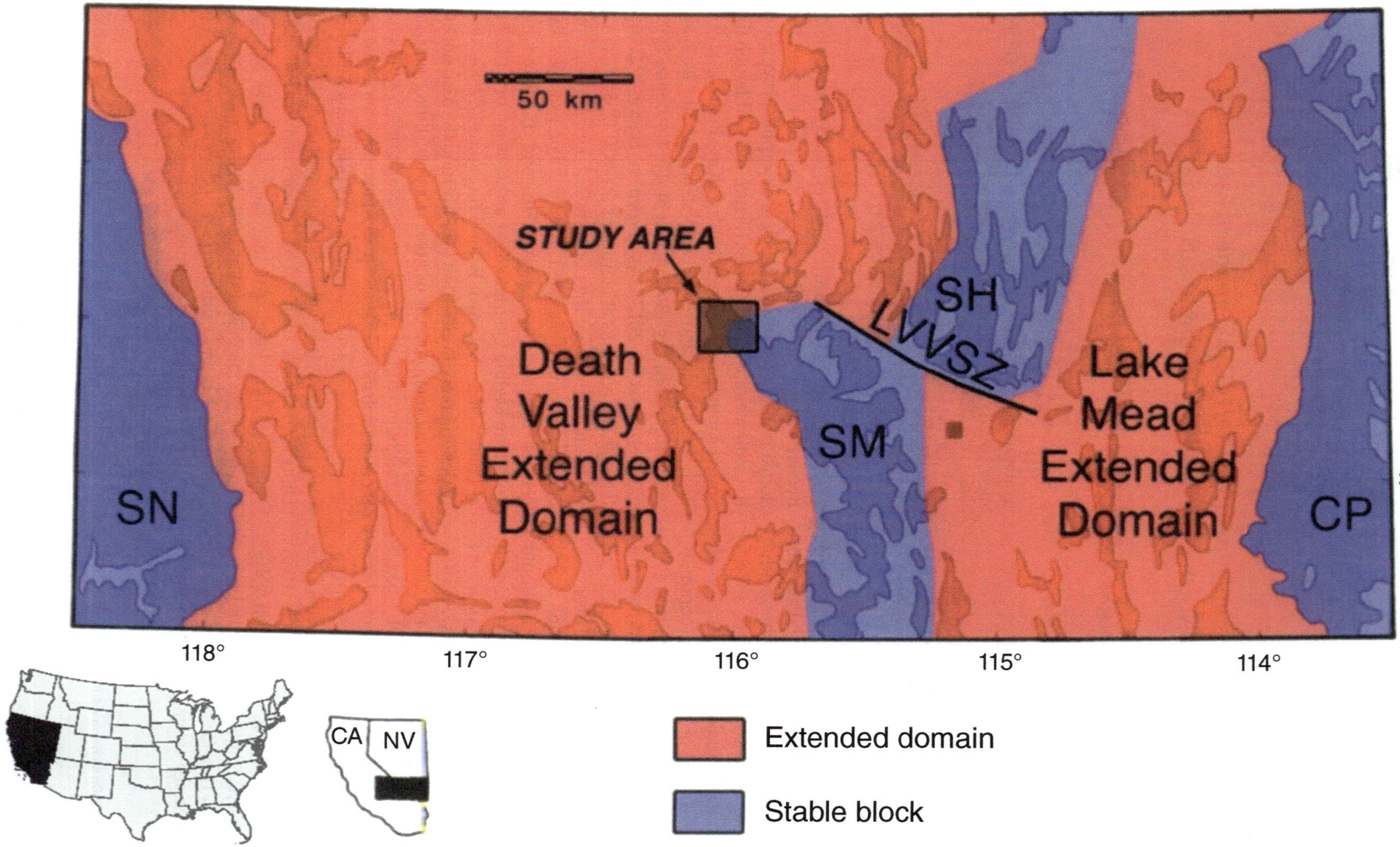

Figure 17.1 Location of the study area in the northwestern Spring Mountains (SM) in relation to the Las Vegas Valley Shear Zone (LVVSZ) and the relatively unextended Sierra Nevada Mountains (SN), Colorado Plateau (CP), and Sheep Range (SH). Source: Adapted from an illustration in Wernicke et al. (1988).

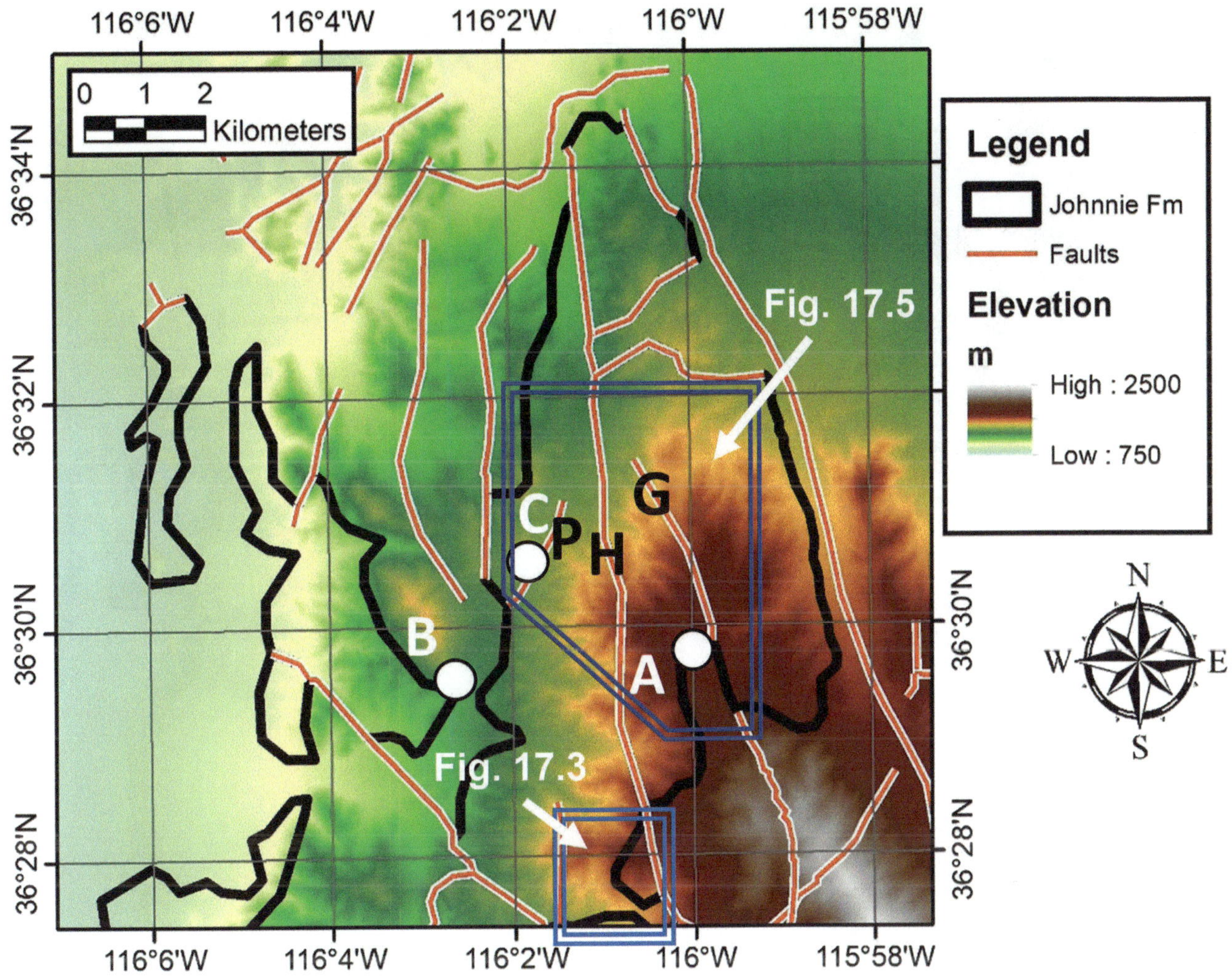

Figure 17.2 Highly generalized map of faults and Johnnie Formation outcrop in the northwestern Spring Mountains, NV. P-Patrick's Fault, H-Hill 5482 Fault, G-Gold Springs Fault. Source: From Stewart and Carlson (1978). See Figure 17.1 for location.

17.3 Using Remote Sensing to Subdivide the Johnnie Formation

Comparison of a 1996 AVIRIS Classic image (Figure 17.3a) with a geologic map (Figure 17.3b) shows that Johnnie Fm. siltstone, quartzite, and carbonate intervals can be imaged south of Point A. Geology mapped on the ground (Abolins1999) provides ground truth. Of particular interest are the regionally widespread Upper Johnnie Fm. (Zju) Rainstorm Member and locally widespread Middle Johnnie cherty dolostone (Zjm1a). The Rainstorm Member has a distinctive ~2 m thick oolitic dolostone near its base and consists mostly of siltstone, carbonate-cemented sandstone, sandy limestone, and intraformational limestone breccia (Figure 17.4). Within the Johnnie Formation in the northwestern Spring Mountains, the intraformational limestone breccia is significant because it is largely confined to the Rainstorm Member. Marker beds are not as readily apparent in the siltstone and sandstone of the Lower Johnnie Formation (Zjl).

Figure 17.3a is a simulated Landsat 5 Thematic Mapper (TM) band ratio image. The rationale for processing data in this way is that there is a large literature devoted to the interpretation of Landsat TM band ratio images (e.g.

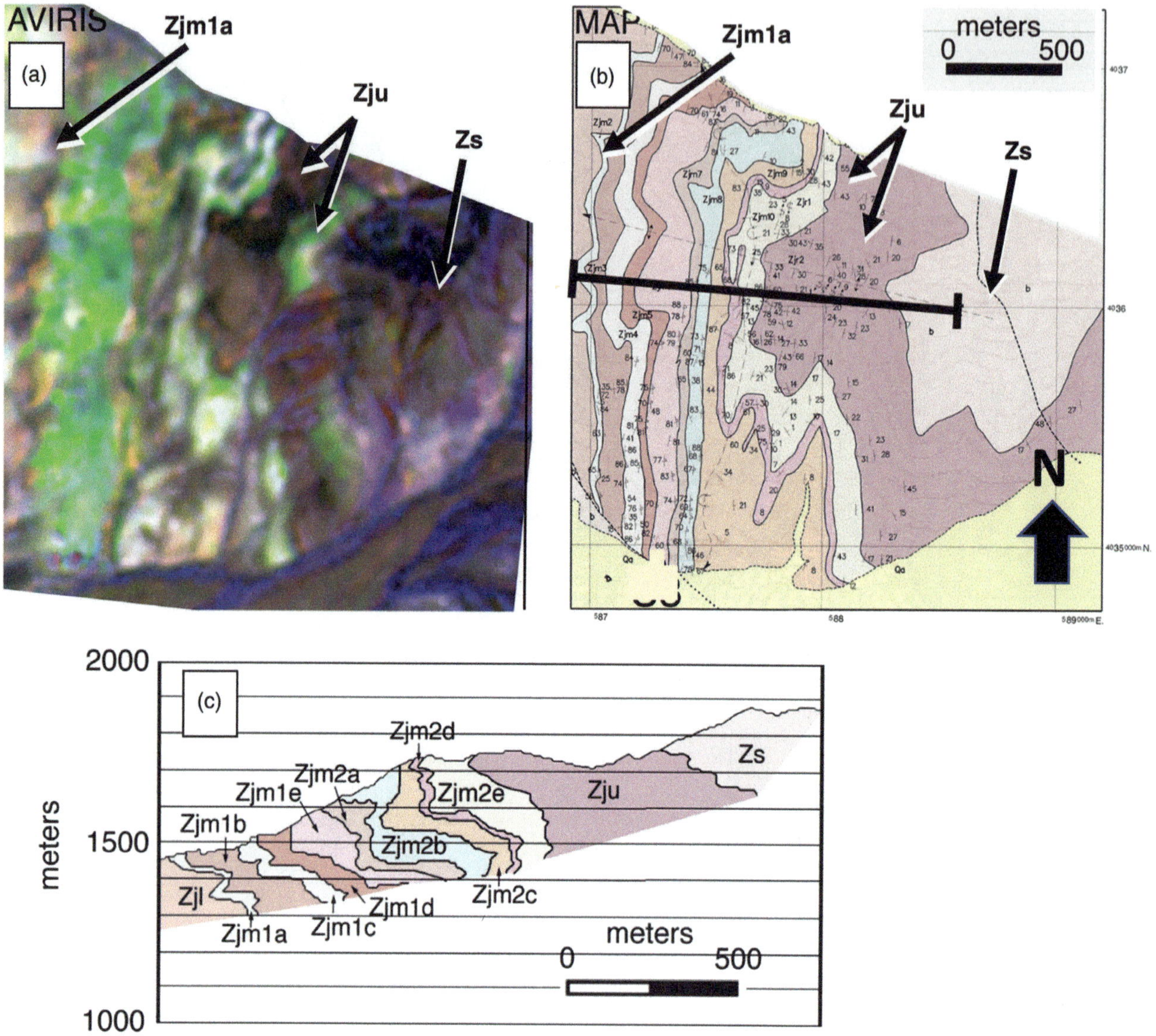

Figure 17.3 (a) AVIRIS Classic simulated Landsat Thematic Mapper band ratio image (Flight line ID: f960411t01p02_r03); (b) geologic map (Abolins 1999); and (c) cross-section (Abolins 1999) of the Middle (Zjm) and Upper (Zju) Johnnie Formation and part of the Stirling Quartzite (Zs). See Figure 17.2 for location.

Figure 17.4 Upper Johnnie Formation (Zju) Rainstorm Member intraformational limestone breccia.

Crippen 1989; Sadek et al. 2020). AVIRIS radiance values are summed over the wavelength ranges of each of the Landsat bands (e.g. AVIRIS bands 10–24 for Landsat 5 Thematic Mapper Band 1), and minimum radiance values are then subtracted to reduce haze. In terms of Landsat 5 TM bands, colors are assigned in the following way: red = band 5/band 7, green = band 5/band 4, and blue = band 3/band 1. To incorporate overall brightness (e.g. shadows), each color is multiplied by the average of the six simulated and haze-reduced visible/near-infrared TM bands.

17.4 Imaging Faults Within the Johnnie Formation

Principal component analysis (PCA) is used to image three faults in the center of the study area (Figures 17.5 and 17.6). The rationale for using PCA is that it has been used in carbonate remote sensing studies (e.g. Bellian et al. 2007). Two faults, the Hill 5482 Fault and the Gold Springs Fault, offset the cherty dolostone (Zjm1a), but both offset it by relatively modest amounts because they place Lower Johnnie Fm. (Zjl) strata against Lower Johnnie Fm. strata. For the Hill 5482 Fault, the hanging wall cutoff is ~60 m lower than the footwall cutoff, and the hanging wall cutoff is ~125 m lower than the footwall cutoff for the Gold Springs Fault. In

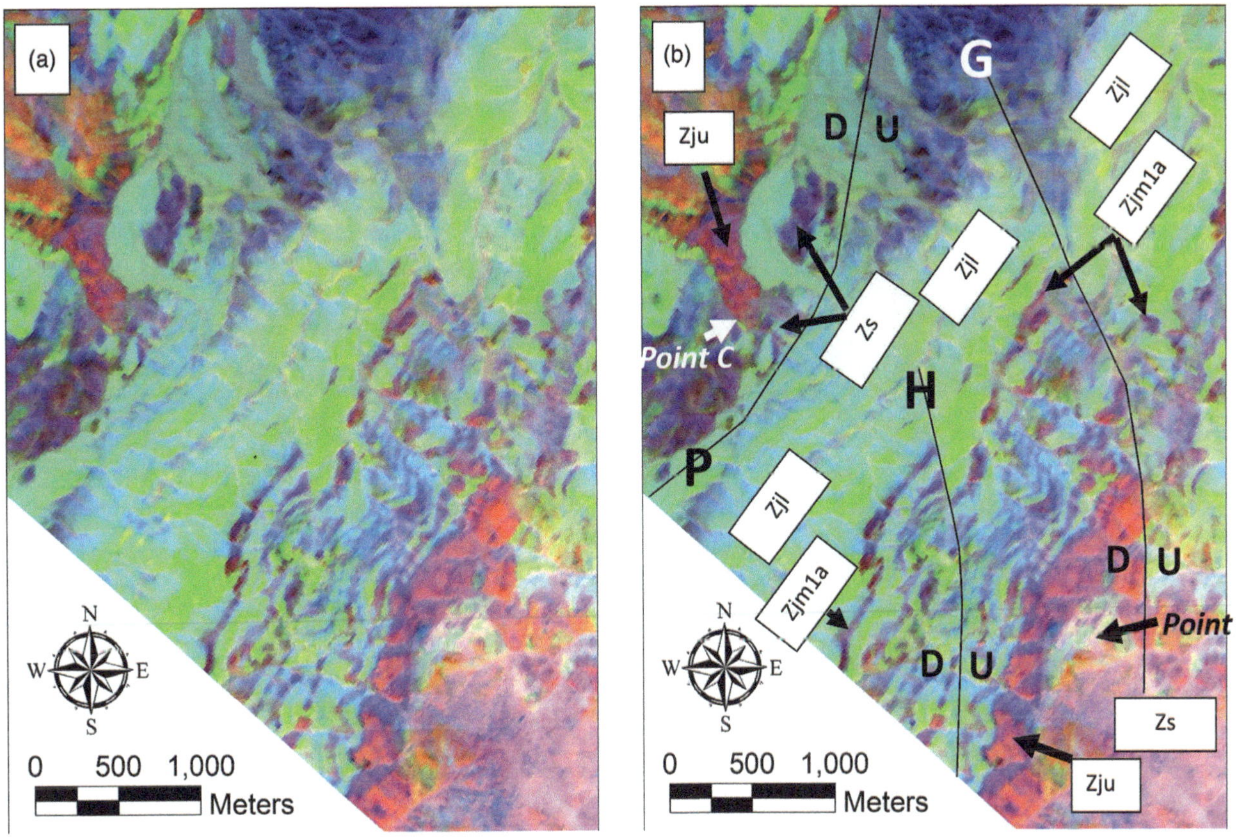

Figure 17.5 (a) AVIRIS Classic principal component image (red = PC3, green = PC2, and blue = PC1) and (b) interpretation. Principal component analysis based on 164 AVIRIS bands from parts of the spectrum least affected by atmospheric absorption (AVIRIS bands 1–107, 114–152, 171, 176–177, 180–192, and 195–196). To reduce haze, minimum radiance values were subtracted from each band. P-Patrick's Fault, H-Hill 5482 Fault, G-Gold Springs Fault. Flight line ID: f960411t01p02_r04.

contrast, the third fault, Patrick's Fault,[1] has a much larger stratigraphic throw because it juxtaposes the Stirling Quartzite (Zs) and Upper Johnnie Formation (Zju) on the northwest with the Lower Johnnie Formation (Zjl) on the southeast. Geologic mapping (Figure 17.7) shows that the Upper Johnnie Fm. (Zju) and Stirling Quartzite (Zs) are at an elevation of only ~1350 m at Point C on the northwest side of the fault, which is ~470 m lower than the elevation of the contact between the two at Point A. These key structural relationships clarify the location of a major normal fault that separates the northwestern Spring Mountains from the adjoining extended domains (Figure 17.8).

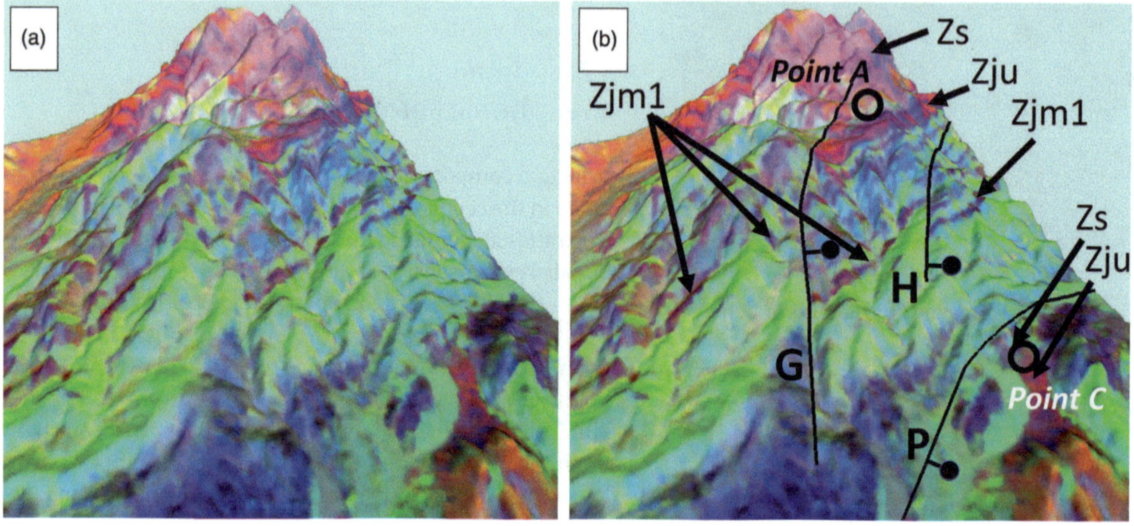

Figure 17.6 (a) Visualization (southeast view) of the footwall of Patrick's Fault and (b) interpretation. See Figure 17.5 and text for explanation of principal component image. Bar-and-ball ornament is on down-thrown side of each normal fault. P-Patrick's Fault, H-Hill 5482 Fault, G-Gold Springs Fault. Vertical exaggeration: 2×.

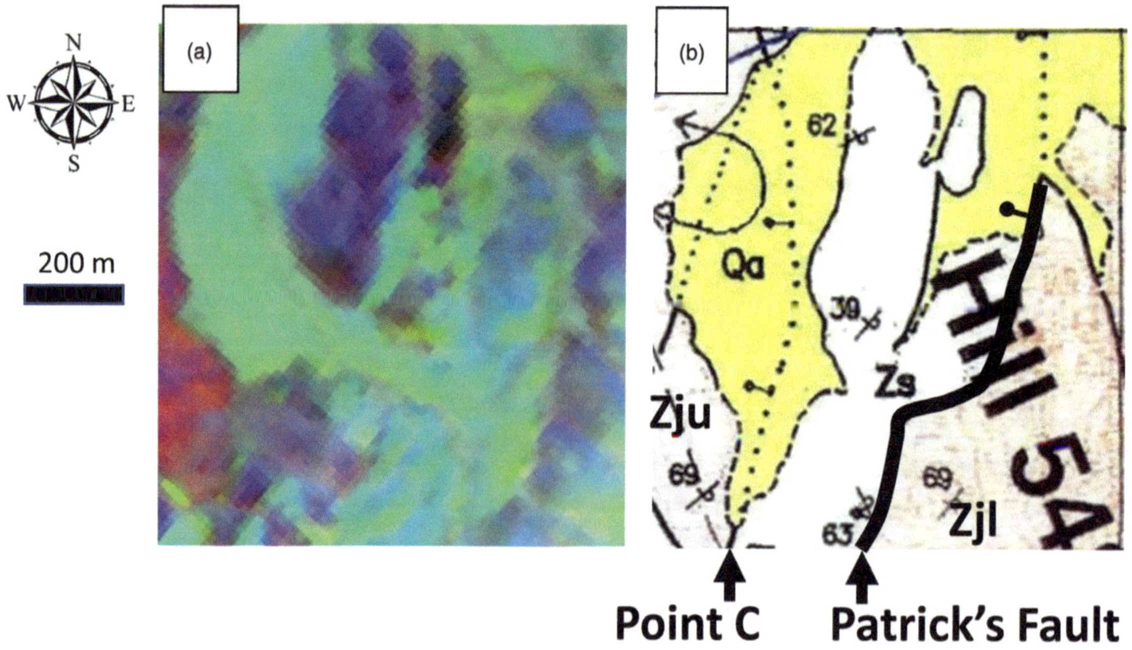

Figure 17.7 (a) AVIRIS Classic principal component image and (b) geologic map (Abolins 1999) of units outcropping along Patrick's Fault. Zjl-Lower Johnnie Formation, Zju-Upper Johnnie Formation, Zs-Stirling Quartzite. Bar-and-ball ornament is on down-thrown side of each normal fault. See Figure 17.5b for location of Point C.

[1] This fault is "Paddy's Fault" in Abolins (1999). As is customary, Paddy's Fault was named after a location named on a U.S. Geological Survey 1 : 24 000 topographic map. "Patrick's Fault" is used here out of cultural sensitivity.

Figure 17.8 Major normal faults separating the northwestern Spring Mountains from adjoining extended domains. See Figure 17.1 for location.

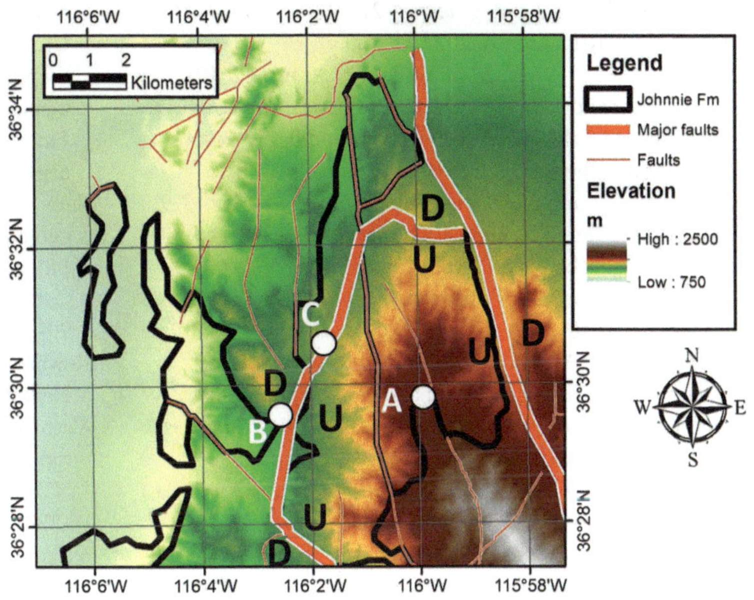

Acknowledgements

Geologic mapping and image processing and interpretation were supported by a 1993 National Science Foundation Graduate Research Fellowship Award to Mark Abolins and NSF-EAR 96-14780 (Brian Wernicke, PI). Ron Blom and Bob Crippen (NASA-JPL) provided image processing and interpretation expertise, and Ron Blom requested the 1996 AVIRIS flight.

The AVIRIS simulated band ratio image (Fig. 17.3a) was processed with ER Mapper software, and the AVIRIS PCA image (Figures 17.5, 17.6, and 17.7a) was processed with ENVI software. Maps and 3D visualizations were prepared with ESRI GIS software. Wiley Blackwell Acquisition Editors (present: Frank Weinreich, past: Andrew Harrison), Stacey Woods (Handling Editor), and proofreading team. Summarized in Misra and Mukherjee (2022).

References

Abolins, M.J. (1999). I. Stratigraphic Constraints on the Number of Discrete Neoproterozoic Glaciations and the Relationship between Glaciation and Ediacaran Evolution; II. The Kwichup Spring Thrust in the Northwest Spring Mountains, Nevada: Implications for Large-Magnitude Extension and the Structure of the Cordilleran Thrust Belt. Ph.D. thesis. Pasadena, California, California Institute of Technology, 341 p.

Bellian, J.A., Beck, R., and Kerans, C. (2007). Analysis of hyperspectral and lidar data: remote optical mineralogy and fracture identification. *Geosphere* 3: 491–500.

Burchfiel, B.C., Fleck, R.J., Secor, D.T. et al. (1974). Geology of the Spring Mountains. *Nevada: Geological Society of America Bulletin* 85 (7): 1013–1022.

Crippen, R.E. (1989). Selection of Landsat TM band and band-ratio combinations to maximize lithologic information in color composite displays, 7th Thematic Conference, Remote Sensing Exploration Geology, Calgary, Alberta, p. 917–921.

Dasgupta, S. and Mukherjee, S. (2017). Brittle shear tectonics in a narrow continental rift: asymmetric non-volcanic Barmer basin (Rajasthan, India). *The Journal of Geology* 125: 561–591.

Dasgupta, S. and Mukherjee, S. (2019). Remote sensing in lineament identification: examples from western India. In: *Problems and Solutions in Structural Geology and Tectonics*, Developments in Structural Geology and Tectonics Book Series, vol. 5. Series Editor: Mukherjee, S (ed. A. Billi and A. Fagereng), 205–221. Elsevier. ISSN: 2542-9000. ISBN: 9780128140482.

Dasgupta, B.M., Mukherjee, S., and Chatterjee, R. (2022). Depositional system, morphological signatures, tectonics and sedimentation pattern along the transform margin- Palar-Pennar basin, Indian east coast. *Journal of*

Petroleum Science & Engineering 211: 110155. https://doi.org/10.1016/j.petrol.2022.110155.

Giallorenzo, M.A., Wells, M.L., Yonkee, W.A. et al. (2017). Timing of exhumation, wheeler pass thrust sheet, southern Nevada and California: late Jurassic to middle cretaceous evolution of the southern Sevier fold-and-thrust belt. *Geological Society of America Bulletin* 130 (3–4): 558–579.

Gogoi, M.P., Gogoi, B., and Mukherjee, S. (2022). Tectonic instability of the petroliferous upper Assam valley (NE India): a geomorphic approach. *Journal of Earth System Science* 131: 18.

Misra, A.A. and Mukherjee, S. (2022). Introduction to "atlas of structural geological and geomorphological interpretation of remote sensing images". In: *Atlas of Structural Geological and Geomorphological Interpretation of Remote Sensing Images* (ed. A.A. Misra and S. Mukherjee). Wiley Blackwell.

Misra, A.A., Bhattacharya, G., Mukherjee, S., and Bose, N. (2014). Near N-S paleo-extension in the western Deccan region in India: does it link strike-slip tectonics with India-Seychelles rifting? *International Journal of Earth Sciences* 103: 1645–1680.

Sadek, M.F., El-kalioubi, B.A., Ali-Bik, M.W. et al. (2020). Utilizing Landsat-8 and ASTER data in geologic mapping of hyper-arid mountainous region: case of Gabal Batoga area. *South Eastern Desert of Egypt: Environmental Earth Sciences* 79 (5): 101. https://doi.org/10.1007/s12665-020-8845-4.

Slate, J.L., Berry, M.E., Rowley, P.D., et al. (1999). Digital geologic map of the Nevada Test Site and vicinity, Nye, Lincoln, and Clark Counties, Nevada, and Inyo County, California: U.S. Geological Survey Open-File Report 99–554–A, 54p., 2 plates.

Stewart, J.H. and Carlson, J.E. (1978). Geologic Map of Nevada: U.S. Geological Survey, 1:500,000.

Wernicke, B.P., Axen, G.J., and Snow, J.K. (1988). Basin and range extensional tectonics at the latitude of Las Vegas, NV. *Geological Society of America Bulletin* 100 (11): 1738–1757.

Witkosky, R. and Wernicke, B.P. (2018). Subsidence history of the Ediacaran Johnnie formation and related strata of southwest Laurentia: implications for the age and duration of the Shuram isotopic excursion and animal evolution. *Geosphere* 14 (5): 2245–2276.

18

Coseismic Surface Rupture and Related Disaster During the 2018 Mw 7.5 Palu Earthquake, Sulawesi Island, Indonesia

Jinrui Liu[1], Dengyun Wu[1], Zhikun Ren[1], Jie Chen[2], Peng Guo[1], Gongming Yin[2], Hongliu Ran[1], Chuanyou Li[2], and Gang Su[3]*

[1] *Key Laboratory of Seismic and Volcanic Hazards, China Earthquake Administration, Beijing, China*
[2] *State Key Laboratory of Earthquake Dynamics, Institute of Geology, China Earthquake Administration, Beijing, China*
[3] *China earthquake Disaster Prevention Center, China Earthquake Administration, Beijing, China*

18.1 The 2018 Mw 7.5 Palu Earthquake

Remote sensing has been a powerful tool to interpret structural geology and geomorphology of terrains (e.g. Misra et al. 2014; Kaplay et al. 2017; Dasgupta and Mukherjee 2017, 2019; Dasgupta et al. 2022; Gogoi et al. 2022). At 6:02 p.m. on 28 September 2018 (10:02 a.m., 28 September Coordinated Universal Time [UTC]), a magnitude Mw 7.5 earthquake struck the district around Palu City, Sulawesi Island, Indonesia. More than 2000 people died and more than 70 000 houses were reported to be damaged by the earthquake due to the associated strong ground motion and related disaster. Before the occurrence of the 2018 Palu earthquake, we arrived at Palu City on 27 September and experienced the earthquake. We conducted investigations of surface ruptures and related disasters in Palu City beginning one day after the occurrence of the earthquake. A number of remote-sensing, unmanned aerial vehicle (UAV), and photography images were obtained.

18.2 Coseismic Surface Rupture and Related Disaster Produced by the 2018 Palu Earthquake

The causative fault of the Palu earthquake was the Palu-Koro fault, which is a sinistral strike-slip fault cutting through Sulawesi Island (e.g. Stevens et al. 1999; Socquet et al. 2006, 2019; Wu et al. 2020). Along this fault within the Palu basin, the coseismic surface rupture developed mainly within a deformation zone several meters to several tens of meters wide, causing serious damage and casualties, and was dominantly characterized by left-lateral dislocation (Song et al., 2019; Wu et al., 2020; Figures 18.1, 18.2). Using the deformed roads as linear surface markers, coseismic displacements were measured (Figures 18.3–18.5). Meanwhile, the Palu earthquake also generated cascading hazards including tsunami and sand liquefactions, considered to be the main cause of catastrophic loss. Cyclics hearing of loosely compacted, fluid-saturated sediments during earthquake-induced ground motion results in reduced shear strength in the affected media and the occurrence of liquefaction (Quigley et al. 2013). During the 2018 Palu earthquake, west of the rupture zone, the topography as well as the buildings were seriously disrupted by mud flows due to sand liquefaction caused by the amplified ground motion (Figures 18.6–18.8). The strike-slip Palu earthquake also generated tsunamis, which hit the coast lines around the narrow Palu bay (Figures 18.9–18.11), especially the eastern coast where the celebration of the foundation of Palu City was being held during the earthquake. Palu City was awarded the status of administrative city on 27 September 1978, and 2018 marked the 30th anniversary of the city, so the celebration lasted until September 28. The continental crust moved northward by the strike-slip Palu fault and landslide along the coast from the Palu earthquake; these were also observed and reported during the earthquake, which may be the cause of the tsunami generated by the strike-slip Palu earthquake (Figure 18.1a; Socquet et al. 2019).

*Corresponding Author: rzk@ies.ac.cn

Atlas of Structural Geological and Geomorphological Interpretation of Remote Sensing Images, First Edition.
Edited by Achyuta Ayan Misra and Soumyajit Mukherjee.

Companion website: www.wiley.com/go/misra/RemoteSensingImages

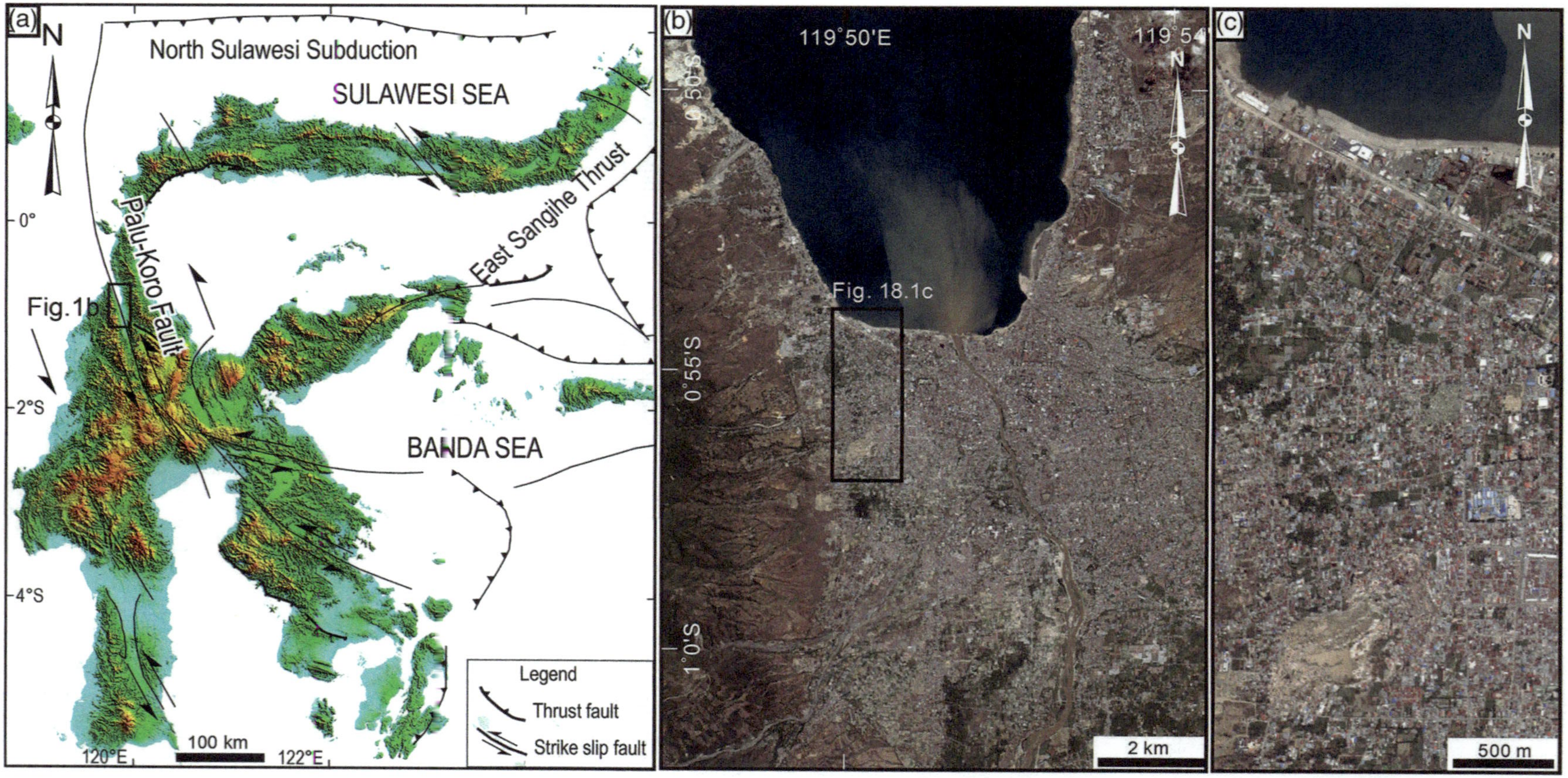

Figure 18.1 Uninterpreted Pléiades image with a resolution of 0.5 m. (a) Map of Sulawesi Island; the base map is from the 90 m Shuttle Radar Topography Mission (SRTM) digital elevation model (DEM), obtained from International Scientific & Technical Data Mirror Site, Computer Network Information Center, Chinese Academy of Sciences (www.gscloud.cn). (b) Pléiades PHR1A image of Palu City and its surrounding area acquired on 5 Oct 2018. Pleiades is a constellation of two very-high-resolution satellites capable of acquiring imagery of any point on the globe in under 24 hours for civil and military users. The satellite was launched in December 2011 with a sensor that provides Multi Spectral (MS) images. (c) Pléiades PHR1A image of part of Palu City, which is the enlarged view of the black box in (b). The image was configured into four bands with different solar irradiance (watt m^{-2} μm^{-1}): 1594.0 for B2, 1831.0 for B1, 1915.0 for B0, and 1060.0 for B3, respectively.

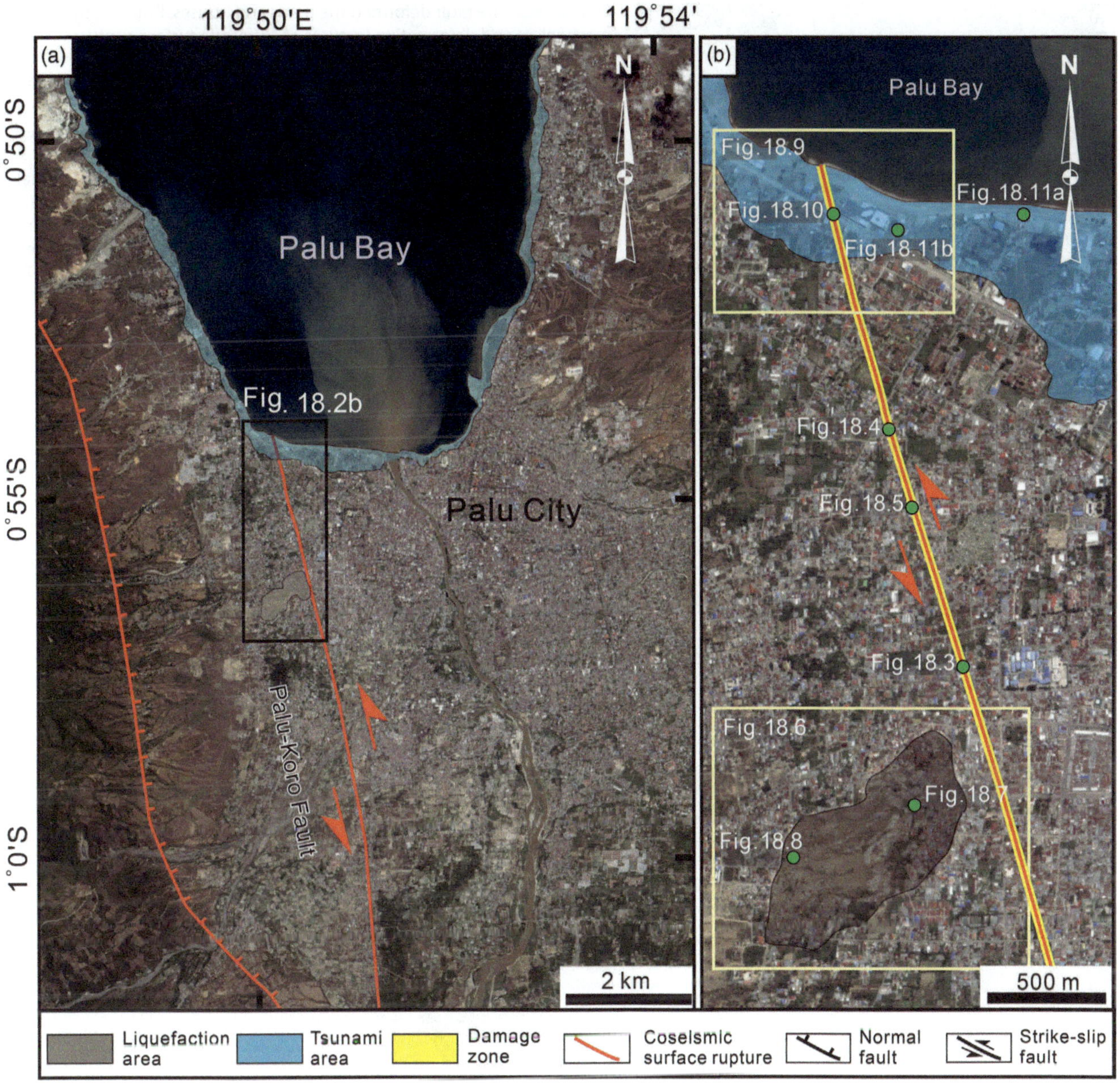

Figure 18.2 Interpreted Pléiades image shown in Figure 18.1. (a) Pléiades PHR1A image of Palu City and its surrounding area. Red lines represent Palu-Koro fault with coseismic rupture. Blue translucent area indicates the tsunami area. (b) Pléiades PHR1A image of part of Palu City. The red-yellow line indicates coseismic surface rupture and damage zone by strike-slip fault motion. The fault trace and coseismic rupture can be clearly identified in the subsequent figures. The blue translucent area indicates the tsunami area. See the black rectangle in (a) for location.

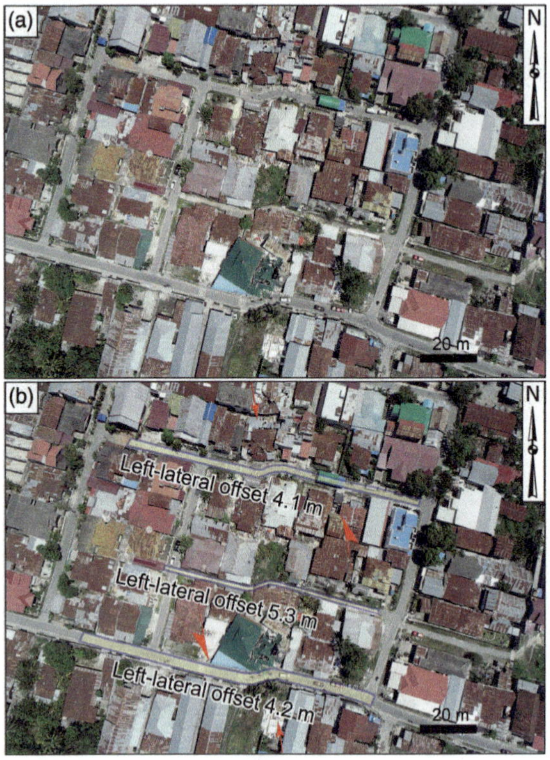

Figure 18.3 The fault deformed the roads and houses. Both the focal mechanism and deformation features of coseismic surface ruptures indicated that the displacements were dominated by almost pure left-lateral strike-slip, with minor vertical displacements, and therefore they could be directly measured in the field (Wu et al. 2020). (a) Uninterpreted UAV image in Palu City. The consumer grade UAV DJI Phantom 4 was used to collect images. This UAV is equipped with a 20-megapixel, 1-in. Complementary Metal-Oxide-Semiconductor (CMOS) sensor and internal GPS, which can fly for ~30 minutes. We set the flying height at 100 m in the UAV Application and then produced a target image object space resolution of ~3.5 cm. See Figure 18.2b for location. (b) Interpretation of UAV image shown in Figure 18.3a. The surface rupture mainly showed left-lateral offset without vertical offset, and the deformation width was only ~20 m, indicating that it is caused by a pure strike-slip motion. Small red arrows indicate the location of the fault that deformed the houses and roads. Based on the dislocated roads, the surface coseismic sinistral offsets were 4.1, 5.3, and 4.2 m, respectively, at the three locations.

Figure 18.4 Left-lateral offset road. (a) Uninterpreted field photograph taken in Palu City. See Figure 18.2b for location. (b) Interpretation of field photography shown in Figure 18.4a. The red line depicts the location of the fault. The road was left-laterally displaced by 3.5 m. The different degree of damage to structures caused by coseismic rupture is due to the different stability of structures. On the other hand, the damage zone of coseismic surface rupture is very narrow, and only the buildings just passing through the coseismic rupture will be seriously damaged, such as the structures on the right side of the road in the figure. The large-scale structures in the following figures were razed to the ground mainly due to sand liquefaction.

Figure 18.5 Left-lateral offset road. (a) Uninterpreted field photograph taken in Palu City. See Figure 18.2b for location. (b) Interpretation of field photography shown in (a). The red line depicts the location of the fault. The road was left-laterally displaced by 1.8 m.

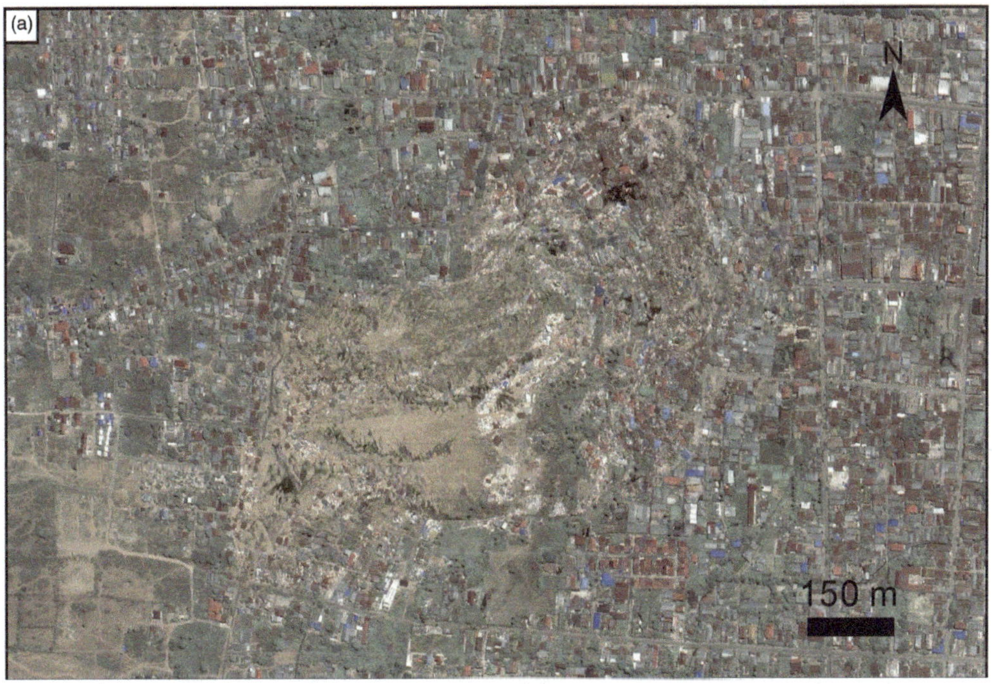

Figure 18.6 Liquefaction area in Palu city. (a) Uninterpreted Pléiades PHR1A image of liquefaction area. See Figure 18.2b for location. (b) Interpretation of Pléiades PHR1A image of liquefaction area. The yellow dotted line is the boundary of sand liquefaction area. During the 2018 Palu earthquake, sand liquefaction and site amplification damaged this area more seriously than other regions of Palu City (Watkinson and Hall 2019), with ~5000 people lost and buried in the liquefied area covering several square kilometers. At this site, the sand liquefaction was more destructive than the left-lateral shear caused by coseismic surface rupture constrained in a narrow zone less than 20 m across.

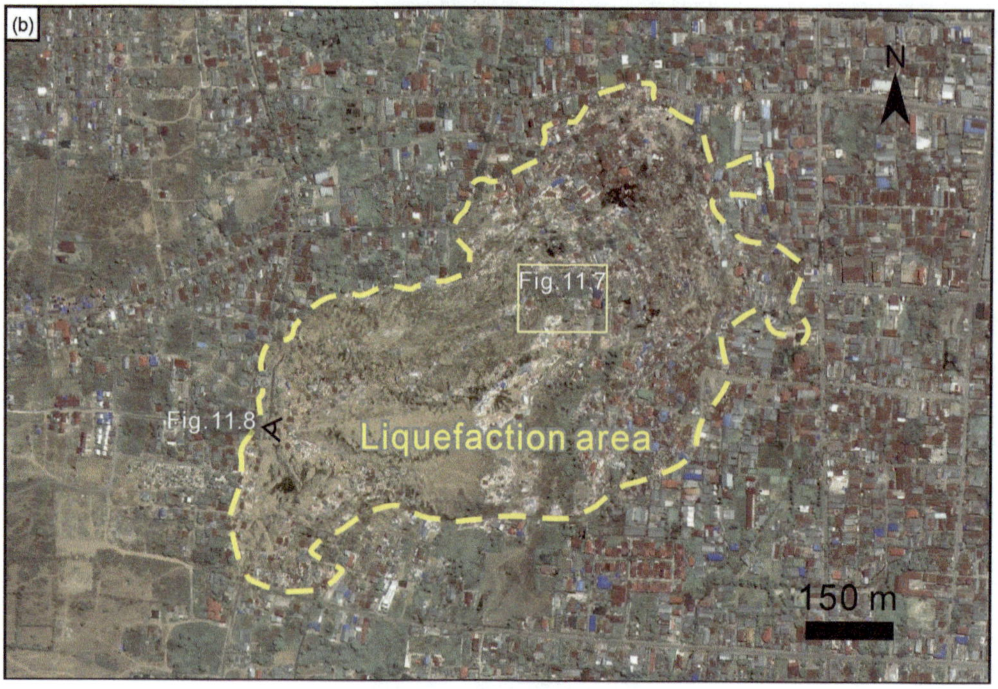

Figure 18.6 (*Continued*)

Figure 18.7 UAV image in Palu city showing sand liquefaction damaged lots of buildings. See Figure 18.6b for location.

Figure 18.8 Houses damaged by sand liquefaction. See Figure 18.6b for location.

Figure 18.9 Coastline around the Palu bay hit by tsunami. (a) Uninterpreted Pléiades PHR1A image of tsunami area. See Figure 18.2b for location. (b) Interpretation of Pléiades PHR1A image of tsunami area. The blue dotted line is the boundary of sand tsunami area. The earthquake-induced tsunamis were possibly due to the northward motion of the seafloor and coseismic landslides. Strike-slip ruptures may generate some vertical land motion if the rake is not purely horizontal thereby displacing the water column (Socquet et al. 2019). The tsunami caused devastating damage to the coastal areas.

Figure 18.10 A UAV image showing houses, roads, and automobiles damaged by the tsunami. See Figure 18.9 for location.

Figure 18.11 Buildings damaged by the tsunami. (a) Mercure Hotel, where our team stayed, collapsed during the earthquake and then was damaged by tsunami. When the earthquake happened, we were resting on the 1st to 3rd floor of the hotel and immediately ran out of the hotel with the crowd. Fortunately, no one was hurt. The east side of the hotel has collapsed, shown in the figure. See Figure 18.2b for location. (b) A church destroyed by the tsunami. See Figure 18.9b for location.

Figure 18.11 (*Continued*)

Acknowledgements

Soumyajit Mukherjee invited to submit an article in the edited book. Achyuta Ayan Misra handled this article. Wiley Blackwell Acquisition Editors (present: Frank Weinreich, past: Andrew Harrison), Stacey Woods (Handling Editor), and proofreading team. Summarized in Misra and Mukherjee (2022).

References

Dasgupta, S. and Mukherjee, S. (2017). Brittle shear tectonics in a narrow continental rift: asymmetric non-volcanic Barmer basin (Rajasthan, India). *The Journal of Geology* 125: 561–591.

Dasgupta, S. and Mukherjee, S. (2019). Remote sensing in lineament identification: examples from western India. In: *Problems and Solutions in Structural Geology and Tectonics*, Developments in Structural Geology and Tectonics Book Series, vol. 5. Series Editor: Mukherjee, S. (ed. A. Billi and A. Fagereng), 205–221. Elsevier. ISSN: 2542-9000. ISBN: 9780128140482.

Dasgupta, B.M., Mukherjee, S., and Chatterjee, R. (2022). Depositional system, morphological signatures, tectonics and sedimentation pattern along the transform margin- Palar-Pennar basin, Indian east coast. *Journal of Petroleum Science & Engineering* 211: 110155. https://doi.org/10.1016/j.petrol.2022.110155.

Gogoi, M.P., Gogoi, B., and Mukherjee, S. (2022). Tectonic instability of the petroliferous upper Assam valley (NE India): a geomorphic approach. *Journal of Earth System Science* 131: 18. https://doi.org/10.1007/s12040-021-01752-6.

Kaplay, R.D., Md, B., Mukherjee, S., and Kumar, T.V. (2017). Morphotectonic expression of geological structures in eastern part of south East Deccan volcanic province (around Nanded, Maharashtra, India). In: *Tectonics of the Deccan Large Igneous Province* (ed. S. Mukherjee, A.A. Misra, G. Calvès and M. Nemčok), 317–335. London, Special Publications 445: Geological Society.

Misra, A.A. and Mukherjee, S. (2022). Introduction to "atlas of structural geological and geomorphological interpretation of remote sensing images". In: *Atlas of Structural Geological and Geomorphological Interpretation of Remote Sensing Images* (ed. A.A. Misra and S. Mukherjee). Wiley Blackwell. ISBN: 9781119813354.

Misra, A.A., Bhattacharya, G., Mukherjee, S., and Bose, N. (2014). Near N-S paleo-extension in the western Deccan region in India: does it link strike-slip tectonics with India-Seychelles rifting? *International Journal of Earth Sciences* 103: 1645–1680.

Quigley, M.C., Bastin, S., and Bradley, B.A. (2013). Recurrent liquefaction in Christchurch, New Zealand, during the Canterbury earthquake sequence. *Geology* 41 (4): 419–422. https://doi.org/10.1130/G33944.1.

Socquet, A., Simons, W., Vigny, C. et al. (2006). Micro block rotations and fault coupling in SE Asia triple junction (Sulawesi, Indonesia) from GPS and earthquake slip vector data. *Journal of Geophysical Research* 111 (B8): B08409. https://doi.org/10.1029/2005JB003963.

Socquet, A., Hollingsworth, J., Pathier, E., and Bouchon, M. (2019). Evidence of super shear during the 2018 magnitude 7.5 Palu earthquake from space geodesy. *Nature Geoscience* 12 (3): 192–199. https://doi.org/10.1038/s41561-018-0296-0.

Song, X., Zhang, Y., Shan, X. et al. (2019). Geodetic observations of the 2018 Mw 7.5Sulawesi earthquake and implications for the kinematics of the Palu fault. *Geophysical Research Letters* 46 (8): 4212–4220. https://doi.org/10.1029/2019GL082045.

Stevens, C., McCaffrey, R., Bock, Y. et al. (1999). Rapid rotations about a vertical axis in a collisional setting revealed by the Palu fault, Sulawesi, Indonesia. *Geophysical Research Letters* 26 (17): 2677–2680. https://doi.org/10.1029/1999GL008344.

Watkinson, I.M. and Hall, R. (2019). Impact of communal irrigation on the 2018 Palu earthquake-triggered landslides. *Nature Geoscience* 12 (11): 940–945. https://doi.org/10.1038/s41561-019-0448-x.

Wu, D., Ren, Z., Liu, J. et al. (2020). Coseismic surface rupture during the 2018 Mw 7.5 Palu earthquake, Sulawesi Island, Indonesia. *GSA Bulletin* https://doi.org/10.1130/B35597.1.

19

Structural and Alteration Mapping Using ASTER Imagery and DEM for Gold Mineralization in the Gadag Schist Belt of Karnataka, India

Nisha Rani[*1], *Venkata R. Mandla*[2], *and Tejpal Singh*[3]

[1] *Geological Survey of India, Hyderabad, Telangana, India*
[2] *CGARD, National Institute of Rural Development and Panchayati Raj, Ministry of Rural Development, Government of India, Hyderabad, Telangana, India*
[3] *CSIR-Central Scientific Instruments Organisation, Chandigarh, India*

19.1 Introduction

Remote sensing has found versatile use in geosciences such as economic geology, structural geology and geomorphology (Misra et al. 2014; Dasgupta and Mukherjee 2017, 2019; Kaplay et al. 2017; Dasgupta et al. 2022; Gogoi et al. 2022). Mineralization is basically controlled by lithology, structures, and hydrothermal minerals. Satellite imageries are useful for mineral exploration in mapping geology, faults and fractures that localize ore deposits, and hydrothermally altered rocks/minerals (Sabins 1999). The purpose of this study was to map hydrothermally altered rocks/minerals and structures that may be associated with gold mineralization in the Gagad Schist Belt (GSB). Hydrothermal alteration zones are considered important guides for mineral exploration and can act as potential zones of gold mineralization (Rani et al. 2020). Structures play an important role in mineral exploration and structural mapping includes faults, folds, synclines and anticlines, and lineaments. The formation of hydrothermal mineral deposit is usually the consequence of interaction between numerous important processes including structural, hydrological, thermal, and geochemical. Recognition of key structural features controlling mineralization can lead to the determination of exploration targets and the discovery of ore bodies (Zhang et al. 2011). The GSB is the northern extremity of Chitradurga schist belt of Western Dharwar Craton (Figure 19.1), mostly occupied by variants of metabasalts, metavolcanics, granite gneiss, conglomerate, greywacke, argillic, and banded iron formation rocks. In the GSB, gold mineralization is associated with several prominent shears, striking NNW–SSE (Vasudev 2009; Rani et al. 2016) and is controlled by shears and marked by wall rock alterations, like chloritization, biotitisation, sericitization, carbonatisation, and silification (Beeraiah and Kumar 2010). Structures and pressure shadow zones wherever they have intersected different lithounits have played an important role in hosting gold mineralization in the GSB. It appears that in the GSB, the gold mineralization is not hosted by any particular lithology, instead distributed in the diversified host rocks (Ugarkar and Deshpande 1999).

For a long time, remote sensing techniques have been used for mineral exploration as indirect tools to map the lithology, structures, and alteration minerals, and they are very helpful in mapping altered minerals (Rajesh 2004; Waldhoff et al. 2008). The alteration minerals have characteristic absorption features from 0.3 to 1.0 μm in visible–near infrared (VNIR) and from 2.0 to 2.5 μm in shortwave infrared (SWIR) wavelength regions (Rani et al., 2017a, 2017b, 2020). The Advanced Spaceborne Thermal Emission and Reflection Radiometer (ASTER) multispectral data have the potential to identify mineral assemblages in different hydrothermal alteration zones because of its 14 spectral bands, which cover a wide spectral range. Therefore, to map the alteration zones and structures in the area, ASTER L1B, data acquired from https://earthdata.nasa.gov, and the Digital Elevation Model (DEM) from https://earthexplorer.usgs.gov were processed. False color composite (FCC) band ratioing, and minimum noise fraction (MNF) techniques were applied to enhance alterations. Lineament, basin, and drainage were derived from the DEM data.

Figure 19.2 shows unprocessed ASTER L1B data. To map the alteration zones, FCC and different band ratio techniques were applied (Figures 19.3–19.5). Figure 19.6

*Corresponding Author: nishargsi@gmail.com

Atlas of Structural Geological and Geomorphological Interpretation of Remote Sensing Images, First Edition.
Edited by Achyuta Ayan Misra and Soumyajit Mukherjee.

Companion website: www.wiley.com/go/misra/RemoteSensingImages

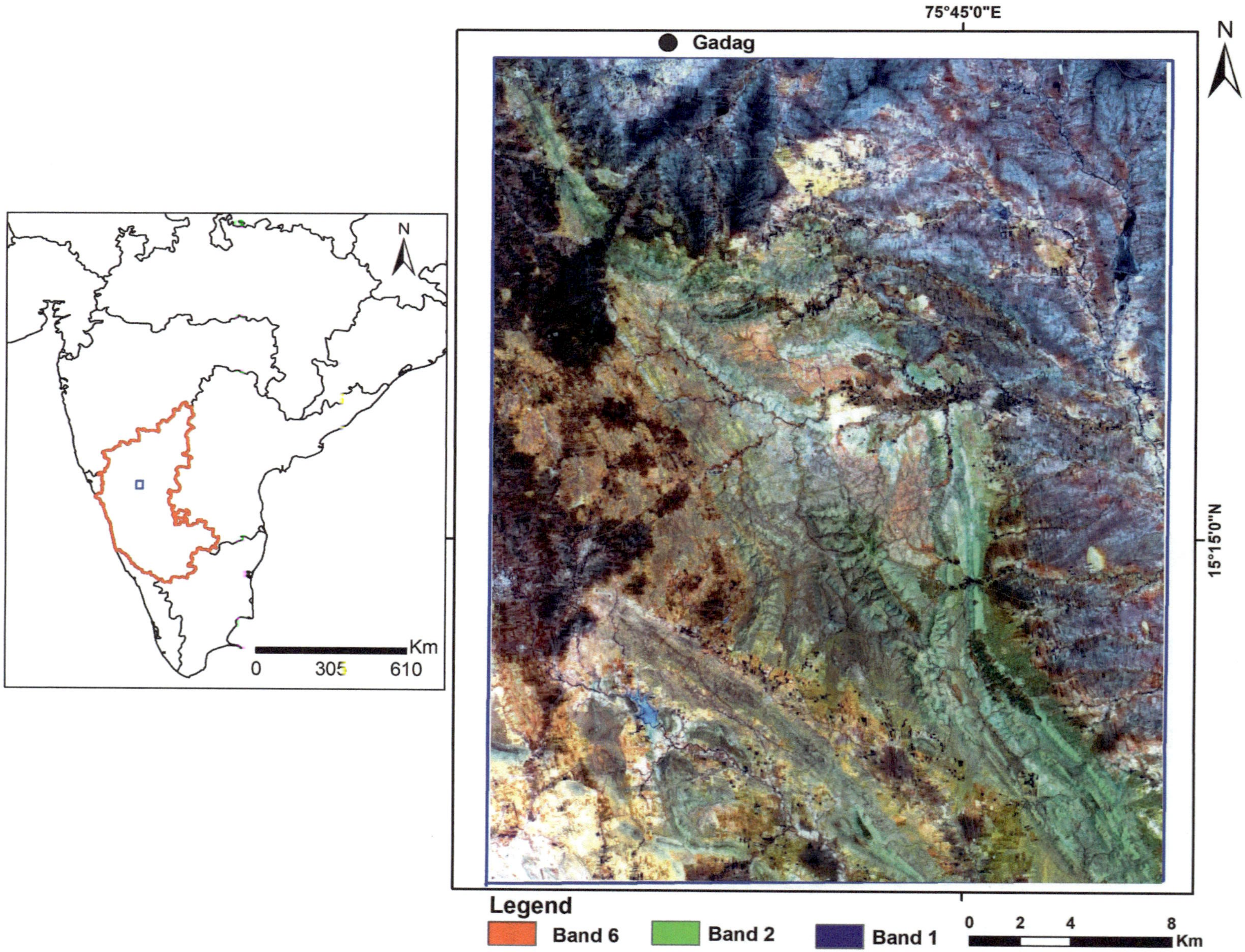

Figure 19.1 Location of Gadag Schist belt of Western Dharwar Craton shown on georeferenced ASTER image (RGB-B6: B2: B1).

Figure 19.2 Unprocessed ASTER satellite imagery (RGB-B1: B2: B3) of Gadag Schist belt of Western Dharwar Craton, India.

Figure 19.3 FCC of ASTER bands 4 : 6 : 8 (RGB) shows argillic (pink, green), phyllic (yellowish-green), and propylitic alteration (light blue). Al-OH minerals (kaolinite, muscovite, and montmorillonite) have the most reflection in band 4 of the SWIR region. The phyllic zone (muscovite/sericite) shows an intense Al–OH absorption feature at 2.20 μm (ASTER band 6), and a less intense feature near 2.38 μm (ASTER band 8). In the propylitic zone, reflectance spectra of the mineral assemblages are characterized by Fe, Mg–OH absorption features, and CO_3 (epidote, chlorite, and carbonate) at 2.35 μm (ASTER band 8) region.

Figure 19.4 The band ratio 4/6 enhances the argillic, band 4/7 to phyllic, and band 3/1 to Fe-oxide minerals. FCC of band ratios 4/6 : 4/7 : 3/1(RGB) indicate argillic and phyllic hydrothermal alterations in yellowish and orange to pink, and iron oxides in light blue.

Figure 19.5 The band ratio 4/6 enhances the argillic, band 5/6 enhances the phyllic alteration, and band 5/8 enhances the propylitic alteration. In FCC of band ratios 4/6 : 5/6 : 5/8 (RGB), argillic to phyllic alterations are highlighted in rust and orange and propylitic alteration inlight blue to magenta.

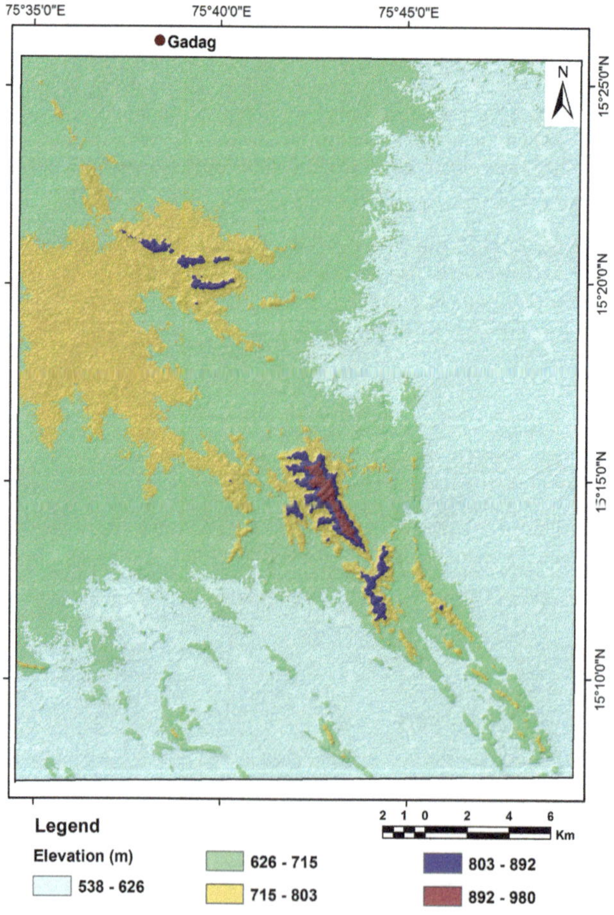

Figure 19.6 DEM data superimposed on the hillshade map for better visualization and interpretation of structures in the area.

is the DEM of the area, superimposed on hillshade map. Figures 19.7 and 19.8 are the MNF images of the ASTER highlighting structures. Figure 19.9 is a DEM derived drainage & lineament map. Figure 19.10 shows correlation structures, alteration, and gold mineralization. Figure 19.11 shows statistical value of the distance of gold mineralization from the lineaments calculated at the interval of 50 m. Three gold mineralizations occur within 50 m and four are in the 100 m range.

Figure 19.7 FCC image of MNF bands 4, 2, and 3 in RGB, enhancing the structures, lithological contact, and geology of the GSB.

Figure 19.8 FCC image of MNF bands 2, 4, and 5 in RGB, enhancing the structures, lithological contact, and geology of the GSB.

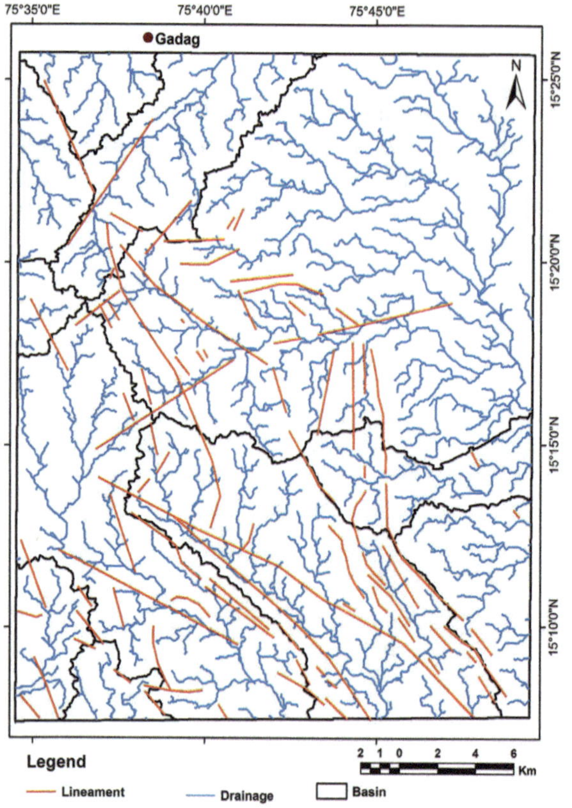

Figure 19.9 Drainage pattern and lineaments of GSB, derived from DEM and ASTER images. Lineaments (lithological contact, linear to sub-linear features, ridge of the hill top, fault and fracture, etc.) are identified by interpreting FCC image 4/6 : 4/7 : 3/1 (RGB); 4/6 : 5/6 : 5/8 (RGB), and FCC image of MNF bands 4 : 2 : 3 and 2 : 4 : 5 (RGB). Where the lineaments correspond to displacement of rock units or sudden change in drainage pattern or rock type they may be related to faulting.

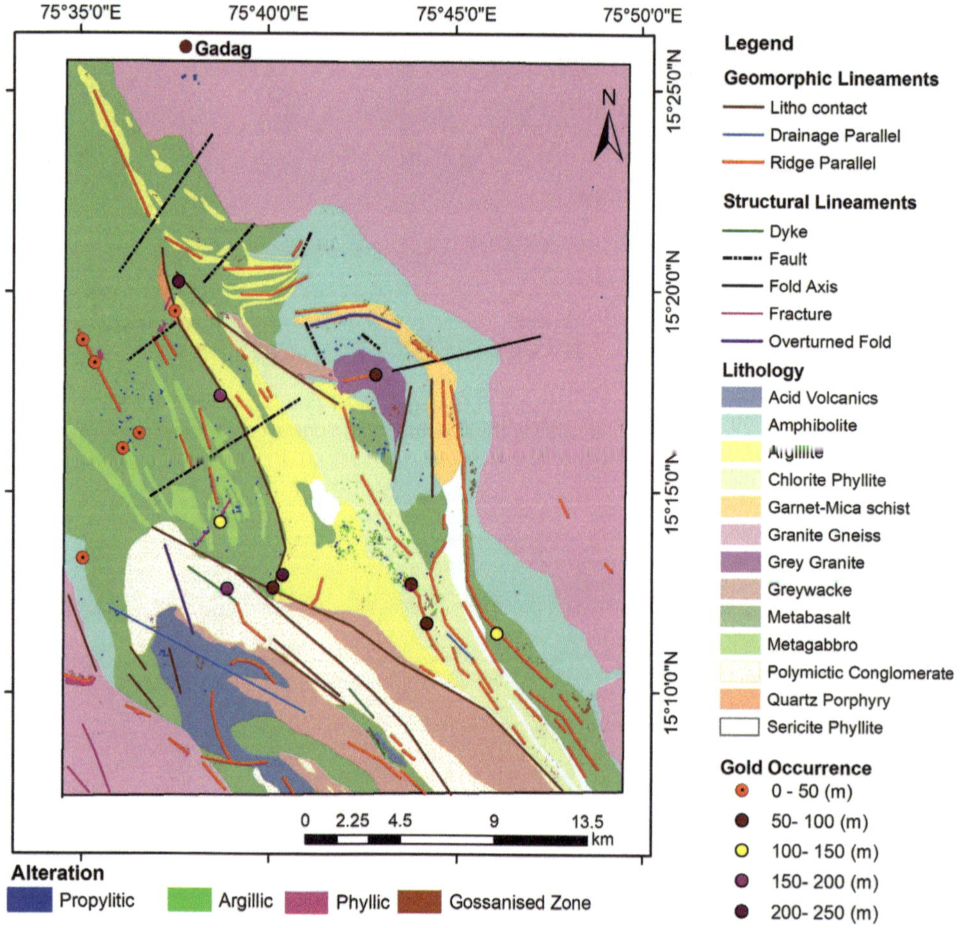

Figure 19.10 The derived lineaments (geomorphic, structural), drainage, alteration zones were superimposed on a geological map (by Geological Survey of India) in a GIS environment, in order to study the relation of the structures, alteration with gold mineralization of the study area. Map shows the gold mineralization is controlled by structures associated with alteration. Occurrence of gold is defined in five classes based on their relation with respect to distance of lineaments. Source: The map is modified from the URL: http://GSI.gov.in/bhukosh.

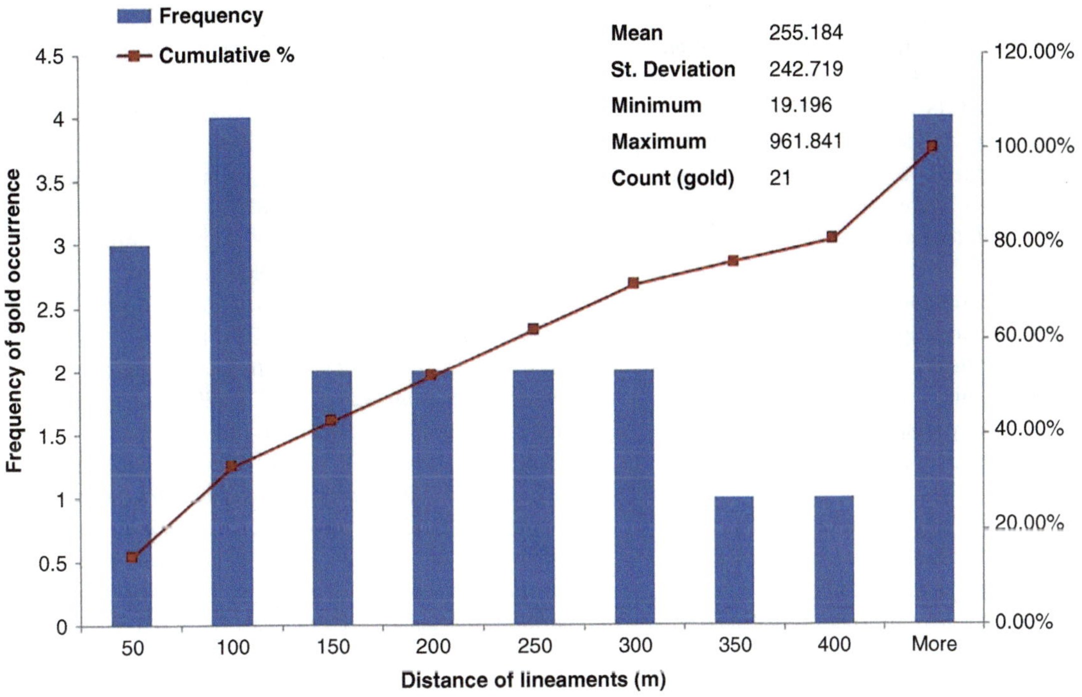

Figure 19.11 Statistical value of the distance of gold mineralization from the lineaments calculated at the interval of 50 m. The histogram graph shows lineament distance (in m) on x-axis and frequency of gold occurrence on y-axis. Three gold mineralizations are occurring within 50 m and four in 100 m range.

Acknowledgements

Soumyajit Mukherjee invited to submit an article in the edited book. Achyuta Ayan Misra handled this article. Wiley Blackwell Acquisition Editors (present: Frank Weinreich, past: Andrew Harrison), Stacey Woods (Handling Editor), and proofreading team. Summarized in Misra and Mukherjee (2022).

References

Beeraiah, M.B. and Kumar, C.G.H. (2010). Report on exploration for gold in Nagavi sector, Gadag schist belt, Gadag district, Karnataka (E-I SATGE). Geological Survey of India, Bangalore (Unpublished Report).

Dasgupta, S. and Mukherjee, S. (2017). Brittle shear tectonics in a narrow continental rift: asymmetric non-volcanic Barmer basin (Rajasthan, India). *The Journal of Geology* 125: 561–591.

Dasgupta, S. and Mukherjee, S. (2019). Remote sensing in lineament identification: examples from western India. In: *Problems and Solutions in Structural Geology and Tectonics*, Developments in Structural Geology and Tectonics Book Series, vol. 5. Series Editor: Mukherjee, S (ed. A. Billi and A. Fagereng), 205–221. Elsevier. ISSN: 2542-9000. ISBN: 9780128140482.

Dasgupta, B.M., Mukherjee, S., and Chatterjee, R. (2022). Depositional system, morphological signatures, tectonics and sedimentation pattern along the transform margin – Palar-Pennar basin, Indian east coast. *Journal of Petroleum Science & Engineering* 211: 110155. https://doi.org/10.1016/j.petrol.2022.110155.

Gogoi, M.P., Gogoi, B., and Mukherjee, S. (2022). Tectonic instability of the petroliferous upper Assam valley (NE India): a geomorphic approach. *Journal of Earth System Science* 131: 18. https://doi.org/10.1007/s12040-021-01752-6.

Kaplay, R.D., Md, B., Mukherjee, S., and Kumar, T.V. (2017). Morphotectonic expression of geological structures in eastern part of south East Deccan volcanic province (around Nanded, Maharashtra, India). In: *Tectonics of the Deccan Large Igneous Province* (ed. S. Mukherjee, A.A. Misra, G. Calvès and M. Nemčok), 317–335. London, Special Publications 445: Geological Society.

Misra, A.A. and Mukherjee, S. (2022). Introduction to "atlas of structural geological and geomorphological interpretation of remote sensing images". In: *Atlas of Structural Geological and Geomorphological Interpretation of Remote Sensing Images* (ed. A.A. Misra and S. Mukherjee). Wiley Blackwell. ISBN: 9781119813354.

Misra, A.A., Bhattacharya, G., Mukherjee, S., and Bose, N. (2014). Near N-S paleo-extension in the western Deccan region in India: does it link strike-slip tectonics with India-Seychelles rifting? *International Journal of Earth Sciences* 103: 1645–1680.

Rajesh, H.M. (2004). Application of remote sensing and GIS in mineral resource mapping – an overview. *Journal of Mineralogical and Petrologicl Sciences* 99: 83–103.

Rani, N., Mandla, V.R., and Singh, T. (2016). Performance of image classification on hyperspectral imagery for lithological mapping. *Journal of the Geological Society of India* **88** (4): 440–448.

Rani, N., Mandla, V.R., and Singh, T. (2017a). *Spatial Distribution of Altered Minerals in the Gadag Schist Belt (GSB) of Karnataka, Southern India Using Hyperspectral Remote Sensing Data*. Geocarto International https://doi.org/10.1080/10106049.2015.1132484.

Rani, N., Mandla, V.R., and Singh, T. (2017b). Evaluation of atmospheric corrections on hyperspectral data with special reference to mineral mapping. *Geoscience Frontiers* 8 (4): 797–808. https://doi.org/10.1016/j.gsf.2016.06.004.

Rani, N., Singh, T., and Mandla, V.R. (2020). Mapping hydrothermal alteration zone through aster data in Gadag Schist Belt of Western Dharwar Craton of Karnataka, India. *Environmental Earth Sciences* 79: 526. https://doi.org/10.1007/s12665-020-09269-9.

Sabins, F.F. (1999). Remote sensing for mineral exploration. *Ore Geology Reviews* 14: 157–183.

Ugarkar, A.G. and Deshpande, M.P. (1999). Lithology and gold mineralisation of Gadag gold field, Dharwar Craton; evidences for epigenesis of gold in diversified host rocks. *Indian Mineralogist* 33 (1): 37–52.

Vasudev, V.N. (2009). *Field Guide to Selected Gold Prospects in Karnataka and Andhra Pradesh*, 1–56. Bangalore: Geological Society of India.

Waldhoff, G., Bubenzer, O., Bolten, A. et al. (2008). Spectral analysis of ASTER, Hyperion, and Quickbird data for geomorphological and geological research in Egypt (DakhlaAsis, Western Desert). *The International Archives of the Photogrammetry, Remote Sensing and Spatial Information Sciences* XXXVII (Part B8): 1201–1206.

Zhang, Y., Robinson, J., and Schaubs, P.M. (2011). Numerical modelling of structural controls on fluid flow and mineralization. *Geoscience Frontiers* 2 (3): 449–461. https://doi.org/10.1016/j.gsf.2011.05.011.

20

Identifying Subtle Deformation Structures from Satellite Images in Parts of the Mesozoic Kachchh (Kutch) Basin, Kachchh District, Gujarat, India

Achyuta Ayan Misra[*1], Arijit Ghosh[1], and Atul Kumar Patidar[2]*

[1] Reliance Industries Ltd., Reliance Corporate Park, Mumbai, Maharashtra, India
[2] Department of Petroleum Engineering and Earth Sciences, University of Petroleum and Energy Studies, Dehradun, Uttarakhand, India

20.1 Introduction

Remote sensing has been an excellent tool to interpret structural geology and geomorphology (e.g. Misra et al. 2014; Kaplay et al. 2017; Dasgupta et al. 2022; Gogoi et al. 2022). The Kachchh (Kutch) Basin is a seismically active pericratonic rift basin on the western continental margin in Gujarat, India (Dasgupta and Mukherjee 2017, 2019; Shaikh et al. 2020). It was classified as a volcanic passive margin (Biswas 2014). It is bounded by the Nagar Parkar Hill on the north, Little Rann on the southeast, the Gulf of Kachchh on the south, and the Arabian Sea to the west. This basin has a unique landscape with highlands surrounded by extensive mudflats with salt covered white patches (*salinas*). Most of the highlands (structural uplifts) here expose Jurassic rocks while the surrounding plains (structural lows) are covered by Tertiary, Quaternary, and Rann sediments. During the marine incursion the uplifts were separate islands. There are five such islands viz. Pachchham, Khadir, Bela, Chorar, and Wagad. The first four are collectively called the "Island Belt" since they occur in one E–W linear belt toward the northernmost limit of the basin (Figure 20.1).

20.2 Regional Geology

The Kachchh Basin exposes a thick sequence of Mesozoic and Cenozoic sedimentary rocks with an intervening phase of the Deccan Traps (Biswas 1977, 1978). The tectonic framework comprises domal structures of Mesozoic rocks as a result of several episodes of uplift and bounded by E–W trending intrabasinal faults. The major intrabasinal master faults are the Island Belt Fault (IBF), Kachchh Mainland Fault (KMF), Katrol Hill Fault (KHF), South Wagad Fault (SWF), and Gedi Fault (GF) (Figure 20.1). The Mesozoic sequence comprises the Jhurio (Bajocian to Bathonian), Jumara (Callovian to Oxfordian), Jhuran (Late Jurassic to Early Cretaceous), and Bhuj (Tithonian to Albian) Formation. The key type sections of these sequences are exposed mainly in three areas: the Island Belt, the Kachchh Mainland, and the Wagad Uplift.

The Kachchh rift opened during the Late Triassic–Early Jurassic and witnessed continuous sedimentation until the Santonian. The Meoszoic rift sedimentation in Kachchh Basin was followed by the Deccan Trap eruptive phase (Vanik et al. 2018). All known forms of intrusive bodies are present and are mainly concentrated in the narrow deformation zones accompanying the master faults. The post-Deccan Trap inversion of the rift resulted in the formation of intra-basinal uplift with corresponding structural lows due to movement along faults (Biswas and Khattri 2002).

In this work, we show how remote sensing can help in the interpretation of subtle deformation structures. The interpretation on the remote sensing data must always be accompanied by a field work to support ground evidence to the interpretations. The images presented here in Figures 20.2–20.15 are high-resolution (<1 m) imagery provided by the ESRI ArcMap software package. These images are sourced from Digital Globe, CNES Airbus, GeoEye, Earthstar Geographics, etc. The date of acquisition is not provided with the imagery. There are two Google Earth images used here, which are sourced from Maxar Technologies and/or CNES Airbus. The date of acquisition information is presented with the imagery and is mentioned on individual images here.

*Corresponding Author: achyutaayan@gmail.com

Atlas of Structural Geological and Geomorphological Interpretation of Remote Sensing Images, First Edition.
Edited by Achyuta Ayan Misra and Soumyajit Mukherjee.

Companion website: www.wiley.com/go/misra/RemoteSensingImages

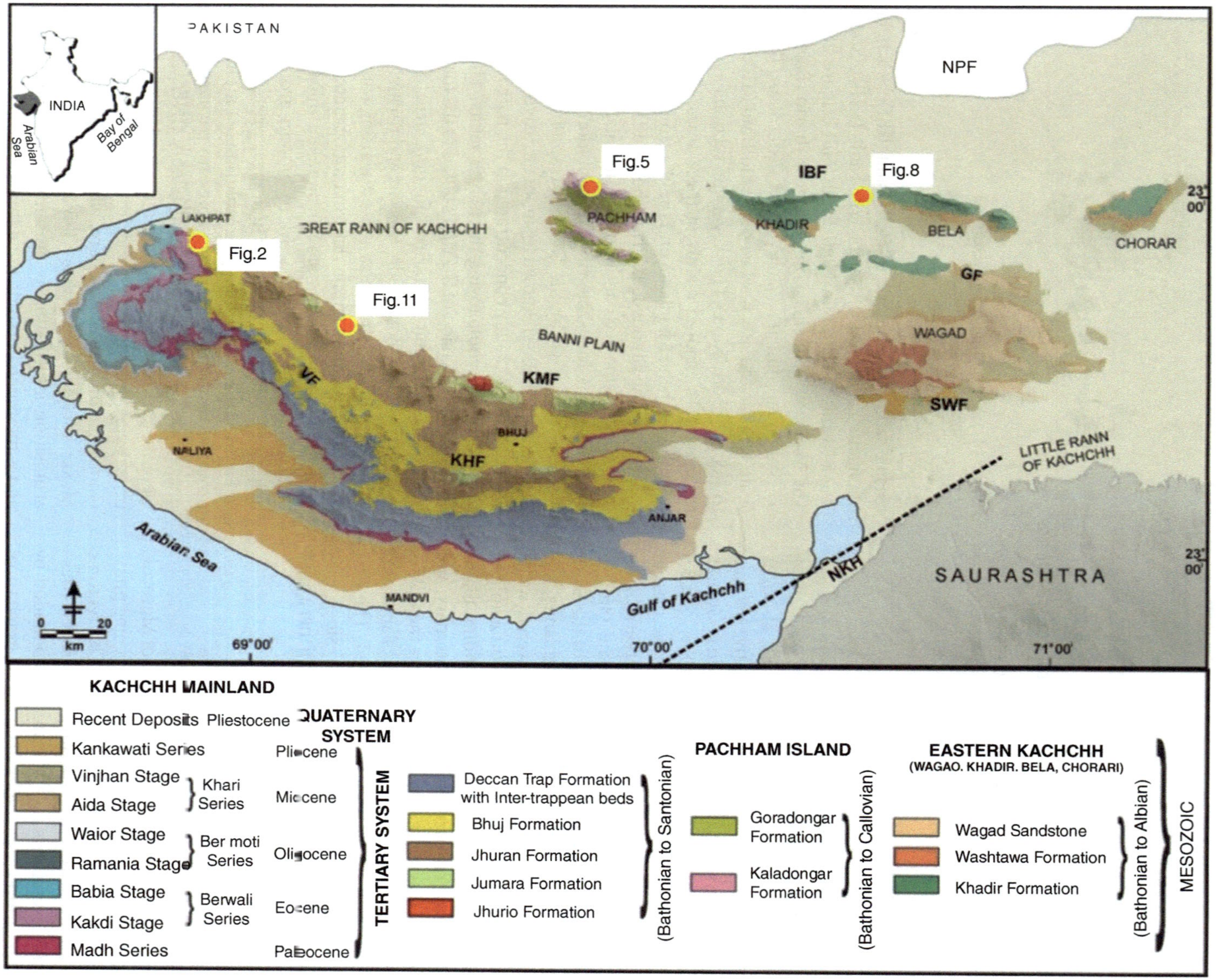

Figure 20.1 Geological map of the Kachchh Basin overlay on Digital Elevation Model. Steep north facing scarp along the Island belt range and highly undulated topography of Mainland area can be seen (Source: reproduced from Maurya et al. (2017)). Red dots: positions of ground truth locations. (Also look at Figure 20.1 in Shaikh et al. 2022 for data compilation).

Figure 20.2 Clean satellite image of Guneri/Ghuneri Dome, westernmost area of Kutch. Source: Image courtesy: ESRI (Environmental Systems Research Institute).

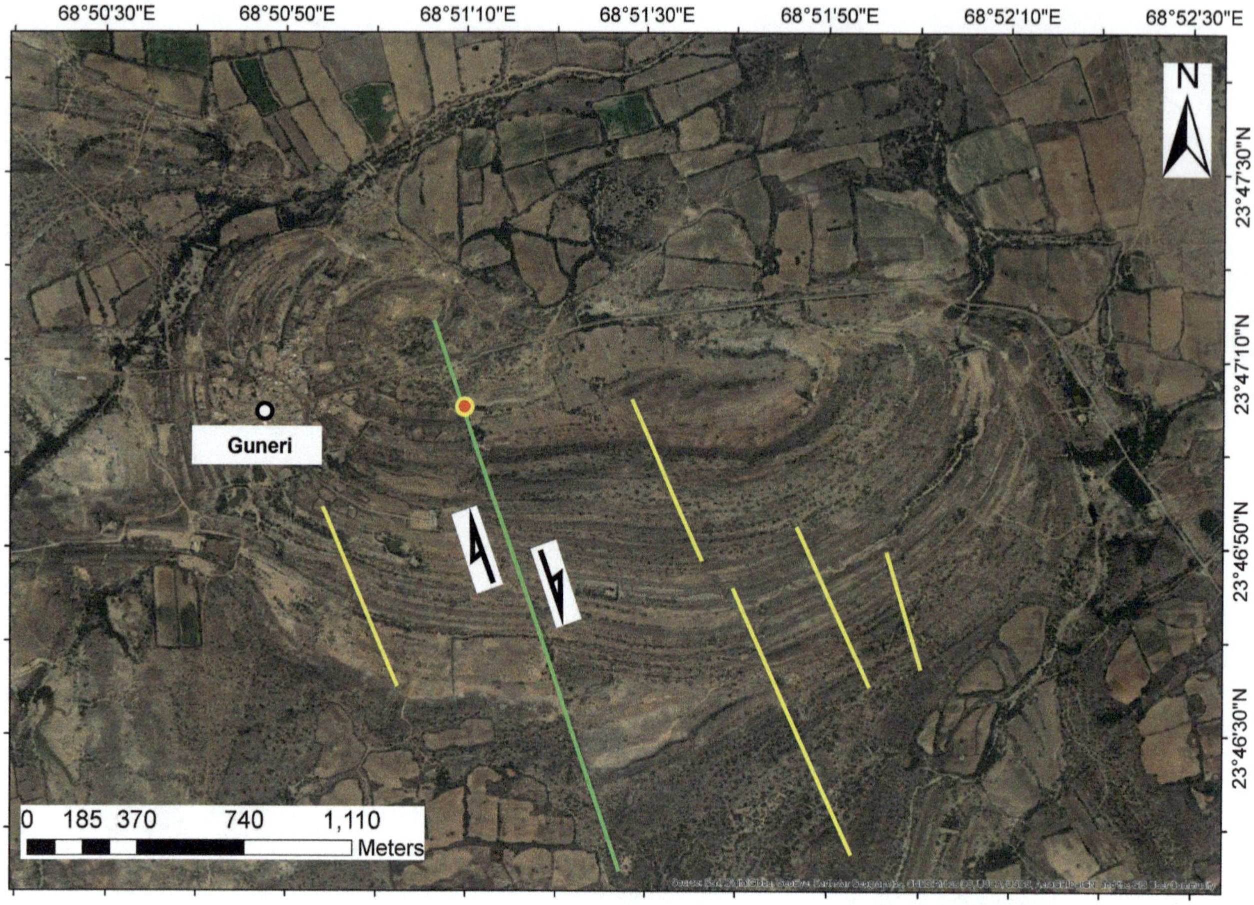

Figure 20.3 Interpretation of the image in Figure 20.2. A number of faults can be interpreted on the image. These faults are dextral, strike-slip to highly oblique-slip in nature. The fault marked in green has the maximum horizontal displacement among the other faults interpreted here (in yellow). The fault marked in green was selected for supporting evidence during field work (Figure 20.4). Red dot: ground truth location.

Figure 20.4 Ground truth location for the fault marked in green in Figure 20.3. (a) Panoramic view of the area and the fault plane. (b) Close up view of the fault plane. There is a clear shear surface on the ground level in the sub-horizontal outcrop and also on the sub-vertical outcrop on the small elevation in the background. The offset is dextral oblique slip. The yellow limestone (Y lst) marker bed is also seen dextrally displaced. Arijit Ghosh stands on the yellow limestone marker bed, which is dextrally displaced.

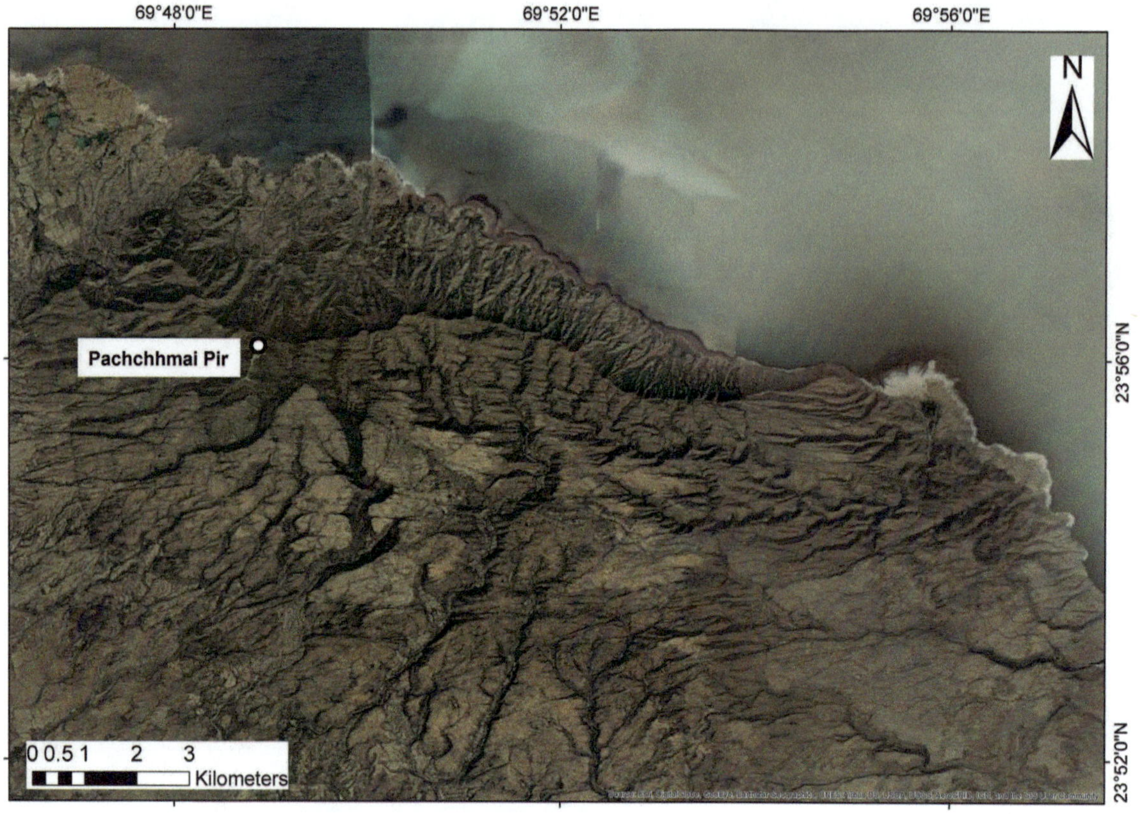

Figure 20.5 Clean satellite image of Pachchham Island, N Kutch. Source: Image courtesy: ESRI.

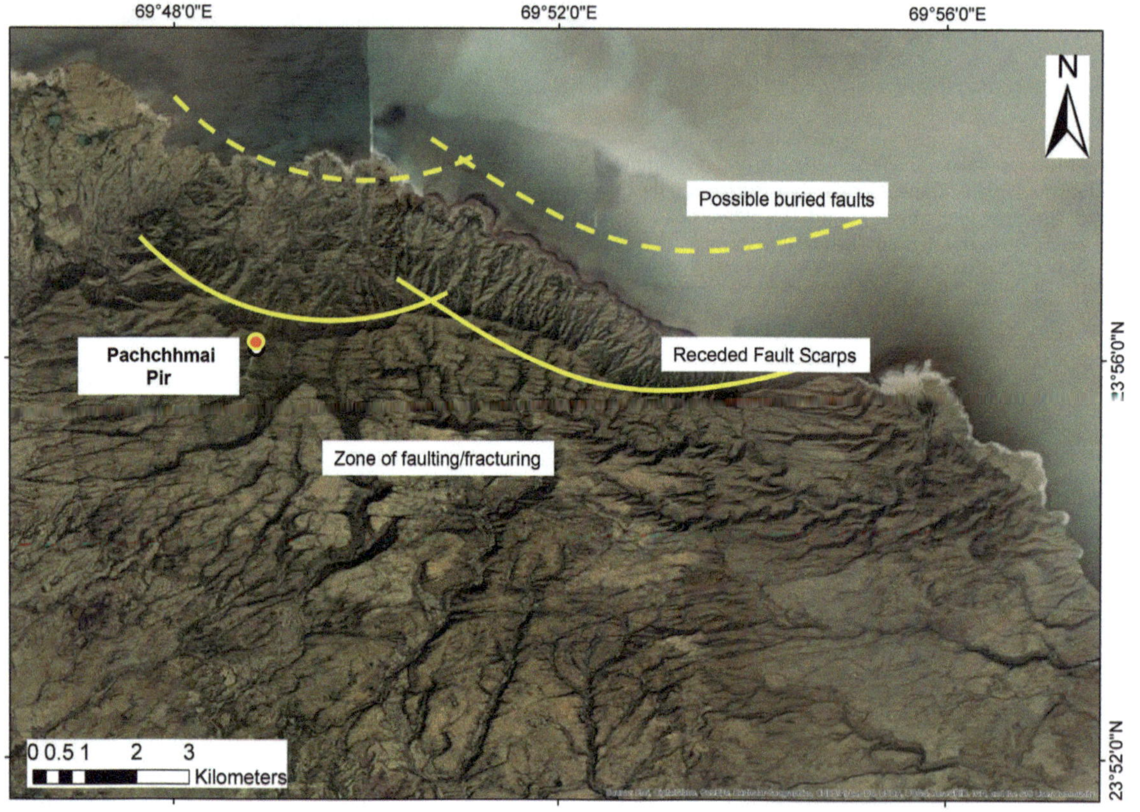

Figure 20.6 Interpretation of the satellite image in Figure 20.5. This cliff is formed by receded fault scarps of the Island Belt Fault, one of the major faults during the earliest rift phase in the Kutch Basin. This area has the highest elevation in the entire Kutch Basin, indicating the strong rift shoulder uplifts during rifting. The buried fault planes may be present in a location in the Great Rann of Kutch, marked here by yellow dotted lines. Such faults can only be demarcated by reflection seismic data (e.g. Misra and Mukherjee 2018). The zone of intense fracturing and faulting can be observed S of the fault scarp, as linear grooves, which deflect streams. Red dot: point from where ground truth photograph was taken. This is the highest point in Kachchh district, known as Pachchhmai Pir, one of the busiest temples in the area.

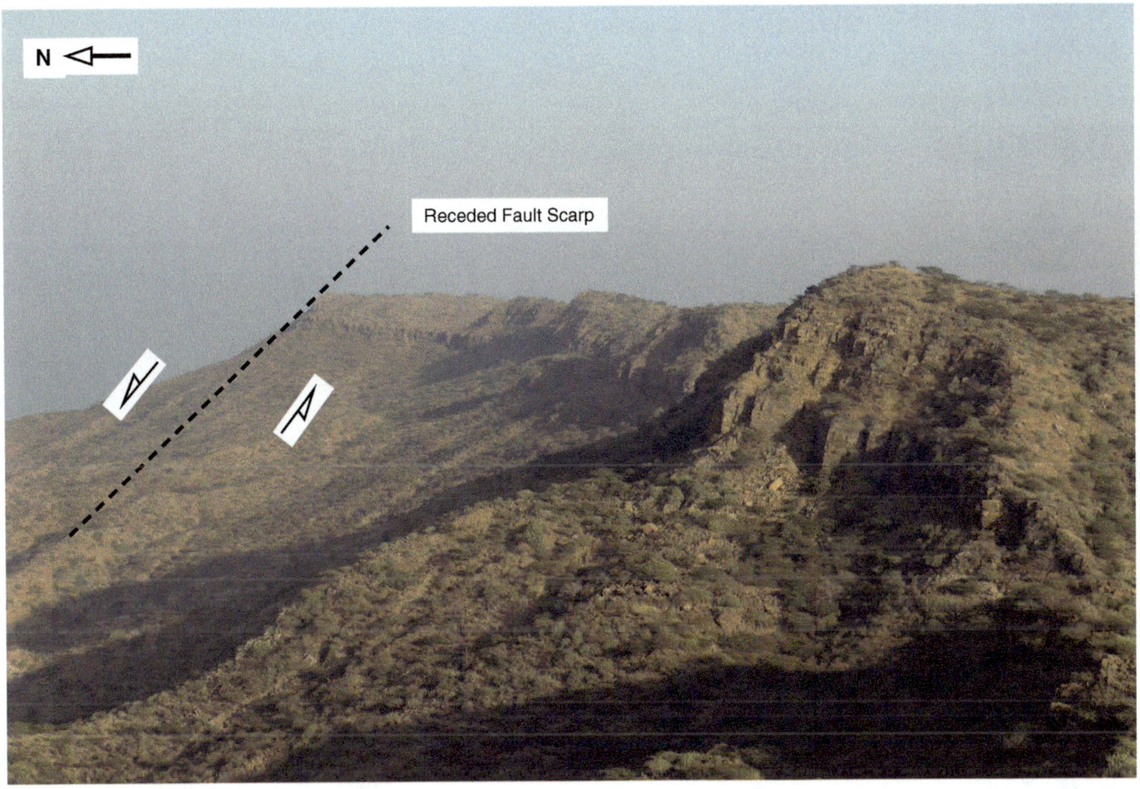

Figure 20.7 Photograph from Pachchhmai Pir viewpoint (red dot in Figure 20.6) looking toward E. The receded fault scarp is marked by a dotted line.

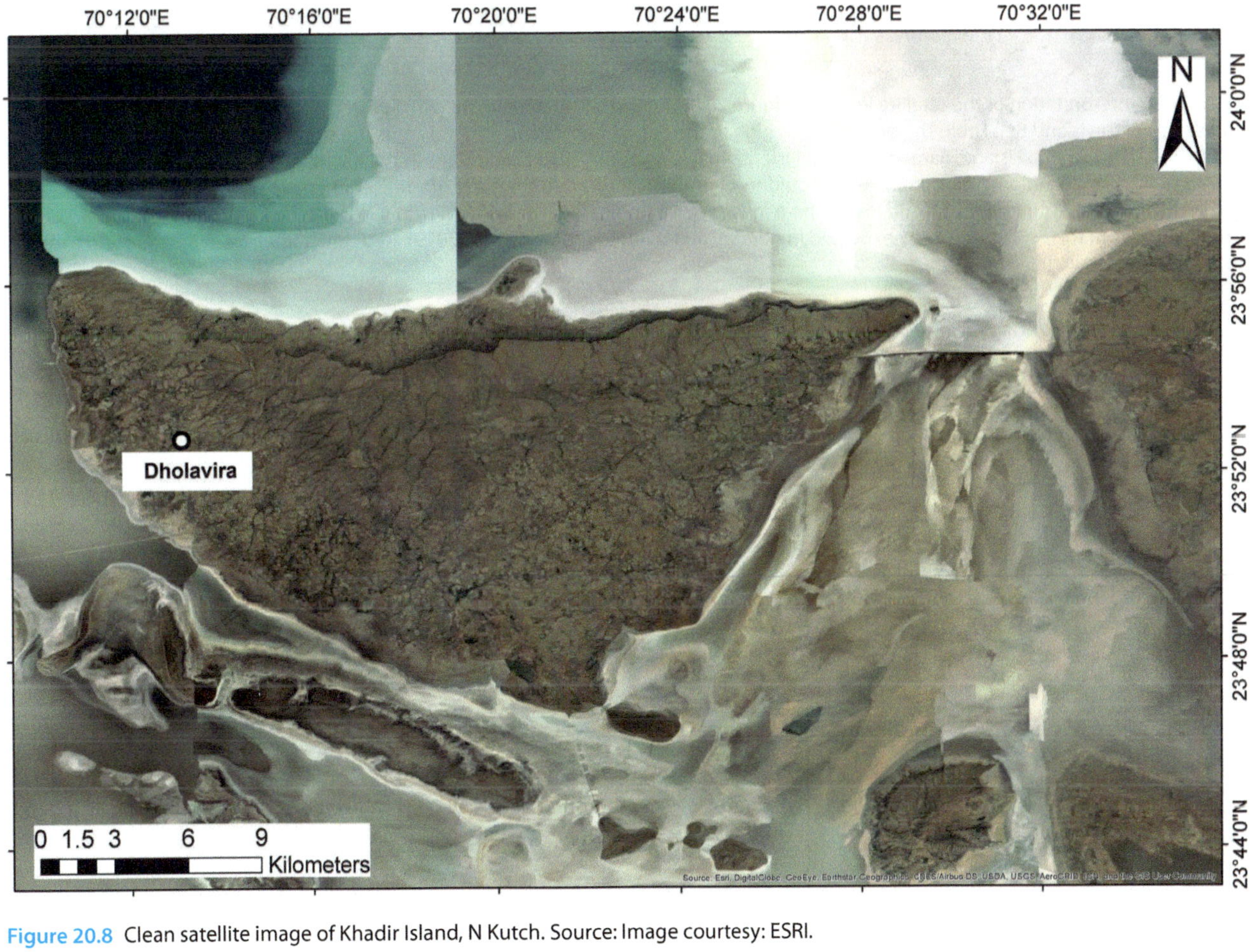

Figure 20.8 Clean satellite image of Khadir Island, N Kutch. Source: Image courtesy: ESRI.

Figure 20.9 Interpretation of the satellite image in Figure 20.8. Similar to the Pachchham cliff (Figure 20.5), this is also formed by receded fault scarps of the Island Belt Fault. This area is not as high as the Pachchham Island, but indicates the significant uplift of the rift shoulders. There is a small island, named Cheriabet Island, that houses the oldest sediments in the Kachchh Basin, the Cheriabet Conglomerate. This conglomerate comprises of boulders of granites, indicating the early syn-rift sediments (Krishna 2017). The cliff of the Khadir Island is highest in the centre and low at the edges, possibly indicating a maximum throw in the centre of the fault and diminishing throws away from the centre, as usually expected in normal fault planes (e.g. Misra et al. 2019). Red dot: point from where ground truth photograph was taken.

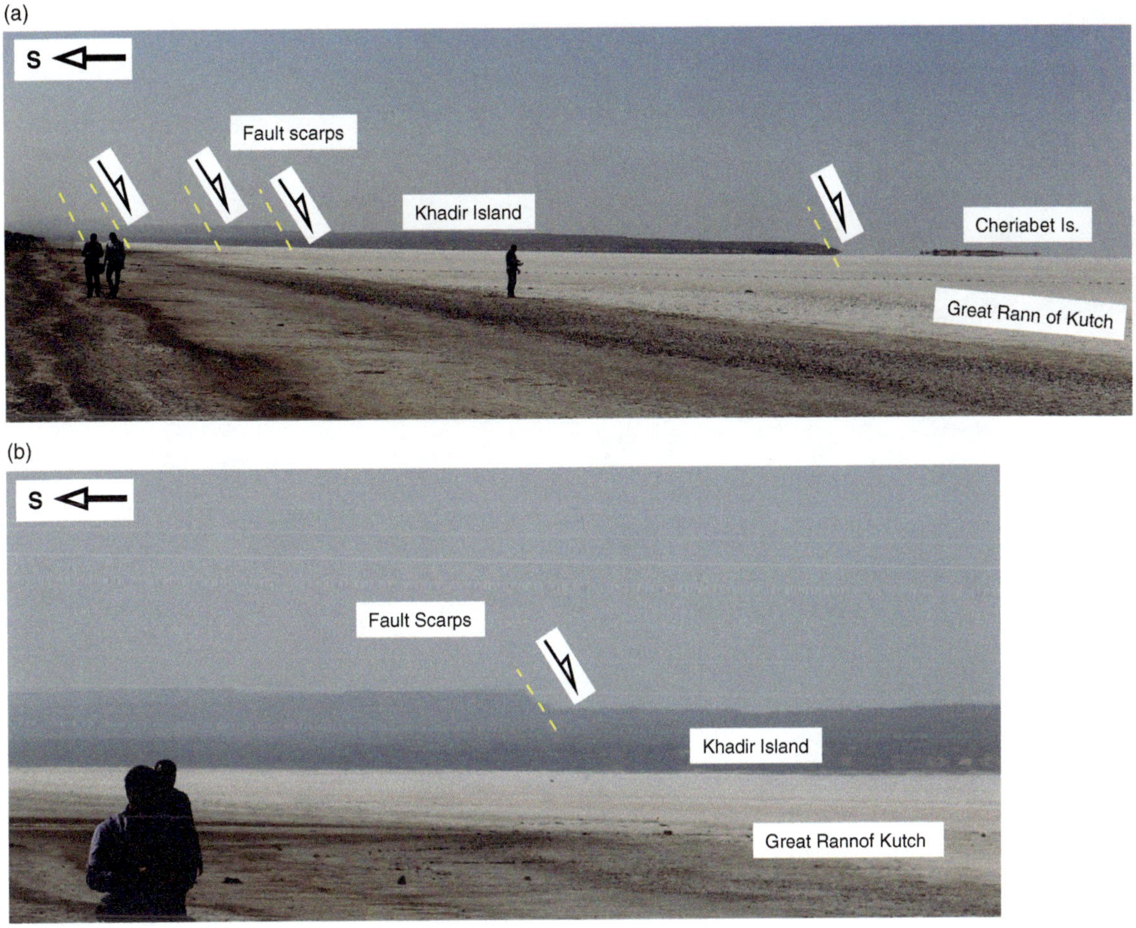

Figure 20.10 Photograph from road connecting Bela Island and Khadir Island (red dot in Figure 20.9) looking toward W. The receded fault scarps are marked by dotted lines. (a) Panoramic view of the Khadir Island and the Cheriabet Island. (b) Closer view of the fault scarps.

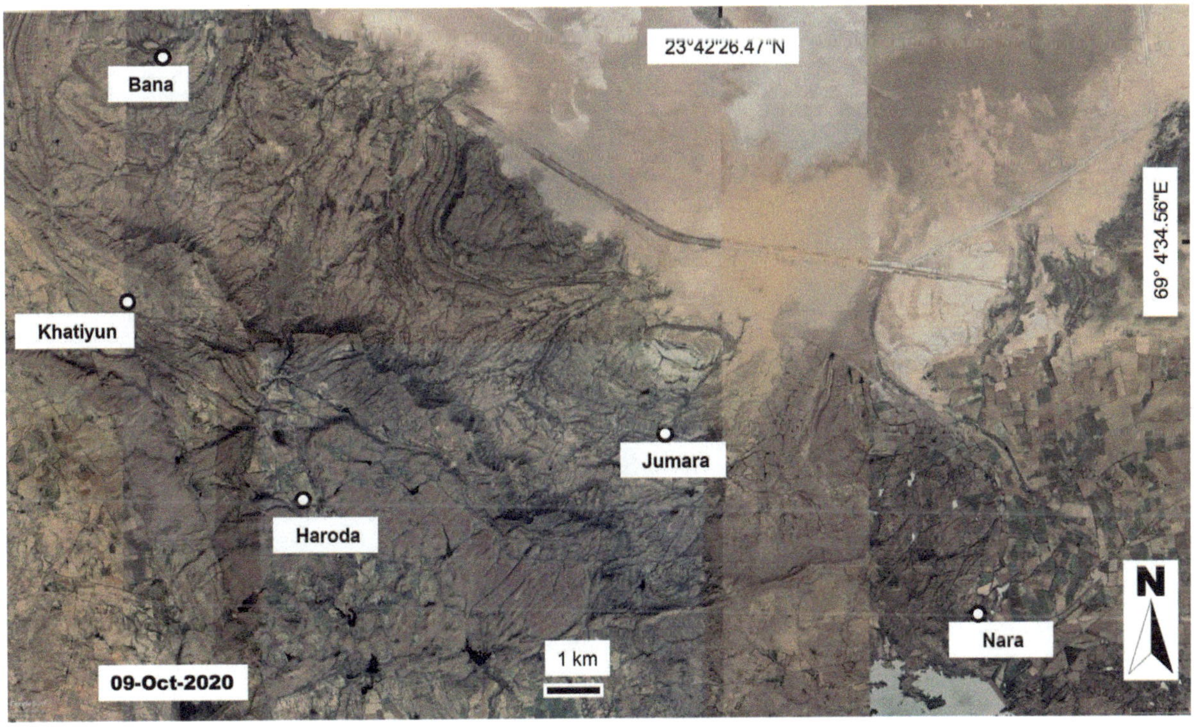

Figure 20.11 Clean satellite image of Jumara Dome area, northern part of Kachchh mainland. Source: Image courtesy: Google Earth.

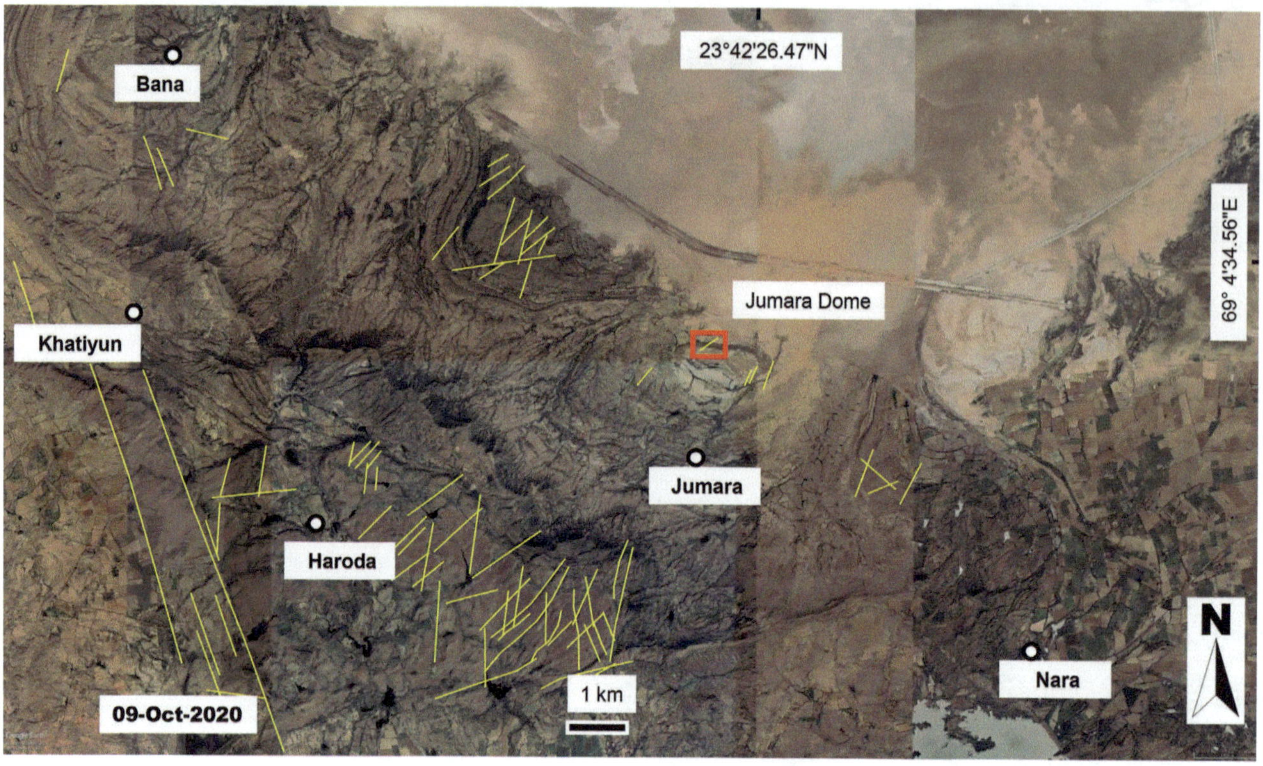

Figure 20.12 Interpretation of the satellite image in Figure 20.11. The possible fault lines on the image are interpreted as lineament by yellow lines. The lineament to the north of the Jumara Dome (marked in red rectangle) was selected for further understanding, shown in Figures 20.13–20.15.

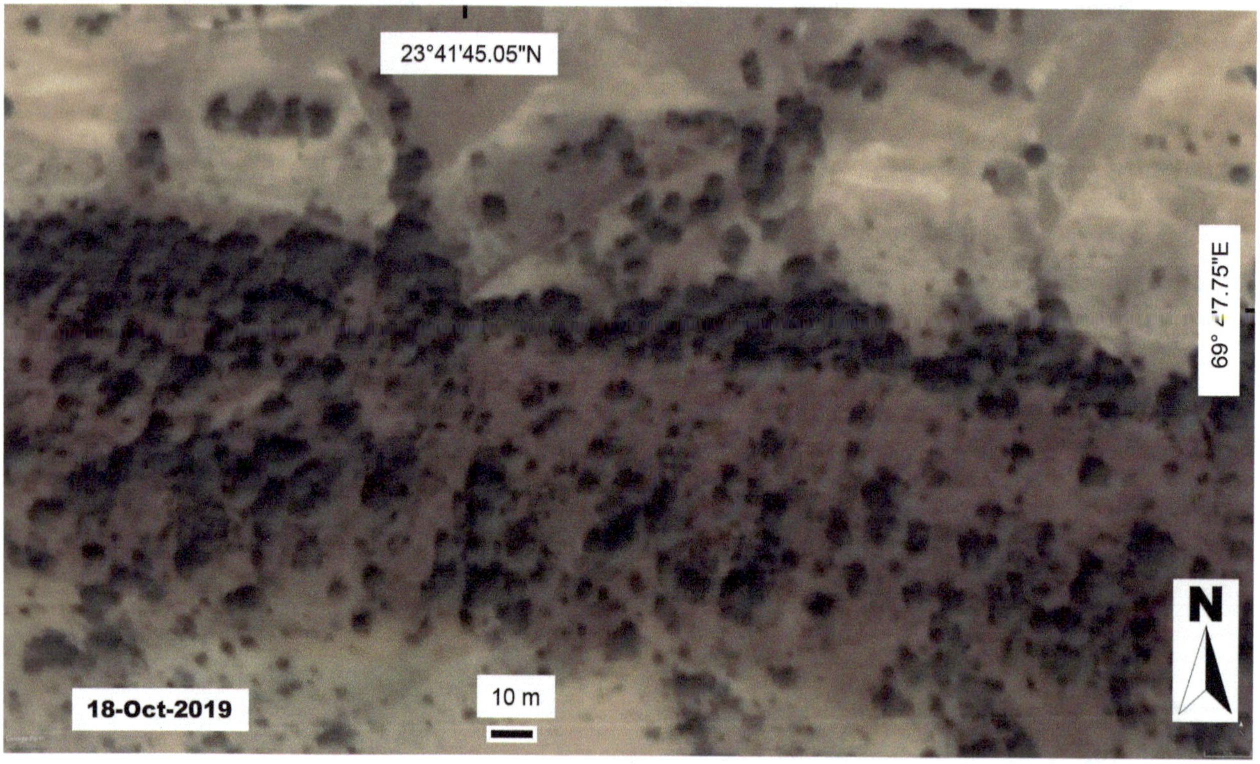

Figure 20.13 Clean satellite image of the northern part of the Jumara Dome, area marked in Figure 20.12 by red rectangle. Source: Image courtesy: Google Earth.

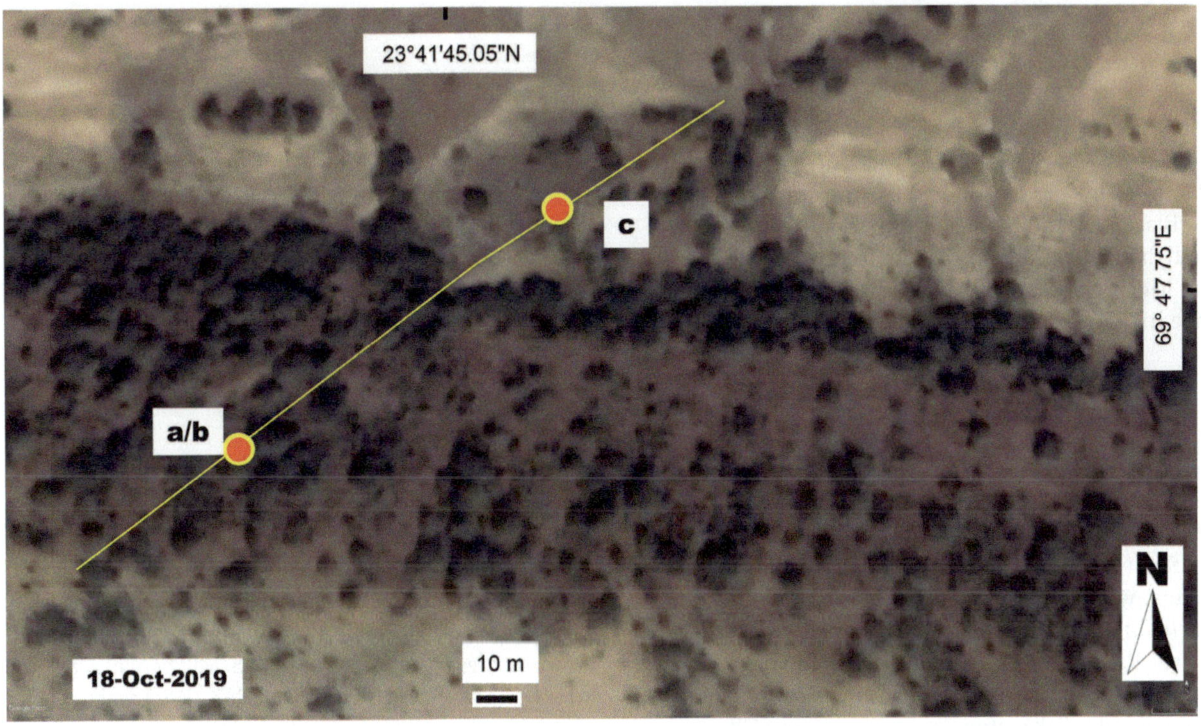

Figure 20.14 Interpretation of the satellite image in Figure 20.13. The possible fault line on the image is interpreted as lineament by yellow line. The lineament appears elevated and could also be a dyke. This was selected to be studied during field work for ground truth support. Red dot: Ground truth locations, corresponding to figure numbers in Figure 20.15.

Figure 20.15 Field photographs for ground truth support for interpretation in Figure 20.14. (a) Slick-en-sided fault plane striking NE–SW, showing sinistral strike-slip shear sense. The dyke can be seen in the background adjacent to the fault plane. Yellow rectangle: area in **b**. (b) Zoomed in view of the fault plane and the slick-en-sides. (c) The dyke, here slightly elevated, to the NE of the area in **a**. The juxtaposition and parallel nature of the fault and the dyke indicated that the dyke may have been emplaced into the fault plane or even that deformation and magmatism were coeval (Misra and Mukherjee 2017).

Acknowledgements

Thanks to Andrew Harrison and the proofreading team (Wiley Blackwell). Soumyajit Mukherjee (IIT Bombay) handled this article and provided minor review comments. Mery Biswas is thanked for her comments. Reliance Industries Ltd. is thanked for funding the fieldworks to Kachchh. Wiley Blackwell Acquisition Editors (present: Frank Weinreich, past: Andrew Harrison), Stacey Woods (Handling Editor), and proofreading team. Summarized in Misra and Mukherjee (2022).

References

Biswas, S.K. (1977). Mesozoic rock-stratigraphy of Kutch, Gujarat. *Quarterly Journal of the Geological Mineralogical and Metallurgical Society of India* 49: 1–51.

Biswas, S.K. (1978). On status of the Bhuj and Umia series of Kutch, W. India. In: *Proceedings of VIIth Indian colloquium micropaleontology and stratigraphy, Madras* (ed. D.A. Rasheed), 6–16. Chennai: Manorama Press.

Biswas, S.K. (2014). Active tectonics of western continental margin of Indo–Pak craton– stress source for SCR earthquakes. *Journal of Earthquake Science and Engineering* 1: 46–58.

Biswas, S.K. and Khattri, K.N. (2002). A geological study of earthquakes in Kutch, Gujarat, India. *Geological Society of India* 60 (2): 131–142.

Dasgupta, S. and Mukherjee, S. (2017). Brittle shear tectonics in a narrow continental rift: asymmetric non-volcanic Barmer basin (Rajasthan, India). *The Journal of Geology* 125: 561–591.

Dasgupta, S. and Mukherjee, S. (2019). Remote sensing in lineament identification: examples from western India. In: *Problems and Solutions in Structural Geology and Tectonics*, Developments in Structural Geology and Tectonics Book Series, vol. 5. Series Editor: Mukherjee, S (ed. A. Billi and A. Fagereng). Elsevier ISSN: 2542-9000.

Dasgupta, S., Biswas, M., Mukherjee, S., and Chatterjee, R. (2022). Depositional system, morphological signatures, tectonics and sedimentation pattern along the transform margin Palar Pennar basin, Indian east coast. *Journal of Petroleum Science & Engineering* 211: 110155. https://doi.org/10.1016/j.petrol.2022.110155.

Gogoi, M.P., Gogoi, B., and Mukherjee, S. (2022). Tectonic instability of the petroliferous upper Assam valley (NE India): a geomorphic approach. *Journal of Earth System Science* 131: 18. https://doi.org/10.1007/s12040-021-01752-6.

Kaplay, R.D., Md, B., Mukherjee, S., and Kumar, T.V. (2017). Morphotectonic expression of geological structures in eastern part of south East Deccan volcanic province (around Nanded, Maharashtra, India). In: *Tectonics of the Deccan Large Igneous Province* (ed. S. Mukherjee, A.A. Misra, G. Calvès and M. Nemčok), 317–335. London, Special Publications 445: Geological Society.

Krishna, J. (2017). Integration of the micro/macro faunal/floral data into Ammonoid stratigraphic framework in the Indian Mesozoics. In: *The Indian Mesozoic Chronicle*, 277–366. Singapore: Springer.

Maurya, D.M., Chowksey, V., Patidar, A.K., and Chamyal, L.S. (2017). A review and new data on neotectonic evolution of active faults in the Kachchh Basin, Western India: legacy of post-Deccan trap tectonic inversion. In: *Tectonics of the Deccan Large Igneous Province* (ed. S. Mukherjee, A.A. Misra, G. Calvés and M. Nemčok), 237–268. London, Special Publications, 445: Geological Society.

Misra, A.A. and Mukherjee, S. (2017). Dyke–brittle shear relationships in the western Deccan strike-slip zone around Mumbai (Maharashtra, India). In: *Tectonics of the Deccan Large Igneous Province* (ed. S. Mukherjee, A.A. Misra, G. Calvés and M. Nemčok) 2017, 269–295. London, Special Publications, 445: Geological Society.

Misra, A.A. and Mukherjee, S. (2022). Introduction to "Atlas of Structural Geological and Geomorphological Interpretation of Remote Sensing Images". In: *Atlas of Structural Geological and Geomorphological Interpretation of Remote Sensing Images* (ed. A.A. Misra and S. Mukherjee). Wiley Blackwell. ISBN: 9781119813354.

Misra, A.A., Bhattacharya, G., Mukherjee, S., and Bose, N. (2014). Near N–S paleo-extension in the western Deccan region, India: does it link strike-slip tectonics with India–Seychelles rifting? *International Journal of Earth Sciences* 103 (6): 1645–1680.

Misra, A.A., Maitra, A., Sinha, N. et al. (2019). Syn-to post-rift fault evolution in a failed rift: a reflection seismic study in Central Cambay Basin (Gujarat), India. *International Journal of Earth Sciences* 108 (4): 1293–1316.

Shaikh, M., Maurya, D.M., Mukherjee, S. et al. (2020). Tectonic evolution of the intra-uplift Vigodi-Gugriana-Khirasra-Netra Fault System in the seismically active Kachchh Rift Basin, India: implications for the western continental margin of the Indian plate. *Journal of Structural Geology* 140: 104124.

Shaikh, M.A., Patidar, A.K., Vanik, N.P. et al. (2022). Building tectonic framework of a blind active fault zone using field and ground-penetrating radar data. *Journal of Structural Geology* https://doi.org/10.1016/j.jsg.2022.104526.

Vanik, N., Shaikh, M.A., Mukherjee, S. et al. (2018). Post-Deccan Trap stress reorientation under transpression: evidence from fault slip analyses from SW Saurashtra, Western India. *Journal of Geodynamics* 121: 9–19.

21

Lineament Analysis in a Part of the Son River Valley, Madhya Pradesh, India

*Achyuta Ayan Misra**

Reliance Industries Ltd., Reliance Corporate Park, Mumbai, Maharashtra, India

21.1 Regional Geology

Remote sensing has been a very effective means to interpret structural geology and geomorphology of terrains (e.g. Misra et al. 2014; Kaplay et al. 2017; Dasgupta and Mukherjee 2017, 2019; Dasgupta et al. 2022; Gogoi et al. 2022). The Permo-Triassic Gondwana rocks crop out in the South Rewa Basin, central India, along with the Late Cretaceous Lameta Beds and Late Cretaceous to Early Paleocene Deccan Traps (Figure 21.1; Chakraborty et al. 2003; Mukherjee et al. 2012). The Deccan igneous rocks occur as sills and dykes while the sedimentary formations are sub-horizontal (Chowdari et al. 2017). The Son River flows through this area. It starts from the Amarkantak hills in Anuppur district (Lat: 22°43′48″N; Long: 82°03′31″E).

21.2 Lineament Analysis

I studied lineaments in the area (Figure 21.1–21.16) on satellite images from Geoeye Satellite sourced from ESRI using the ArcMap software package (ArcGIS Version 10.3.1 of 2015). The images have a resolution of 0.46 m and available in true color composite, i.e. the red, green, and blue bands are merged in red, green, and blue channels, respectively. This gives a true color representation of the area in the satellite image. The dates of acquisition of the images are indeterminate, since ESRI does not include that information for its images. Some interpretation was also done on Google Earth and the dates are mentioned in the respective figures. The lineaments were ground verified to classify them into deformation related lineaments to link them to the regional deformation history. All the lineaments could not be verified and I classified the lineaments into “traffic light” colors to denote confirmed deformation lineament as green and unconfirmed ones as red, whereas the ones that remain doubtful were colored as amber (Figures 21.2, 21.3). The lineaments of the region are mostly tectonic tensile fracturing and some uplift-related fracturing. These lineaments show two major trends: a ~ NE–SW trend, which has higher number of lineaments and a ~ NW–SE trend, which has a lesser number of lineaments. The latter trend, however, has few very long lineaments (Figure 21.3b). This trend matched with the regional fault and fracture trends of the Sohagpur Basin (Mukhopadhyay et al. 2011). There are faults that relate to the ~E–W trend of the Son-Narmada graben. Figures 21.5, 21.8, and 21.11 are unreported in any previous study. The Gondwana coal seams on Raniganj and Barakar formations occur in this region and such studies will help in deciphering the cleat patterns in the coal seams, which are essential analyses in the extraction of coal or coal seam gas.

*Corresponding Author: achyutaayan@gmail.com

Atlas of Structural Geological and Geomorphological Interpretation of Remote Sensing Images, First Edition.
Edited by Achyuta Ayan Misra and Soumyajit Mukherjee.

Companion website: www.wiley.com/go/misra/RemoteSensingImages

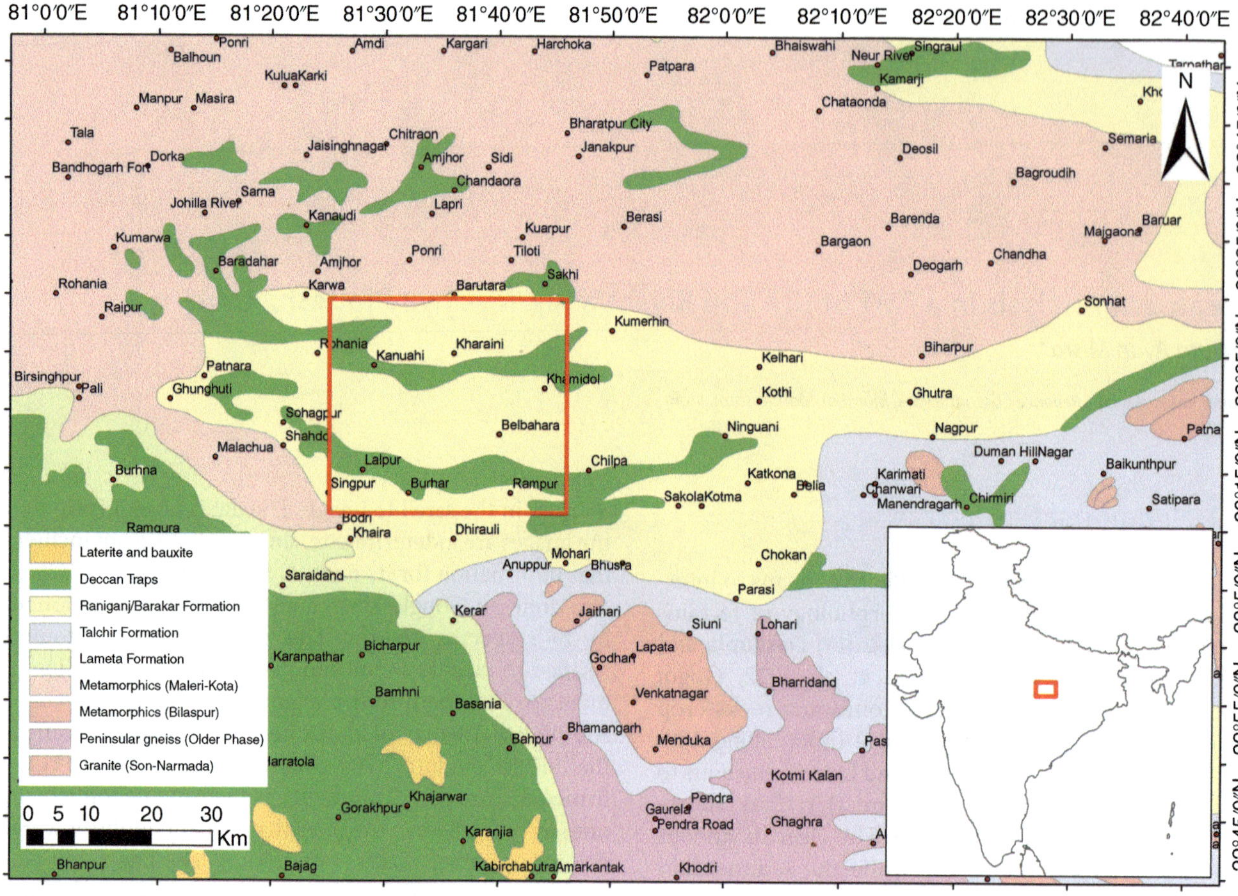

Figure 21.1 Geologic Map of central India. The red square shows the area of the present study. Inset: Location of the area. Source: Modified from Acharyya (1998).

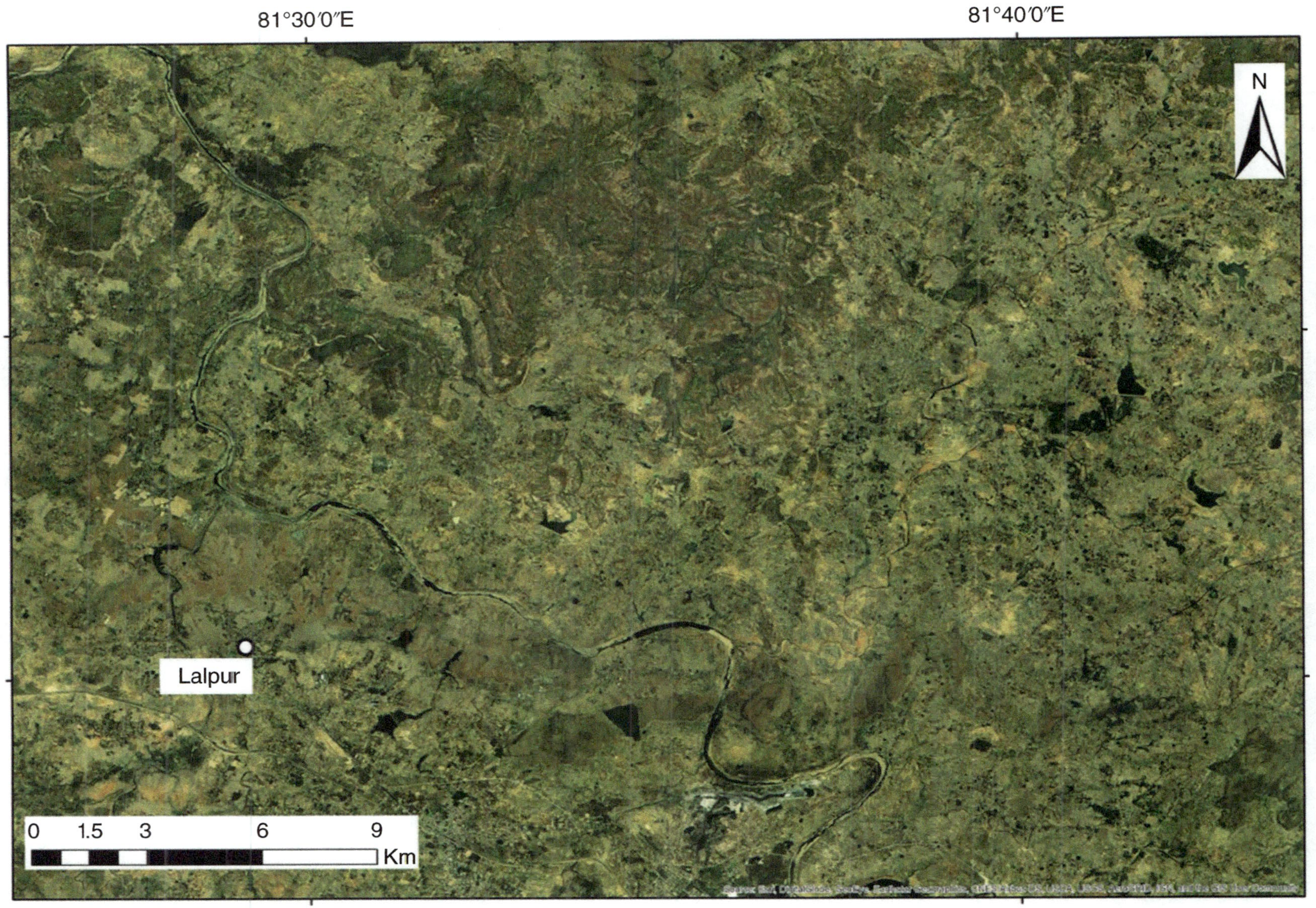

Figure 21.2 Uninterpreted image of the study area. Refer to text for details.

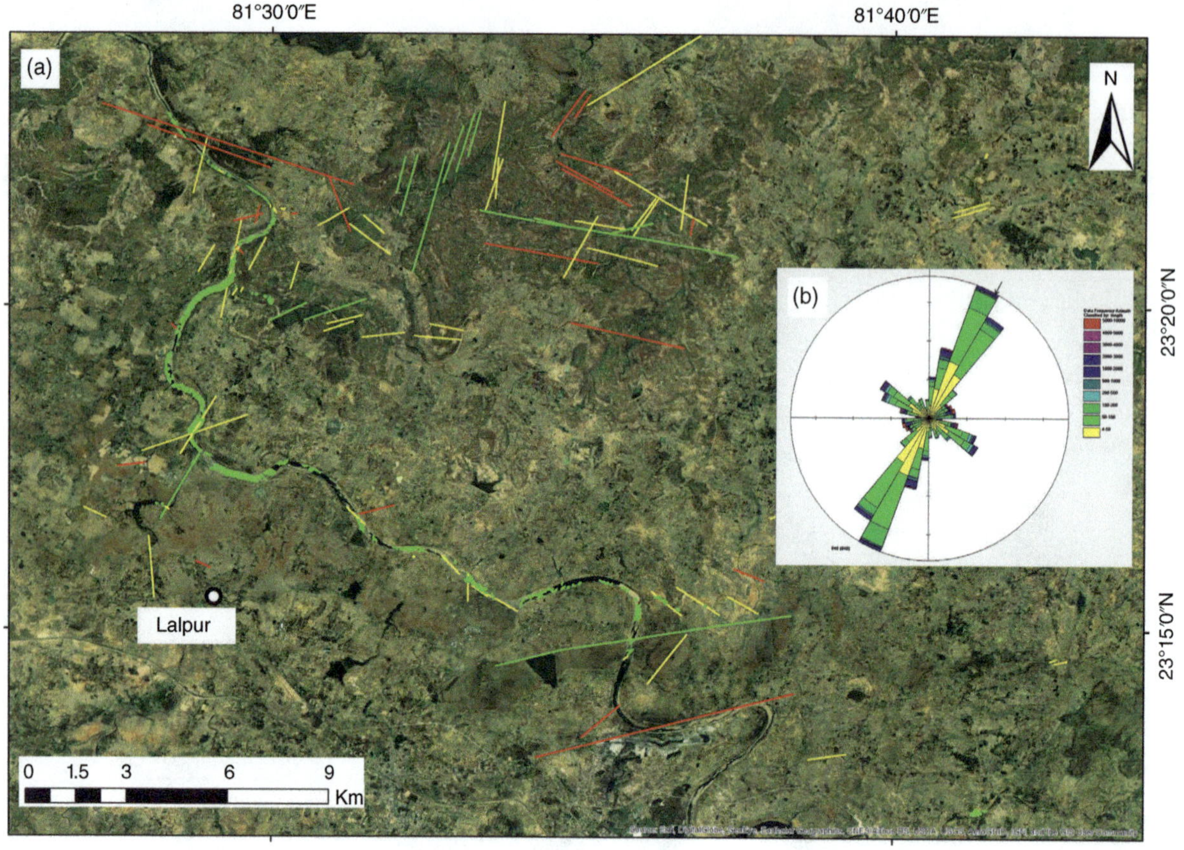

Figure 21.3 (a) Lineaments mapped on the area shown in Figure 21.2. Here most of the lineaments are visible on the river banks, because they are areas devoid of the regolith/soil cover. There are structures in the entire area, however they can be seen neither on satellite images nor on the ground. Careful interpretation is required to infer whether the distribution of the lineaments is uniform or is only along narrow zones throughout the entire area. The lineaments are colored in "traffic light" shades on the confidence of the lineament being related to deformation. (b) Length classified azimuth-frequency rose-plot for the lineaments mapped here. GEOrient 3.2 software (www.holcombe.net.au) used to plot the diagram. See text for details.

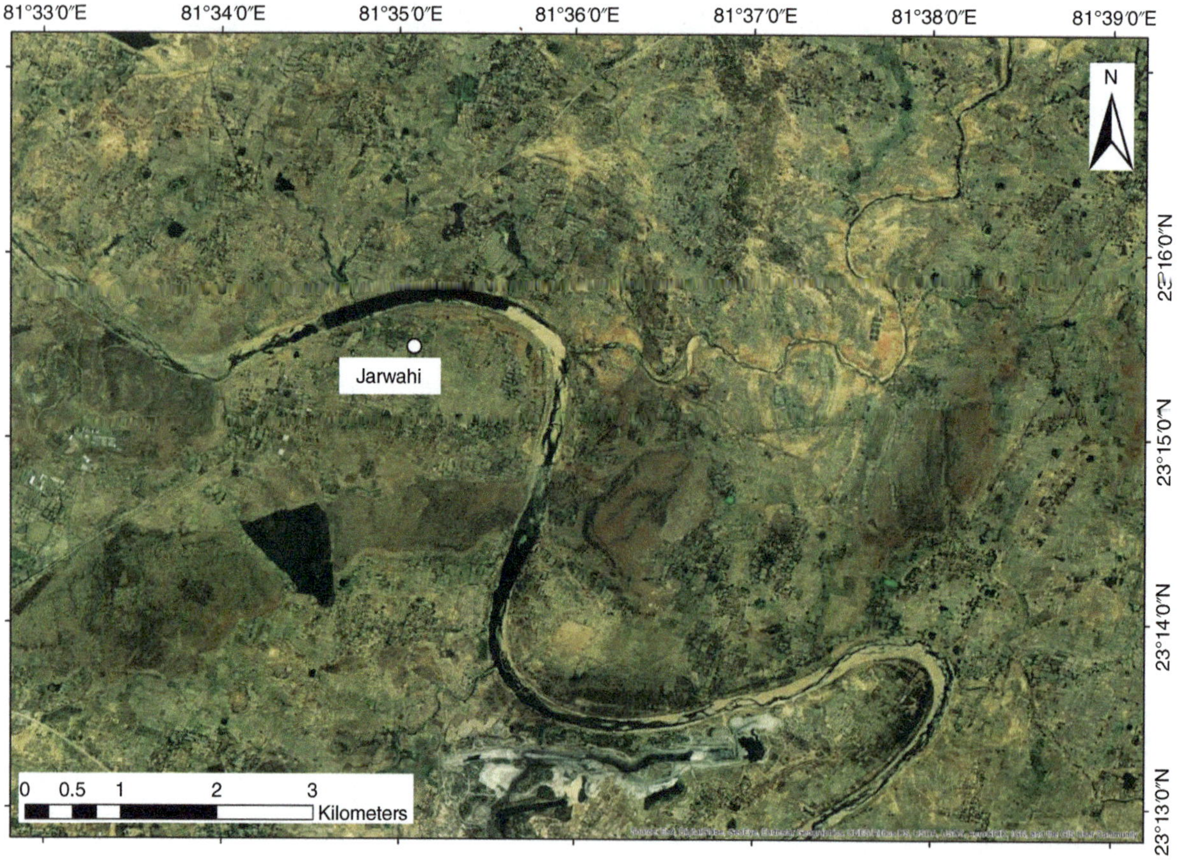

Figure 21.4 Uninterpreted image of eastern part of the study area.

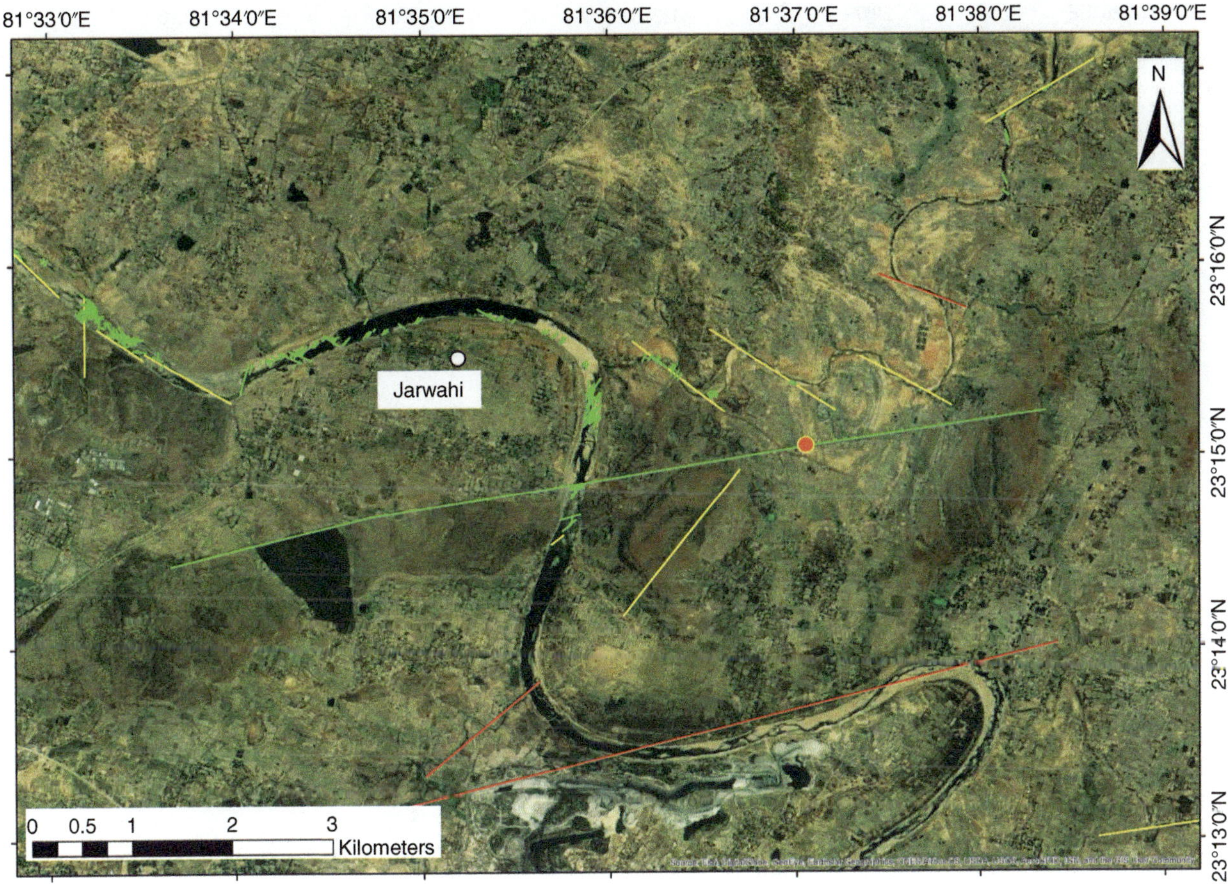

Figure 21.5 Lineaments mapped on the area shown in Figure 21.4.The ~E–W trending lineament clearly interpretable on the satellite image is a N-dipping normal fault. The fault has 5–10 m throw. This was confirmed during ground truthing (Figure 21.6). The incised meander formed here is possibly lineament controlled, and may be related to the uplift of the region due to Tertiary tectonics. There are other linear features visible on the satellite image, which are the lithological contacts between Deccan intrusives and the Gondwana sediments. Red dot: Ground truth location (Figure 21.6). River Son is the major river in this image.

Figure 21.6 Field photos showing ground truth for interpretation in Figure 21.5. (a) Upthrown, S block of the fault, where the Raniganj Formation sandstone occurs on the surface. (b) Downthrown, N block of the fault, where a deep pit has been dug and the Raniganj Formation sandstone is absent. The Raniganj Formation sandstone appears just at the base of the pit, which is ~5 m deep. Thus, the throw of the fault is possibly ~5–10 m.

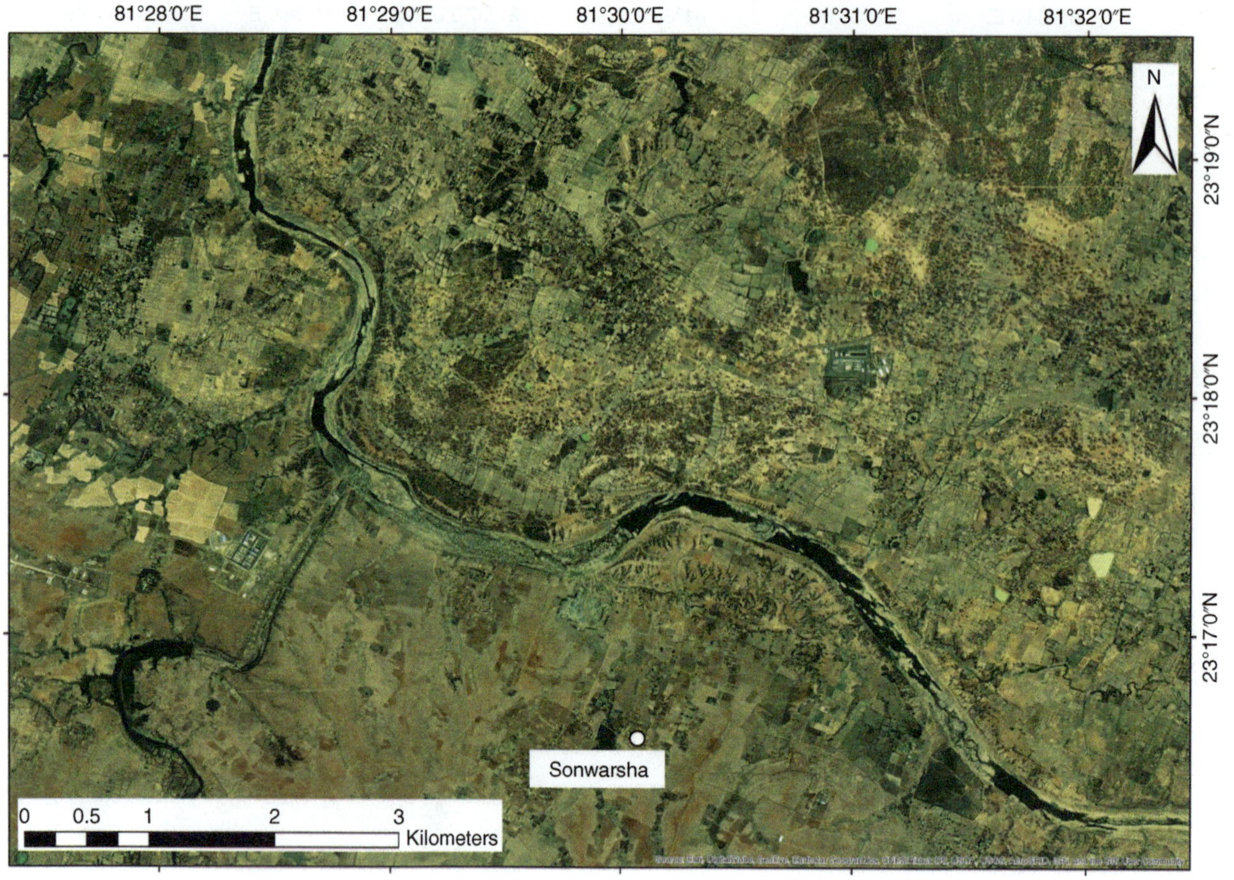

Figure 21.7 Uninterpreted image of central part of the study area.

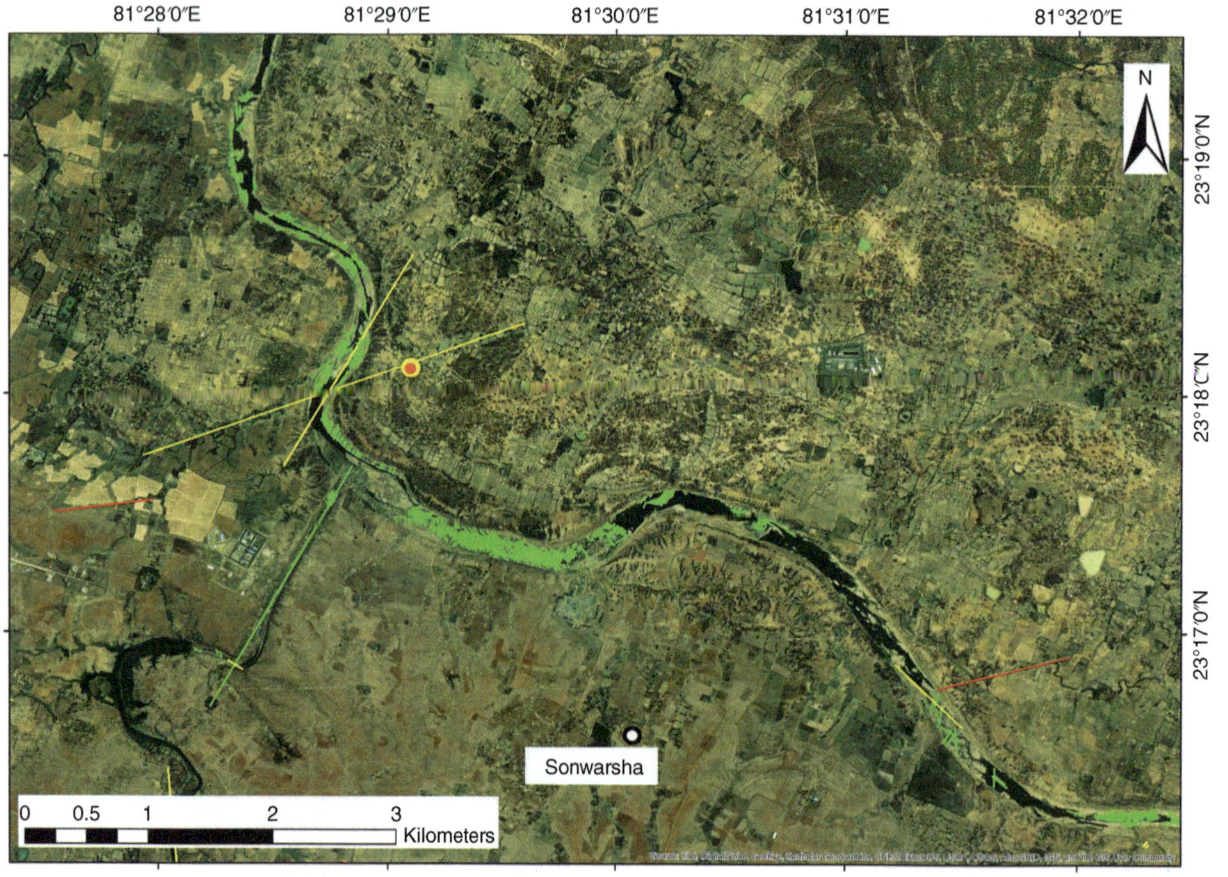

Figure 21.8 Lineaments mapped on the area shown in Figure 21.7. Sarpha Nalla is the river joining the Son River at the left of the image, and it forms a straight channel morphology, indicating to the presence of a lineament, which is a ~W-dipping fault. There are numerous fractures on the N bank of the river at this location, which depict a conjugate morphology besides the presence of tectonic tensile fractures (Figure 21.9). Red dot: Ground truth location (Figure 21.9).

Figure 21.9 Field photos showing ground truth for interpretation in Figure 21.8. Photograph taken pointing toward north showing the m-scale fracturing in the Raniganj Formation sandstone. Note the person in photo for scale.

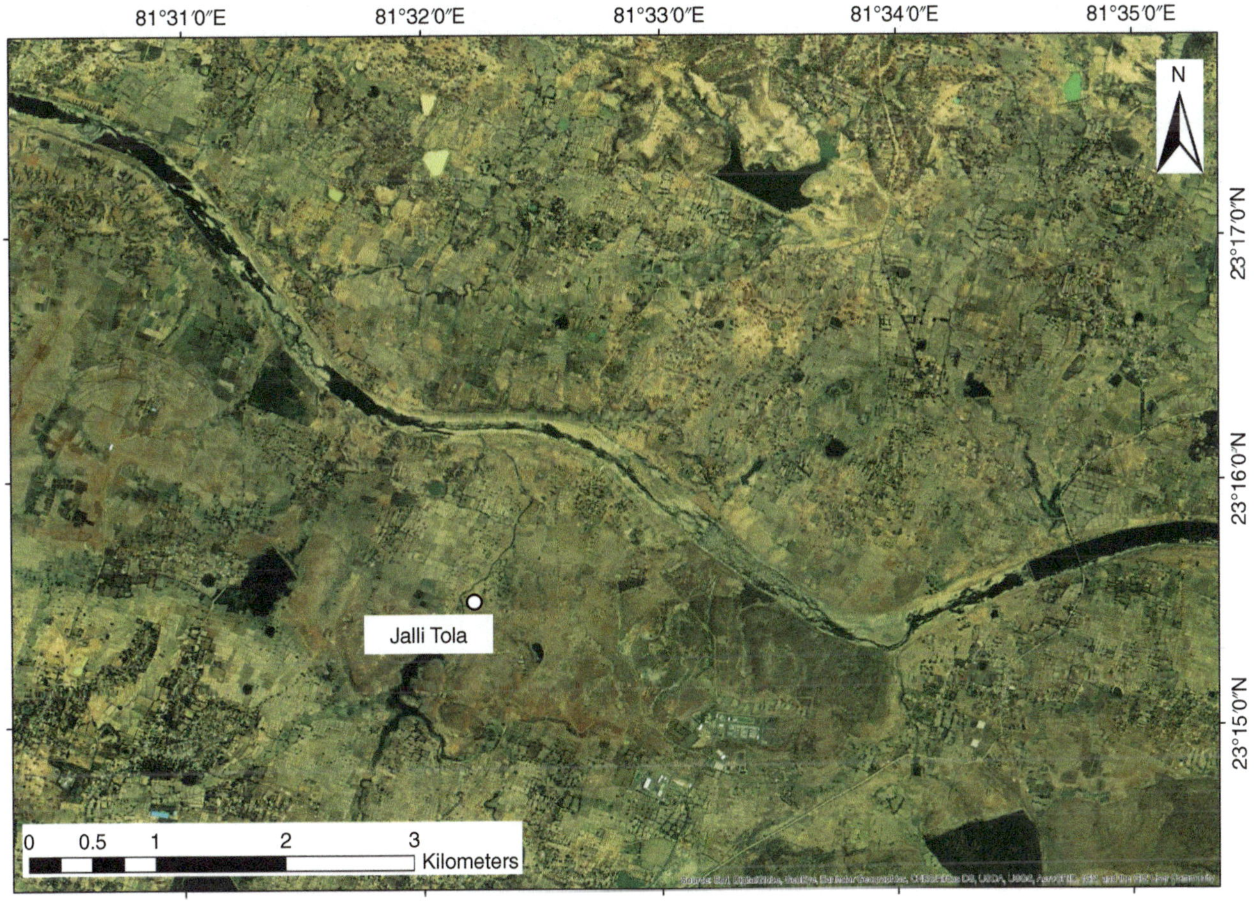

Figure 21.10 Uninterpreted image of central part of the study area.

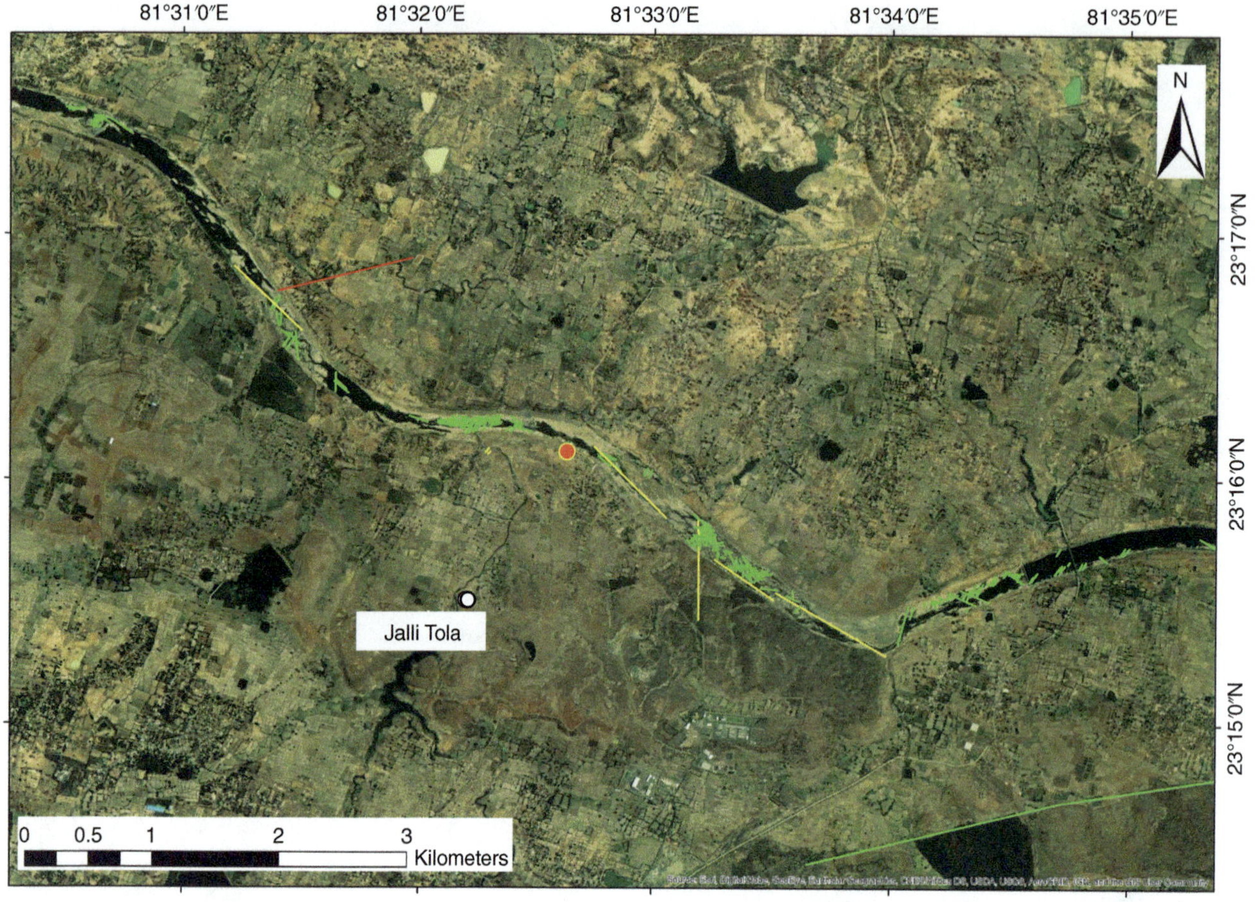

Figure 21.11 Lineaments mapped on the area shown in Figure 21.10. The lineaments classified as green are tensile fractures seen in the sandstones exposed on the river bank. However, the amber ones are possibly faults, but they were not directly observed in the outcrops. However, we observed isolated and clusters of deformation bands (see Fossen et al. 2007) in the sandstones at the riverbank (Figure 21.12). Red dot: Ground truth location (Figure 21.12).

Figure 21.12 Field photos showing ground truth for interpretation in Figure 21.11. (a) bi-directional and (b) unidirectional deformation bands. These are cataclastic deformation bands (see Fossen et al. 2007), 1 mm to 2–3 cm thick, occurring alongside the tensile fracturing. These indicate the presence of a large fault in the area, which may be covered by the river water or the soil cover.

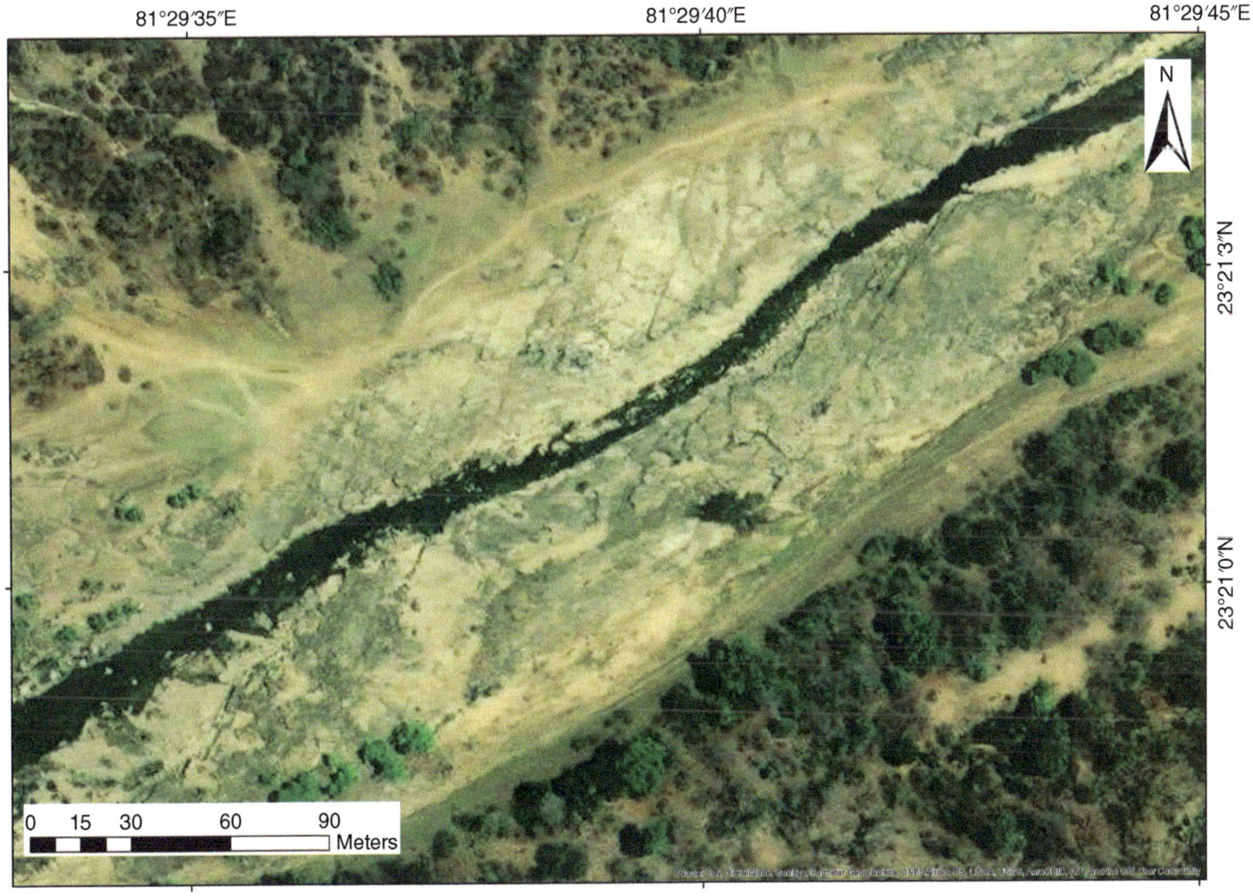

Figure 21.13 Uninterpreted image of northern part of the study area.

Figure 21.14 Lineaments mapped on the area shown in Figure 21.13. These lineaments represent tectonic tensile fracturing. These are nearly equidistant, vertical to the bedding surface and parallel to each other. These were confirmed through ground truthing (Figure 21.15).

Figure 21.15 Field photo showing ground truth for interpretation in Figure 21.14. Note the near parallel and equidistant nature of the fractures. The fracture surfaces are modified by the water action of the river.

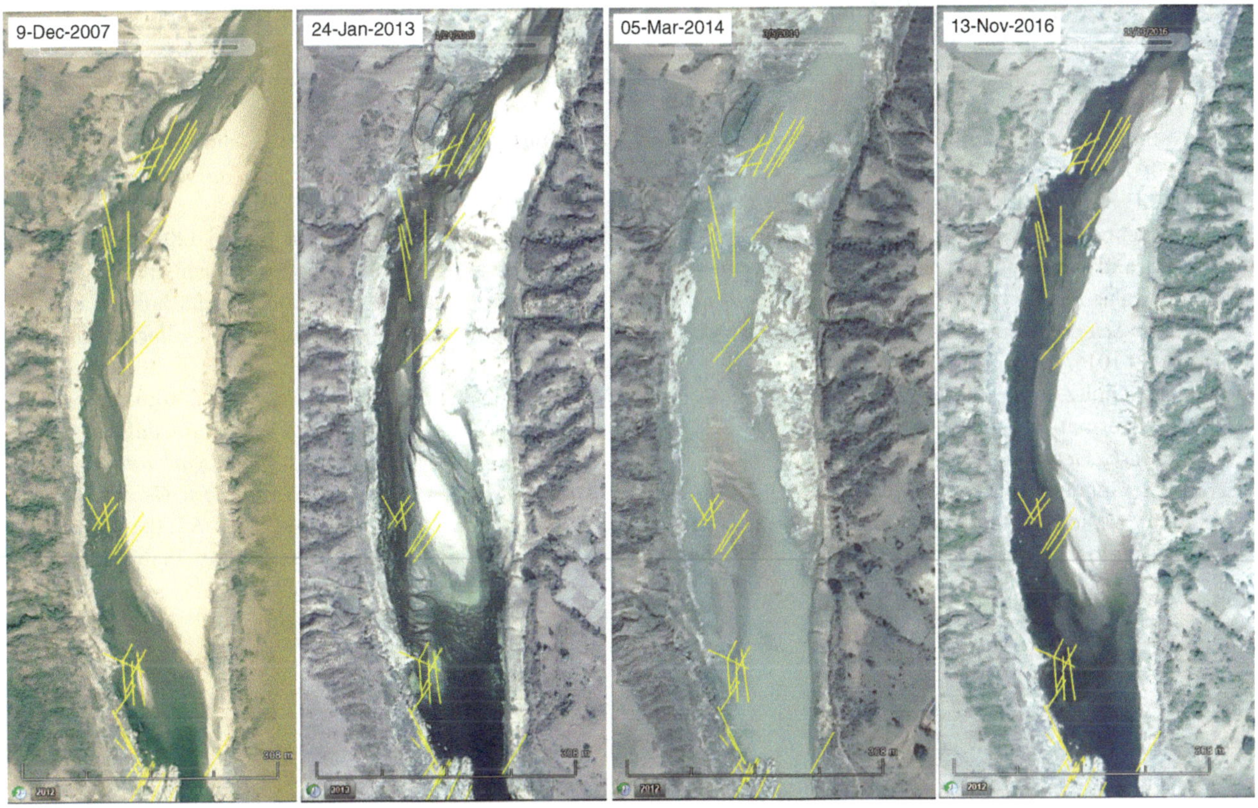

Figure 21.16 Historical imagery obtained from Google Earth showing the importance of comparing images from different seasons and years. The lineaments mapped on each imagery were combined to form the final map. One image does not show all the lineaments because the level of the river water fluctuates and hides the lineaments when it is very high. Similarly, the point bar deposit also hides the lineaments when it covers largest area.

Acknowledgements

Thanks to Andrew Harrison and the proofreading team (Wiley Blackwell). Soumyajit Mukherjee (IIT Bombay) handled this article and provided minor review comments. Mery Biswas is thanked for her comments. Reliance Industries Ltd. is thanked for funding the field-works to the region. Wiley Blackwell Acquisition Editors (present: Frank Weinreich, past: Andrew Harrison), Stacey Woods (Handling Editor), and proofreading team. Summarized in Misra and Mukherjee (2022).

References

Acharyya, S.K. (1998.) Geological Map of India, 7th Edition, 1:2,000,000, Geological Survey of India, Central Head Quarters, Kolkata.

Chakraborty, C., Mandal, N., and Ghosh, S.K. (2003). Kinematics of the Gondwana basins of peninsular India. *Tectonophysics* 377 (3–4), 299–324.

Chowdari, S., Singh, B., Rao, B.N. et al. (2017). Structural mapping based on potential field and remote sensing data, South Rewa Gondwana Basin, India. *Journal of Earth System Science* 126 (6): 1–27.

Dasgupta, S. and Mukherjee, S. (2017). Brittle shear tectonics in a narrow continental rift: asymmetric non-volcanic Barmer basin (Rajasthan, India). *The Journal of Geology* 125: 561–591.

Dasgupta, S. and Mukherjee, S. (2019). Remote sensing in lineament identification: examples from western India. In: *Problems and Solutions in Structural Geology and Tectonics*, Developments in Structural Geology and Tectonics Book Series, vol. 5. Series Editor: Mukherjee, S. (ed. A. Billi and A. Fagereng), 205–221. Elsevier. ISSN: 2542-9000. ISBN: 9780128140482.

Dasgupta, S., Biswas, M., Mukherjee, S., and Chatterjee, R. (2022). Depositional system, morphological signatures, tectonics and sedimentation pattern along the transform margin- Palar-Pennar basin, Indian east coast. *Journal of Petroleum Science & Engineering* 211: 110155. https://doi.org/10.1016/j.petrol.2022.110155.

Fossen, H., Schultz, R.A., Shipton, Z.K., and Mair, K. (2007). Deformation bands in sandstone: a review. *Journal of the Geological Society* 164 (4): 755–769.

Gogoi, M.P., Gogoi, B., and Mukherjee, S. (2022). Tectonic instability of the petroliferous upper Assam valley (NE India): a geomorphic approach. *Journal of Earth System Science* 131: 18. https://doi.org/10.1007/s12040-021-01752-6.

Kaplay, R.D., Md, B., Mukherjee, S., and Kumar, T.V. (2017). Morphotectonic expression of geological structures in eastern part of south East Deccan volcanic province (around Nanded, Maharashtra, India). In: *Tectonics of the Deccan Large Igneous Province* (ed. S. Mukherjee, A.A. Misra, G. Calvès and M. Nemčok), 317–335. London, Special Publications 445: Geological Society.

Misra, A.A. and Mukherjee, S. (2022). Introduction "Atlas of Structural Geological and Geomorphological Interpretation of Remote Sensing Images". In: *Atlas of Structural Geological and Geomorphological Interpretation of Remote Sensing Images* (ed. A.A. Misra and S. Mukherjee). Wiley Blackwell. ISBN: 9781119813354.

Misra, A.A., Bhattacharya, G., Mukherjee, S., and Bose, N. (2014). Near N-S paleo-extension in the western Deccan region in India: does it link strike-slip tectonics with India-Seychelles rifting? *International Journal of Earth Sciences* 103: 1645–1680.

Mukherjee, D., Ray, S., Chandra, S. et al. (2012). Upper Gondwana succession of the Rewa basin, India: understanding the interrelationship of lithologic and stratigraphic variables. *Journal of the Geological Society of India* 79 (6): 563–575.

Mukhopadhyay, A., Mukhopadhyay, S.K., Adhikari, S., and Mondal, A. (2011). *Geology of Sohagpur Coalfield-Bilaspur and Koriya Districts, Chattisgarh and Anuppur and Shahdol Districts, Madhya Pradesh, Geological Survey of India, Bulletin Series A, No. 55*, 104. Kolkata: Geological Survey of India Publication. ISSN 0536-8782.

22

Meso Scale Sinistral Shear, Eastern Dharwar Craton, Telangana, India

*Ankita Biswas**

Geological Survey of India, Hyderabad, Telangana, India

22.1 Overview

Remote sensing has been a very effective means of interpreting structural geology and geomorphology of terrains (e.g. Misra et al. 2014; Kaplay et al. 2017; Dasgupta and Mukherjee 2017, 2019; Dasgupta et al. 2022; Gogoi et al. 2022). The rocks of the Eastern Dharwar craton (EDC) include those of Archaean calc-alkaline granitoid belts, collectively called Dharwar Batholith (2.5–2.7 Ga) (Friend and Nutman 1991; Chadwick et al. 1997; Meert et al. 2010), and accreted linear array of N–S to NNW–SSE trending parallel greenstone belts, which are lithologically similar to the Dharwar Supergroup of the Western Dharwar Carton (WDC) (Chadwick et al. 1997, 2000).

22.2 Description

A meso scale sinistral shear is observed in Sentinel-2A satellite imagery (True color composite, bands 4-3-2) over the granitoid ridges (Figure 22.1) of Peninsular Gneissic Complex-II (PGC-II) (Dashora and Shah 1988–1989, Reddy et al. 1988–1989, Ali and Sumanth 1990–1991; Kazmi 2008–2009), around Fatehpuram village, Telangana, as per inset of Figure 22.2a, b. The PGC is represented by the older migmatitic gneiss, granitic gneiss, and banded gneiss. They are intruded by granitoids of PGC-II represented by granodiorite, hornblende granite, coarse-grained gray/pink granites, alkali feldspar granites, and gray biotite granite. They are later intruded by pegmatite, quartz reef/veins, mafic dykes (pyroxenite, dolerite, and gabbro), and minor granophyre dykes (Dashora and Shah 1989, Reddy et al., 1988–1989, Ali and Sumanth 1990–1991; Kazmi 2008–2009). Four quartzo-feldspathic veins occurred as marker beds and are dragged to the left from the top across the NE–SW sub-vertical plane as observed in the field photographs (horizontal section as seen in Figure 22.2c) representing sinistral sense of shear. The resolution of the Sentinel-2 imagery is 10 m. Bands 4, 3, and 2, i.e. red, green, and blue are used to create a true color composite image of the region to portray the region in true color.

Location: about 4 km south of Fathehpuram, Telangana, India (Lat: 17°51′40.74″N; Long: 79°15′36.54″E). Lithology observed in field: Granite mylonite of PGC-II of Archaean to paleo-Proterozoic with younger intrusives of quartzo-feldspathic veins (After Dashora and Shah 1989, Reddy et al. 1988–1989, Ali and Sumanth 1990–1991; Kazmi 2008–2009).

Acknowledgements

Soumyajit Mukherjee invited to submit an article in the edited book. Achyuta Ayan Misra handled this article. Wiley Blackwell Acquisition Editors (present: Frank Weinreich, past: Andrew Harrison), Stacey Woods (Handling Editor), and proofreading team. Summarized in Misra and Mukherjee (2022). The author is thankful to Shri Subrata Chakraborty, Deputy Director General, SU-Telangana and Shri Biplab Kumar Chakrabarty, Director, SU-Telangana for the opportunity to carry out the field visit in this area.

*Corresponding Author: ankita.iitr09@gmail.com

Atlas of Structural Geological and Geomorphological Interpretation of Remote Sensing Images, First Edition.
Edited by Achyuta Ayan Misra and Soumyajit Mukherjee.

Companion website: www.wiley.com/go/misra/RemoteSensingImages

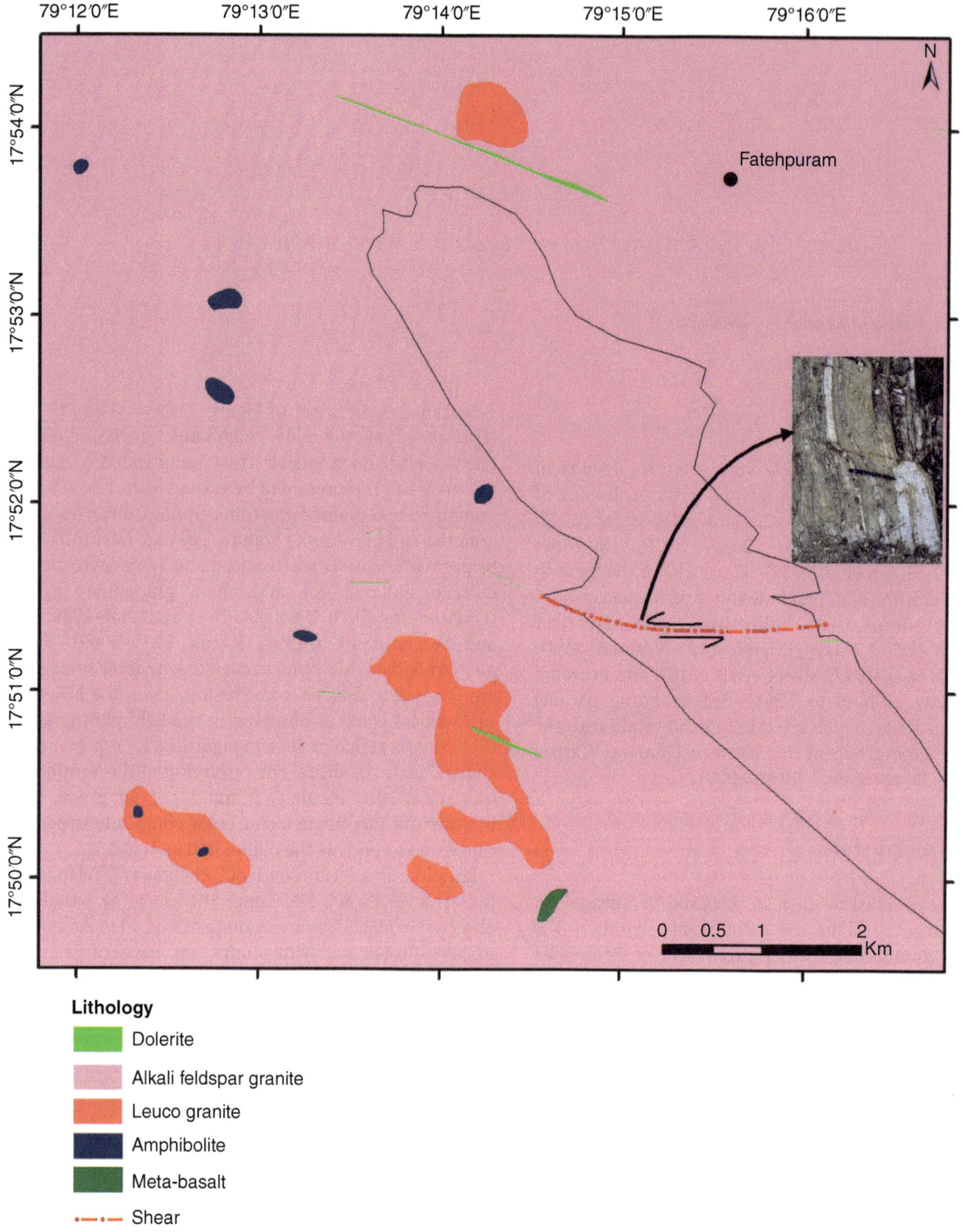

Figure 22.1 Geological map of the study area with field photograph showing sinistrally sheared granite mylonite of PGC II.

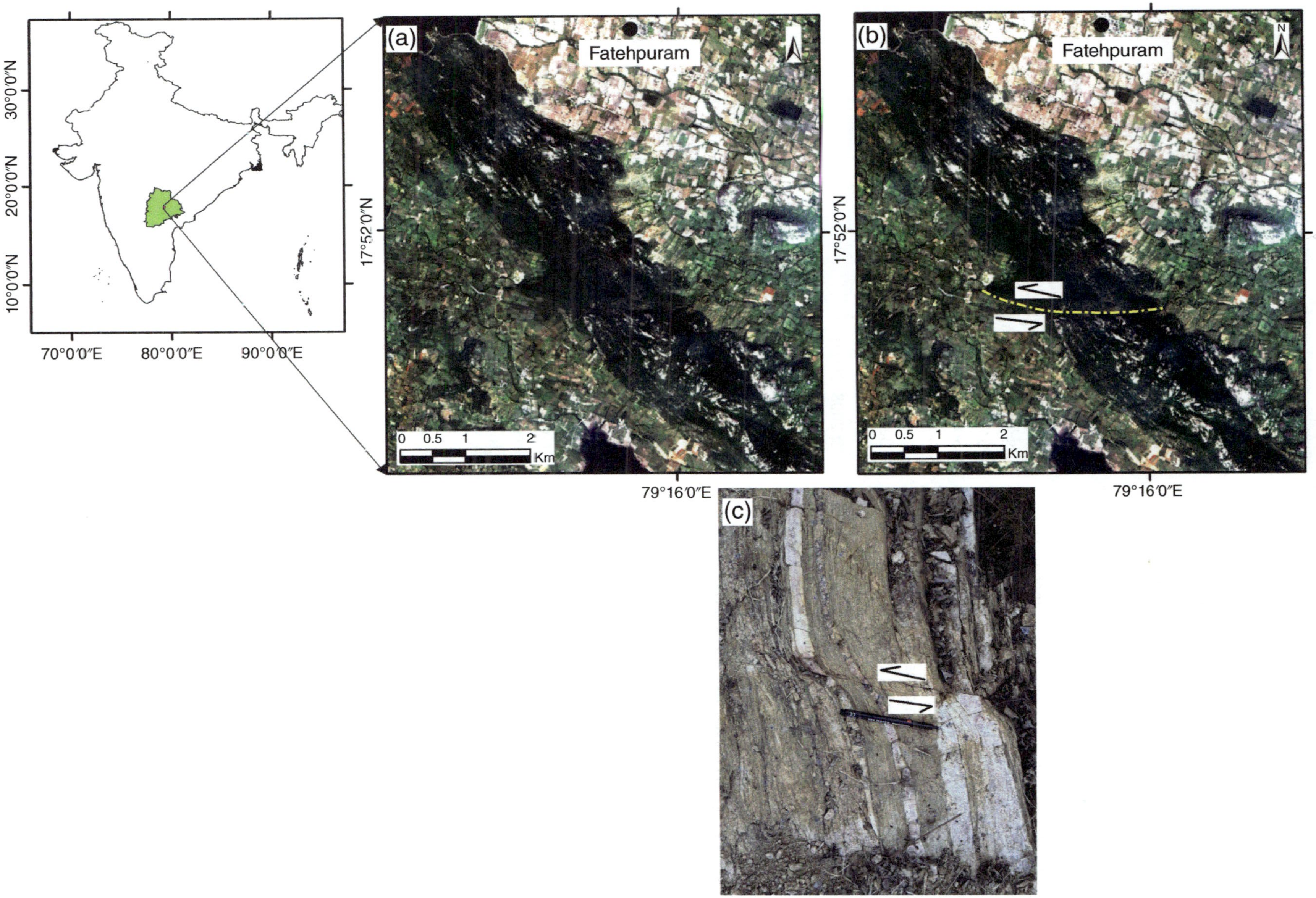

Figure 22.2 Regional Sinistral Shear seen in the granitoid ridges of PGC II group of rocks, from Sentinel-2A satellite imagery in "true color composite" (RGB: Bands 4-3-2) around Fatehpuram. (a) original image; (b) annotated image; (c) field evidence of NE–SW striking sinistral shear on X–Z plane observed in granite mylonites outcrop around Fatehpuram.

References

Ali, A. and Sumanth, N.G. (1990–1991). Report on investigation for sulphides in Ghanpur area, Warangal district, AP, GSI Unpub. Prog. Report for the F.S.: 1990–1991.

Chadwick, B., Vasudev, V.N., and Hegde, G.V. (1997). The Dhawar craton, southern India, and its late Archean plate tectonic setting: current interpretations and controversies. *Journal of Earth System Science* 106 (4): 249–258.

Chadwick, B., Vasudev, V.N., and Hegde, G.V. (2000). The Dharwar craton, southern India, interpreted as the result of late Archaean oblique convergence. *Precambrian Research* 99: 91–11.

Dasgupta, S. and Mukherjee, S. (2017). Brittle shear tectonics in a narrow continental rift: asymmetric non-volcanic Barmer basin (Rajasthan, India). *The Journal of Geology* 125: 561–591.

Dasgupta, S. and Mukherjee, S. (2019). Remote sensing in lineament identification: examples from western India. In: *Problems and Solutions in Structural Geology and Tectonics*, Developments in Structural Geology and Tectonics Book Series, vol. 5. Series Editor: Mukherjee, S (ed. A. Billi and A. Fagereng), 205–221. Elsevier. ISSN: 2542-9000. ISBN: 9780128140482.

Dasgupta, S., Biswas, M., Mukherjee, S., and Chatterjee, R. (2022). Depositional system, morphological signatures, tectonics and sedimentation pattern along the transform margin- Palar-Pennar basin, Indian east coast. *Journal of Petroleum Science & Engineering* 211: 110155. https://doi.org/10.1016/j.petrol.2022.110155.

Dashora, S., Shah, B.M. (1989). Report on systematic geological mapping of the area around Ghanpur and Warangal, falling in toposheet Nos. 56O/5 & 9, GSI Unpub. Prog. Report for the F.S.: 1988–1989.

Friend, C.R.L. and Nutman, A.P. (1991). SHRIMP U–Pb geochronology of the Closepet granite and peninsular gneiss, Karnataka, south 1304 India. *Journal of the Geological Society of India* 38: 357–368.

Gogoi, M.P., Gogoi, B., and Mukherjee, S. (2022). Tectonic instability of the petroliferous upper Assam valley (NE India): a geomorphic approach. *Journal of Earth System Science* 131: 18. https://doi.org/10.1007/s12040-021-01752-6.

Kaplay, R.D., Md, B., Mukherjee, S., and Kumar, T.V. (2017). Morphotectonic expression of geological structures in eastern part of south East Deccan volcanic province (around Nanded, Maharashtra, India). In: *Tectonics of the Deccan Large Igneous Province* (ed. S. Mukherjee, A.A. Misra, G. Calvès and M. Nemčok), 317–335. London, Special Publications 445: Geological Society.

Kazmi, S.M.K. (2008–2009). Preliminary Investigations for gold and other associated elements in Ghanpur Schist Belt, Warangal district, Andhra Pradesh (G4 Stage) GSI Unpub. Prog. Report for the F.S.2007–2008 & 2008–2009).

Meert, J.G., Pandit, M.K., Pradhan, V.R. et al. (2010). The Precambrian tectonic 1449 evolution of India: a 3.0 billion year odyssey. *Journal of Asian Earth Sciences* 39: 483–515.

Misra, A.A. and Mukherjee, S. (2022). Introduction to "Atlas of Structural Geological and Geomorphological Interpretation of Remote Sensing Images". In: *Atlas of Structural Geological and Geomorphological Interpretation of Remote Sensing Images* (ed. A.A. Misra and S. Mukherjee). Wiley Blackwell. ISBN: 9781119813354.

Misra, A.A., Bhattacharya, G., Mukherjee, S., and Bose, N. (2014). Near N-S paleo-extension in the western Deccan region in India: does it link strike-slip tectonics with India-Seychelles rifting? *International Journal of Earth Sciences* 103: 1645–1680.

Reddy, M., Rahmatulla, K., and Ramanaidu, K. (1988–1989). Geology of the area in parts of Warangal district, AP, GSI Unpub. Prog. Report for the F.S.1988–1989.

23

Regional Polyclinal Fold with Faulted Limbs, Rajasthan, India

Ankita Biswas[1] and Priyom Roy[2]*

[1] *Geological Survey of India, Hyderabad, Telangana, India*
[2] *National Remote Sensing Centre, ISRO, Hyderabad, Telangana, India*

23.1 Overview

Remote sensing has been an efficient means of interpreting structural geology and geomorphology of terrains in mega-scale (e.g. Misra et al. 2014; Kaplay et al. 2017; Dasgupta and Mukherjee 2017, 2019; Dasgupta et al. 2022; Gogoi et al. 2022). A regional polyclinal fold with faulted limbs exhibiting observable offsets is seen in Sentinel-2A satellite imagery. The true color composite, bands Red (4): Green (3): Blue (2) shows a synoptic view of the regionally deformed and faulted suite of rocks in the region with the area of interest (Figure 23.1a, marked by black box). The resolution of the Sentinel-2 imagery is 10 m, thus providing significant granularity in identification of the folded geometry and the faults.

23.2 Description

The synoptic view of the region (Figure 23.1a) further shows that the area has undergone regional polyphase deformation with signatures of east west as well as north–south compression. Though the fold appears to be polyclinal with more than one observable axial surfaces, the primary axial trace is oriented east–west. The limbs show some indications of detachment from the associated layers. Both the limbs of the fold are faulted and display offset. We have been able to map at least six co-oriented faults trending northwest–southeast (Figure 23.1b, c). These are possibly one set of shear resulting from the east–west compressive event (Tikoff and Teyssier 1992). The bed thickness has been maintained in the unfaulted hinge zone of the fold. The orthogonal thickness of the layers near the hinge are maintained at ~210 m (Figure 23.1d). The fold may be classified as of Class 1b fold geometry (Ramsay 1967). It is worthwhile to mention that the folded layers show indistinct signatures of inter-layer slip. However, the same cannot be conclusively substantiated from satellite imagery. The location is 2 km SE of Barundani, Rajasthan, India (Lat: 25° 8′43.02″N; Long: 74°57′9.93″E). The rocks are of Lower Bhander (Lakheri) limestone, Samaria Shale with dolomitic limestone of Bhander Group of Vindhyan Supergroup. The age of the rocks are ~750 Ma (Ray 2006).

*Corresponding Author: ankita.iitr09@gmail.com

Atlas of Structural Geological and Geomorphological Interpretation of Remote Sensing Images, First Edition.
Edited by Achyuta Ayan Misra and Soumyajit Mukherjee.

Companion website: www.wiley.com/go/misra/RemoteSensingImages

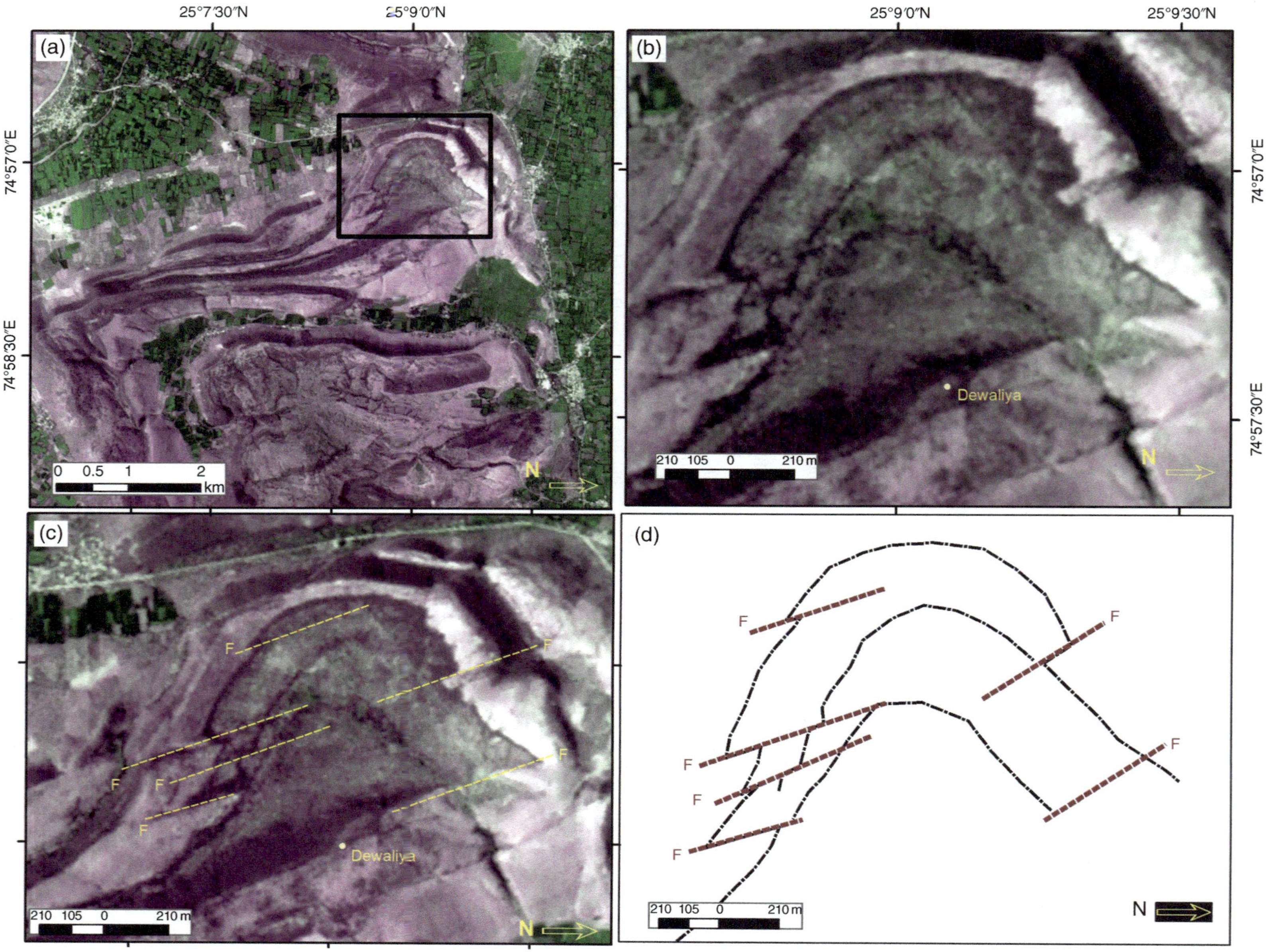

Figure 23.1 Regional polyclinal fold with faulted limbs exhibiting observable offsets, as seen in Sentinel-2A satellite imagery. (a) synoptic view; (b) close-up view showing faulted limbs; (c) annotated image; (d) the diagram of the structure.

Acknowledgements

Soumyajit Mukherjee invited to submit an article in the edited book. Achyuta Ayan Misra handled this article. An anonymous reviewer provided comments. Wiley Blackwell Acquisition Editors (present: Frank Weinreich, past: Andrew Harrison), Stacey Woods (Handling Editor), and proofreading team. Summarized in Misra and Mukherjee (2022).

References

Dasgupta, S. and Mukherjee, S. (2017). Brittle shear tectonics in a narrow continental rift: asymmetric non-volcanic Barmer basin (Rajasthan, India). *The Journal of Geology* 125: 561–591.

Dasgupta, S. and Mukherjee, S. (2019). Remote sensing in lineament identification: examples from western India. In: *Problems and Solutions in Structural Geology and Tectonics*, Developments in Structural Geology and Tectonics Book Series, vol. 5. Series Editor: Mukherjee, S. (ed. A. Billi and A. Fagereng), 205–221. Elsevier ISSN: 2542 9000. ISBN: 9780128140482.

Dasgupta, S., Biswas, M., Mukherjee, S., and Chatterjee, R. (2022). Depositional system, morphological signatures, tectonics and sedimentation pattern along the transform margin- Palar-Pennar basin, Indian east coast. *Journal of Petroleum Science & Engineering* 211: 110155. https://doi.org/10.1016/j.petrol.2022.110155.

Gogoi, M.P., Gogoi, B., and Mukherjee, S. (2022). Tectonic instability of the petroliferous upper Assam valley (NE India): a geomorphic approach. *Journal of Earth System Science* 131: 18. https://doi.org/10.1007/s12040-021-01752-6.

Kaplay, R.D., Babar, M., Mukherjee, S., and Kumar, T.V. (2017). Morphotectonic expression of geological structures in eastern part of south East Deccan volcanic province (around Nanded, Maharashtra, India). In: *Tectonics of the Deccan Large Igneous Province* (ed. S. Mukherjee, A.A. Misra, G. Calvès and M. Nemčok), 317–335. London, Special Publications 445: Geological Society.

Misra, A.A. and Mukherjee, S. (2022). Introduction to "Atlas of Structural Geological and Geomorphological Interpretation of Remote Sensing Images". In: *Atlas of Structural Geological and Geomorphological Interpretation of Remote Sensing Images* (ed. A.A. Misra and S. Mukherjee). Wiley Blackwell ISBN: 9781119813354.

Misra, A.A., Bhattacharya, G., Mukherjee, S., and Bose, N. (2014). Near N-S paleo-extension in the western Deccan region in India: does it link strike-slip tectonics with India-Seychelles rifting? *International Journal of Earth Sciences* 103: 1645–1680.

Ramsay, J.G. (1967). *Folding and Fracturing of Rocks, Structural Geology of Rocks*, 2e. New York: McGraw-Hill.

Ray, J.S. (2006). Age of the Vindhyan Supergroup: a review of recent findings. *Journal of Earth System Science* 115 (1): 149–160.

Tikoff, B. and Teyssier, C. (1992). Crustal-scale, en echelon "P-shear" tensional bridges: a possible solution to the batholithic room problem. *Geology* 20 (10): 927–930.

24

Drainage Architecture and Bar Formation of the Rangit Tributaries, Darjeeling-Sikkim Himalaya, India

*Tanwita Deb**

Indian Institute of Science Education and Research, Bhopal, Bhauri, India (former)

24.1 Introduction

Remote sensing interpretation for structural geology and geomorphology of terrains is widespread (e.g. Misra et al. 2014; Kaplay et al. 2017; Dasgupta and Mukherjee 2017, 2019; Dasgupta et al. 2022; Gogoi et al. 2022). Neotectonics, geology, and climate control the drainage architecture and evolution of associated channel bars in the early stage of a fluvial system (Keller and Pinter 1996; Zhang et al. 2018; Kothyari et al. 2020). Neotectonic activity along the drainage segments of the lower Tista sub-basin and the Rangit tributaries, Eastern Himalaya, has been documented (Ghosh and Sivakumar 2018, 2019; Sarkar et al. 2021). The drainage architecture and bar formation of the Rangit tributaries, India, are controlled by the tectonism and topography of the area. The sinuosity of the Chhota Rangit river is topography-controlled while the zigzag drainage pattern of the Jhepi Khola river is dominantly fault-controlled. The presence of the "pseudo-plain land" in this Himalayan region influences the channel bar formation in Chhota Rangit river, while the eye-shaped bar in the Rammam river is formed due to the local subsidence along the lineaments and is tectonically controlled. Presented here are the drainage morphology and channel bar evolution of the four tributaries (Rammam, Chhota Rangit, Jhapi Khola, and Rongdong Khola) of the Rangit River (in the study area as marked in Figure 24.1a), Darjeeling-Sikkim Himalaya.

24.2 Image Analysis

Although all tributaries are within 10 km of one another, the Rammam (sinuosity 1.15), Jhapi Khola (sinuosity 1.45), and Rongdong Khola (1.49) are less sinuous than the Chhota Rangit (sinuosity 1.86) in the study area (Figure 24.1). The last 12 km flow path before meeting the Rangit river, the Chhota Rangit exhibits a highly sinuous nature (1.86, even though the earlier ~7 km flow path of the Chhota Rangit appears less sinuous) (Figure 24.1b). This highly sinuous nature of the Chhota Rangit river resembles the meandering river channel of the middle course of a fluvial system. Semwal and Chauniyal (2018) have postulated that tectonics and lithology are responsible for the meandering of a river channel in the mountainous region. A possible explanation of the highly sinuous nature of the Chhota Rangit River is that the morphology is dominantly controlled by topography and tectonism. Besides tectonism, relatively low elevated topography (average elevation is between 300 and 600 m with 5–10° slope) compared to the surrounding area acts as "pseudo-plain" geomorphology in the Himalaya, enhancing sinuosity of the Chhota Rangit. The change in Chhota Rangit river sinuosity also follows the sharp fall of the gradient of the river basin to the northern side (fall from ~900 to 540 m) (Figure 24.2a). It possesses mostly point bars along its entire channel (Figure 24.2b). What is more, the last 600 m flow path of Chhota Rangit river passes through a pseudo-plain with

*Corresponding Author: tanwitadeb@gmail.com

Atlas of Structural Geological and Geomorphological Interpretation of Remote Sensing Images, First Edition.
Edited by Achyuta Ayan Misra and Soumyajit Mukherjee.

Companion website: www.wiley.com/go/misra/RemoteSensingImages

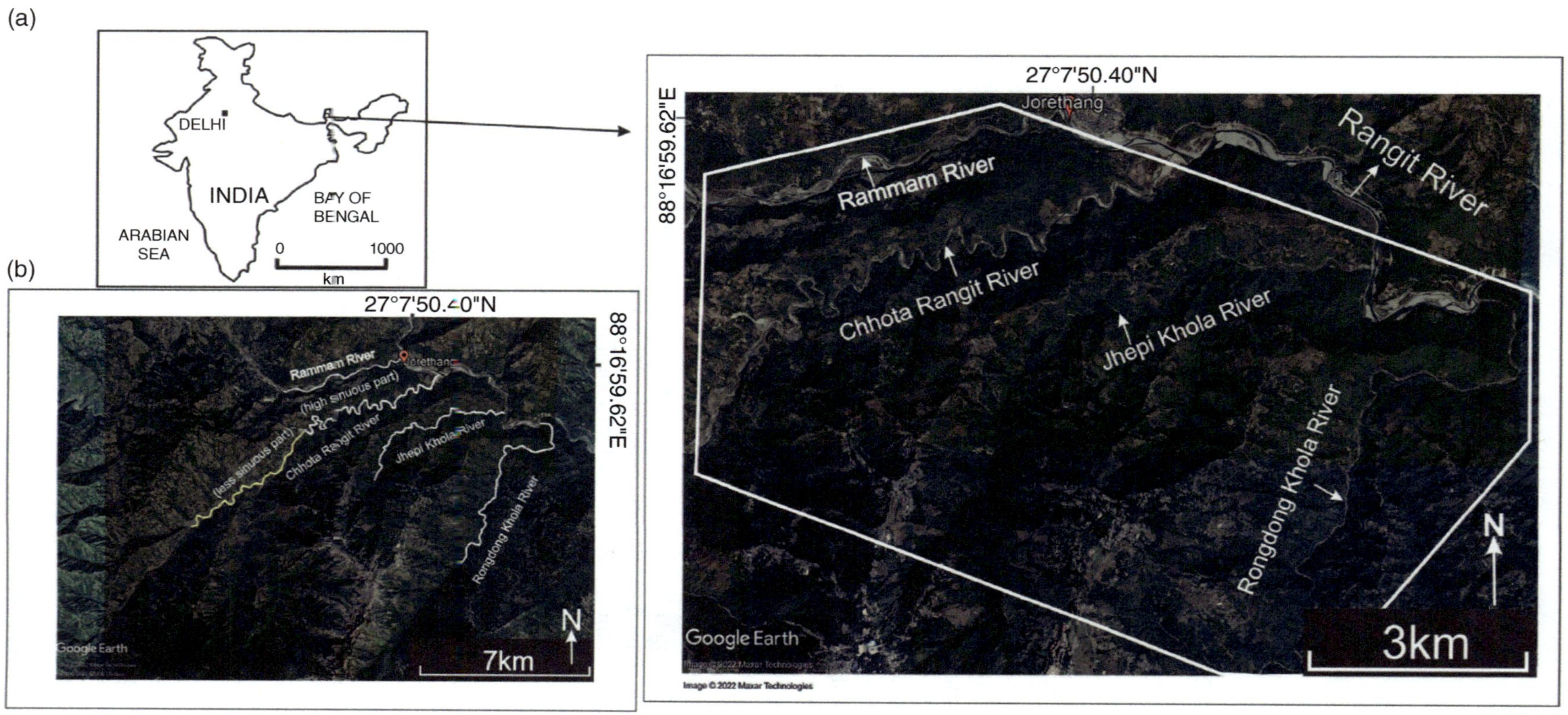

Figure 24.1 (a) Map showing the study area. The satellite image of the study area (marked by the polygon). (b) The four tributaries, Rammam, Chhota Rangit, Jhepi Khola, and Rongdong Khola are present (within 10 km of one another) from the northwest to southeast, respectively, and finally meet the Rangit river. Source: Image courtesy Google Earth.

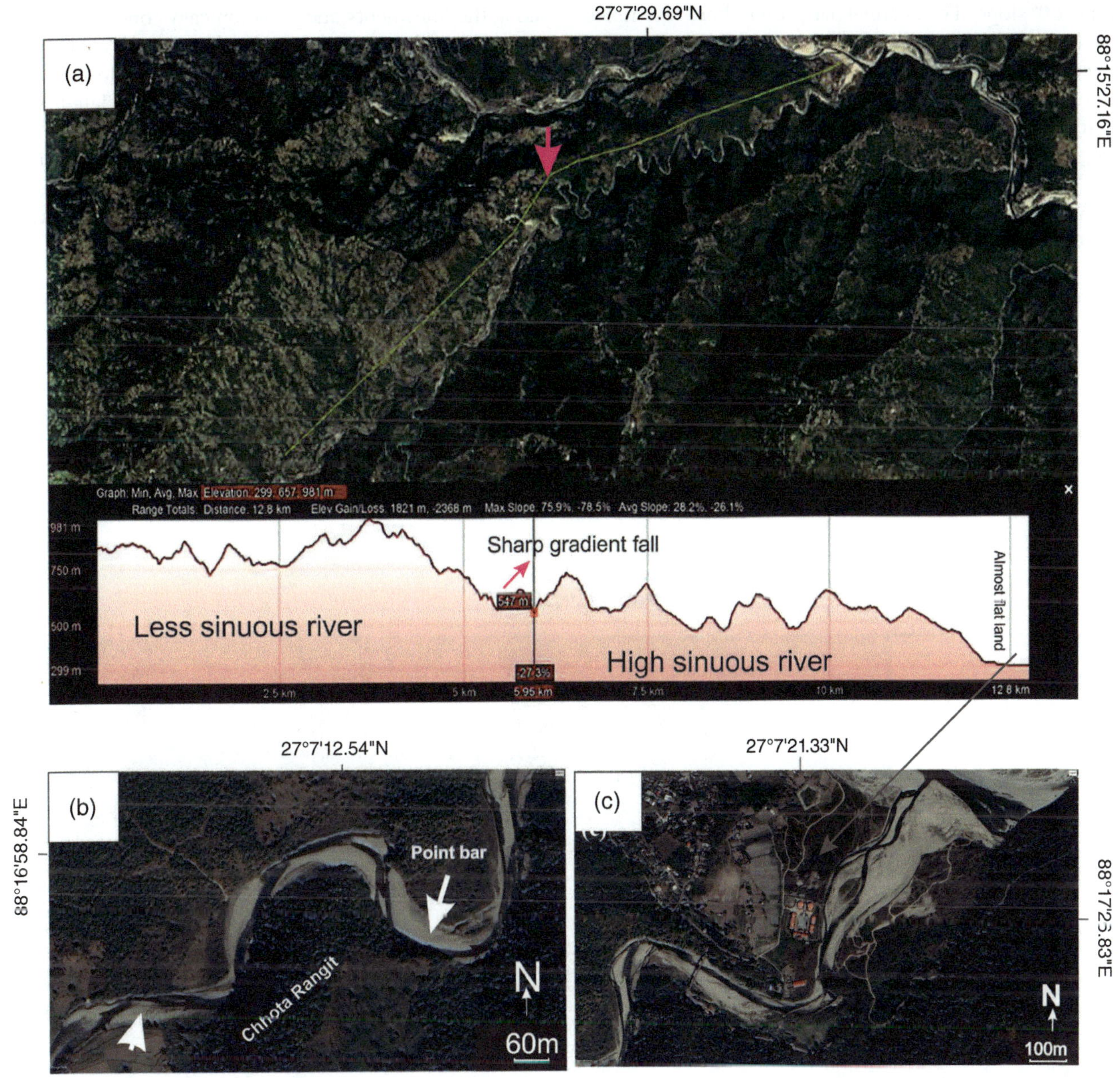

Figure 24.2 (a) Elevation profile shows the sharp change of Chhota Rangit river gradient with the change of sinuosity. The last 600 m flow path acts as a pseudo-plain land. (b) The satellite image of the Chhota Rangit river shows its sinuous nature with the formation of the point bars (marked by white arrows). Flow direction is marked by the black arrows. Extreme east point bar also has a temporary chute channel cutting across (marked by the dotted white arrow). It implies the presence of the periods of flood that lead to the erosion of the point bar. (c) Formation of bar resembles the foothill bar. Source: Image courtesy: Google Earth.

almost 0° slope. The channel bar present here resembles the bar on the foothill area (Figure 24.2c). On the contrary, the Jhepi Khola and Rondong Khola are less sinuous and at places exhibit sharp turns with zigzag morphology, and a few small bifurcations (30–60 m in length) (Figure 24.3). The deflection of the drainages in the Jhepi Khola is due to the presence of the local (NNE–SSW trending) strike-slip faults (Figure 24.3), which are almost parallel to the Gish Transverse fault (see Matin and Mukul 2020). Such drainage morphologies imply the dominance of tectonic significance. The Rammam tributary, on the other hand, is also less sinuous (1.15) compared to the Chhota Rangit (1.86) (Figure 24.4). Furthermore, the Rammam river shows bifurcation of the channel with the formation of the eye-shaped channel bars (250–360 m in length) (Figure 24.5). The eye-shaped channel bar may form due to the local subsidence along the lineaments and is tectonically controlled. Such eyed channel bars are previously reported from the Rispana river, Doon valley, western Himalaya (Joshi 2017). The Palar river, Tamil Nadu (although associated with rift tectonic), also has an eyed channel bar. However, the mechanism of eyed bar formation in the Rispana river and the Palar river is related to the subsidence along the lineaments (Ramasamy and Kumanan 2000; Ramasamy et al. 2011; Selvakumar and Ramasamy 2014; Joshi 2017; Resmi et al., 2017). Presence of lineaments along the Ramman river and surrounding region has been identified by Sarkar et al. (2021). This may indicate that the eyed channel bar in the Rammam river was possibly formed by the local subsidence along the lineaments and tectonically controlled. Further studies are necessary to delineate the cause of the formation of those bars in the area.

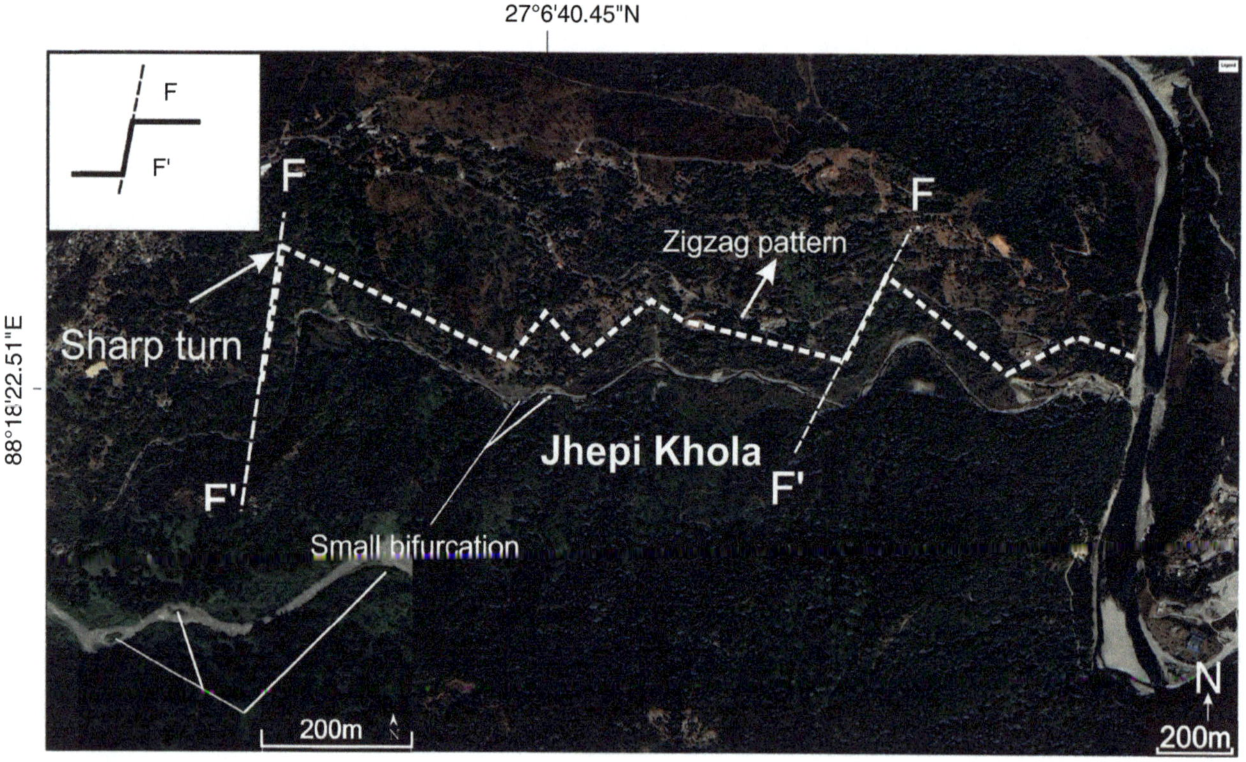

Figure 24.3 Satellite image of the JhepiKhola River exhibits a sharp turn in drainage and zigzag pattern (shown by the dotted line). The sharp drainage deflection (see, inset diagram at top left corner) may be due to the presence of local faults (parallel to the Gish transverse fault). The small bifurcations of the river are due to the tectonic disturbance (zoom image at the bottom left corner). Source: Image courtesy: Google Earth.

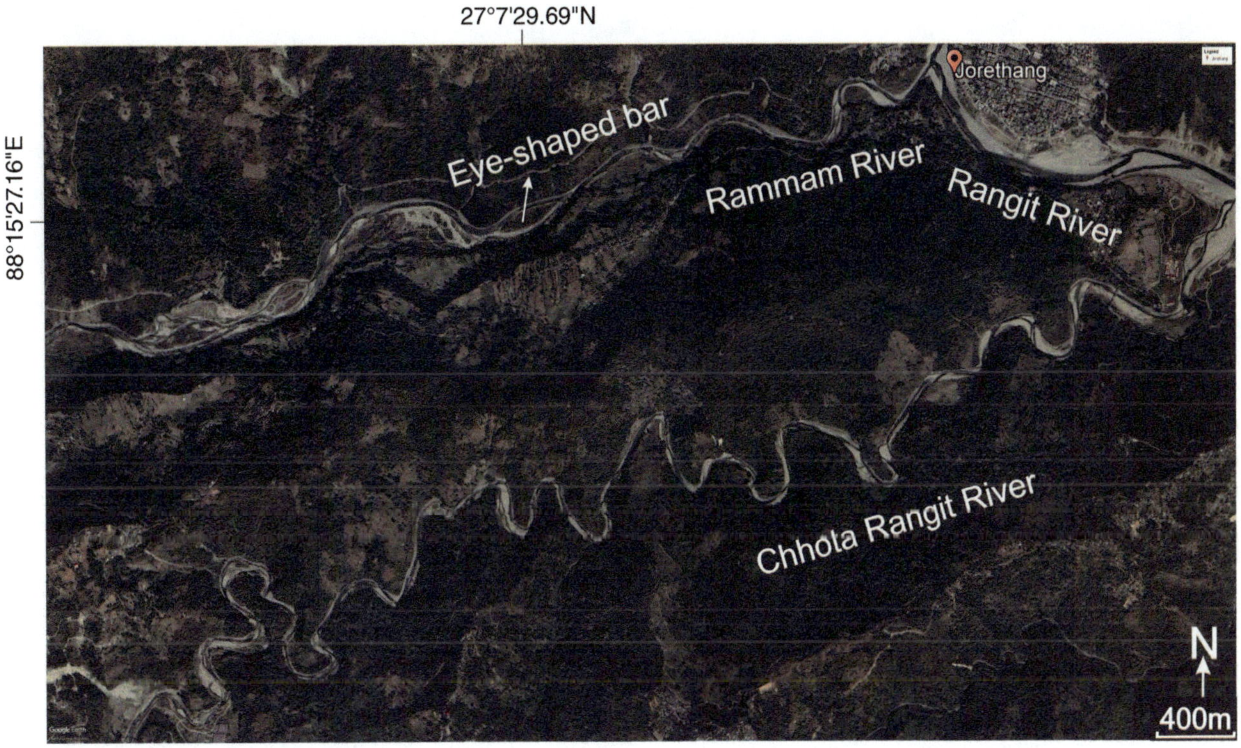

Figure 24.4 The satellite image shows that the Rammam river is less sinuous (1.15) compared to the Chhota Rangit (sinuosity 1.86) river (rivers present 1.5 km apart). The Rammam river has eye-shaped channel bars in the study area (marked by the white arrows). Source: Image courtesy: Google Earth.

Figure 24.5 The satellite image shows an eye-shaped bar in the Rammam river. Source: Image courtesy: Google Earth.

Acknowledgements

I would like to thank Dr. Jyotirmoy Mallik for his support and opinion. I am thankful to Indian Institutes of Science Education and Research, Bhopal for providing the laboratory support. Soumyajit Mukherjee invited to submit an article in the edited book. Achyuta Ayan Misra handled this article. Wiley Blackwell Acquisition Editors (present: Frank Weinreich, past: Andrew Harrison), Stacey Woods (Handling Editor), and proofreading team. Summarized in Misra and Mukherjee (2022).

References

Dasgupta, S. and Mukherjee, S. (2017). Brittle shear tectonics in a narrow continental rift: asymmetric non-volcanic Barmer basin (Rajasthan, India). *The Journal of Geology* 125: 561–591.

Dasgupta, S. and Mukherjee, S. (2019). Remote sensing in lineament identification: examples from western India. In: *Problems and Solutions in Structural Geology and Tectonics*, Developments in Structural Geology and Tectonics Book Series, vol. 5. Series Editor: Mukherjee, S. (ed. A. Billi and A. Fagereng), 205–221. Elsevier. ISSN: 2542-9000. ISBN: 9780128140482.

Dasgupta, B.M., Mukherjee, S., and Chatterjee, R. (2022). Depositional system, morphological signatures, tectonics and sedimentation pattern along the transform margin – Palar-Pennar basin, Indian east coast. *Journal of Petroleum Science & Engineering* 211: 110155. https://doi.org/10.1016/j.petrol.2022.110155.

Ghosh, S. and Sivakumar, R. (2018). Assessment of morphometric parameters for the development of relative active tectonic index and its significant for seismic hazard study, an integrated geoinformatic approach. *Environment and Earth Science* 77: 600.

Ghosh, S. and Sivakumar, R. (2019). An assessment of geomorphometric anomalies and their significance on the seismotectonic activity through geoinformatics. *Journal of Earth System Science* 128: 178.

Gogoi, M.P., Gogoi, B., and Mukherjee, S. (2022). Tectonic instability of the petroliferous upper Assam valley (NE India): a geomorphic approach. *Journal of Earth System Science* 131: 18. https://doi.org/10.1007/s12040-021-01752-6.

Joshi, B.C. (2017). Morphotectonic analysis of Dehradun valley of Garhwal Himalaya, Uttarakhand, India. *International Journal of Emerging Technology and Advanced Engineering* 7: 502–511.

Kaplay, R.D., Md, B., Mukherjee, S., and Kumar, T.V. (2017). Morphotectonic expression of geological structures in eastern part of south East Deccan volcanic province (around Nanded, Maharashtra, India). In: *Tectonics of the Deccan Large Igneous Province* (ed. S. Mukherjee, A.A. Misra, G. Calvès and M. Nemčok), 317–335. London, Special Publications 445: Geological Society.

Keller, E.A. and Pinter, N. (1996). *Active Tectonics: Earthquakes, Uplift and Landscape*, 2e, 359. New Jersey: Prentice Hall.

Kothyari, G.C., Kotlia, B.S., Talukdar, R. et al. (2020). Evidences of neotectonic activity along Goriganga river, higher central Kumaun Himalaya. *Indian Geological Journey* 55: 6123–6146. https://doi.org/10.1002/gj.3791.

Matin, A. and Mukul, M. (2020). Himalayan cross faults affect thrust sheet geometry: an example from the Munsiari thrust sheet near the Gish transverse fault zone, frontal Darjeeling Himalaya, India. *Journal of Asian Earth Science* 199: 104400.

Misra, A.A. and Mukherjee, S. (2022). Introduction to "Atlas of Structural Geological and Geomorphological Interpretation of Remote Sensing Images". In: *Atlas of Structural Geological and Geomorphological Interpretation of Remote Sensing Images* (ed. A.A. Misra and S. Mukherjee). Wiley Blackwell. ISBN: 9781119813354.

Misra, A.A., Bhattacharya, G., Mukherjee, S., and Bose, N. (2014). Near N-S paleo-extension in the western Deccan region in India: does it link strike-slip tectonics with India-Seychelles rifting? *International Journal of Earth Sciences* 103: 1645–1680.

Ramasamy, S.M. and Kumanan, C.J. (2000). Eyed drainages observed in IRS imagery in Tamil Nadu and their geological significance. *International Journal of Remote Sensing* 21: 475–481.

Ramasamy, S.M., Kumanan, C.J., Selvakumar, R., and Saravanavel, J. (2011). Remote sensing revealed drainage anomalies and related tectonics of South India. *Tectonophysics* 501: 41–51.

Resmi, M.R., Achyuthan, H., and Jaiswal, M.K. (2017). Middle to late Holocene paleochannels and migration of the Palar River, Tamil Nadu: implications of neotectonic activity. *Quaternary International* 443: 211–222.

Sarkar, A., Roy, L., Das, S., and Sengupta, S. (2021). Fluvial response to active tectonics: analysis of DEM-derived longitudinal profiles in the Rangit River basin, eastern Himalayas, India. *Environmental Earth Sciences* 80 (7): 1–22.

Selvakumar, R. and Ramasamy, S.M. (2014). Evaluating influence of active tectonics on spatial distribution pattern of floods along eastern Tamil Nadu, India. *Geomorphology* 25–34.

Semwal, S. and Chauniyal, D. (2018). Meanders geometry of Alaknanda River in Srinagar Valley (Garhwal Himalaya). *Indian Journal of Geography and Environment Management* 15–16 (2018): 49–61.

Zhang, J.Y., Liu-Zeng, J., Scherler, D. et al. (2018). Spatiotemporal variation of late quaternary river incision rates in Southeast Tibet, constrained by dating fluvial terraces. *Lithosphere* 10: 662–675.

Index

Atlas of Structural Geological and Geomorphological Interpretation of Remote Sensing Images, First Edition.
Edited by Achyuta Ayan Misra and Soumyajit Mukherjee.

Companion website: www.wiley.com/go/misra/RemoteSensingImages

Printed in the USA/Agawam, MA
September 16, 2026

910661.199